CROP IMPROVEMENT
An Integrated Approach

CROP IMPROVEMENT

An Integrated Approach

Edited By

C P MALIK
GULZAR S SANGHERA
PUSHP SHARMA

MD PUBLICATIONS PVT LTD
NEW DELHI
www.mdppl.com

Published by :

MD Publications Pvt Ltd
"MD House", 11, Darya Ganj,
New Delhi - 110 002
Phone : +91-11-45355555
E-mail : contact@mdppl.com
Website: www.mdppl.com

ISBN : 978-81-7533-456-4

Published and Printed by Mr. Pranav Gupta on behalf of **MD Publications Pvt Ltd** at Mohit Enterprises.

PREFACE

Today crop productivity is experiencing a grim challenge for several reasons, and of prime attention is global climate change, which influences food production all the world over. Concomitantly resources are shrinking but population is increasing. With a view to enhancing the agricultural productivity, it is imperative to develop crop species which could favorably respond to changing environmental conditions. Such plant species should yield enhanced productivity under sub optimal conditions. In a way species should be able to withstand biotic and abiotic stresses. Though the plants have the potential to cope with adverse environments the breeders have hunted for genes which could enhance this potential through cross breeding and selection. Genes for abiotic stress resistances include drought, salinity, temperature, light, etc.

Over the past few decades' research studies concerning physiological, biotechnological approaches were instrumental in enhancing our understanding of the patterns of plant responses to variable stresses and the mechanism operating.

Biotechnological interventions have made well recognized impact and have revolutionized agricultural productivity. Technologies for micro propagation of elite planting materials; molecular breeding for accelerated improvement of specific traits by pyramiding genes from pertinent gene pool; molecular diagnosis of varieties; transgenic organisms incorporating foreign genes of interest into target crops, etc.

Genomics has facilitated the discovery of new genes and DNA markers so important for the application of molecular breeding and transgenic technologies.

It is now feasible to manipulate complete traits controlled by network of genes rather than single genes as in rice (vit A). The possibility of conversion of C3 plant species into C4 with enhanced photosynthesis does not seem to be a distant possibility. Life processes are being increasing unfolded through genomics and bioinformatics. Genome of large number of crop and horticultural species including, maize, rice, plum, tomato was sure to boost our efforts to meet several challenges

Areas like plant genomics and molecular breeding; efficient characterization of genetic resources and search for novel alleles; Climate change and adaptation strategies by different cereals and oil seed crop species; biotic and abiotic stress tolerance are being researched by different groups.

Nanotechnology, which concerns with production and utilization of nanomaterials is an emerging area of great interest and opportunity. It has wider applications in agriculture, food preservation, food packaging and development and utilization of multiplexed diagnostic kits development. Further, bio-selective surfaces are the novel innovations of nanotechnology with enhanced or reduced capability to bind or hold specific molecules. This provides the technique to exploit bioselective surfaces to detect presence of bacteria or viruses. The use of these bio surfaces in the development of biosensors and in purification of mixture of bio molecules can be anticipated.

The present volume titled **CROP IMPROVEMENT: AN INTEGRATED APPROACH** has ten chapters divisible in three disciplines. The first discipline dealing with biotechnology; includes **Crop Improvement in sustainable way through genomic intervention; Advances in hybrid rice technology through application of genetic engineering; Nuclear and organelle specific markers with specific emphasis on chloroplast and its uses in genetic analyses; Assessing morphological, biochemical, utility and molecular diversity in Coriander (*Coriandrum sativum* L.); the status of crop improvement in seed species.** The second discipline of the book deals with **Translocation of photoassimilates, carbon partitioning and crop productivity;** and **Recent advances in photosynthesis for enhancement of crop productivity; Morphological, physiological and biochemical response in plants subjected to salt stress** As stated in the title another chapter summarizes **Utilization of Nanotechnology in crop improvement**. Indeed this constitutes wide and interesting area of research now-a-days. There are many products available in the market which shows their impressive applicability in the field of agriculture. Some of the nanodevices are CNT's , nanobiosensors, nanoparticles, etc. and have contributed in crop improvement, disease detection, disease treatment, pathogen detection, food and packaging, etc. All these nanodevices have helped to increase the crop yield with high quality and also for food safety and biosecurity. Though these nanodevices have given the agriculture a new stature yet they have some side effects on human health and environment and these have to be minimized.

We gratefully appreciate all the learned contributors for their cooperation in reviewing useful information in the selected area of their interest.

The opinions and text contained in their reviews are those of the authors and we have endeavored to honor their ideas.

We are sure that the present volume will prove extremely useful for those interested in exploiting different facets of crop improvement.

The editors record their appreciation and sincere gratitude to Chancellor (JNU, Jaipur), Vice-Chancellor SKUAST (Srinagar) and Vice-Chancellor (PAU, Ludhiana) for encouragement, cooperation and constant support.

We record our sincere thanks to our colleagues Drs. Himakshi Bhati Kushwaha and Amandeep Hora for rendering help from time to time and in the preparation of index.

We are very grateful to Mr Pranav Gupta, Director of MD Publications Pvt Ltd, New Delhi, for his cooperation and support for this book.

Editors

LIST OF CONTRIBUTORS

ANILA BADIYAL

DEPARTMENT OF CROP IMPROVEMENT, CSKHPKV, PALAMPUR (H.P)-176062, INDIA

WASEEM HUSSAIN

DEPARTMENT OF CROP IMPROVEMENT, CSKHPKV, PALAMPUR (H.P)-176062, INDIA

NAVDEEP SINGH JAMWAL

DEPARTMENT OF CROP IMPROVEMENT, CSKHPKV, PALAMPUR (H.P)-176062, INDIA

GULZAR S SANGHERA

SKUAST-KASHMIR, MOUNTAIN RESEARCH CENTRE FOR FIELD CROPS, KHUDWANI, ANANTNAG, 192102, JAMMU & KASHMIR, INDIA

SAJAD HUSSAIN DAR

SKUAST-KASHMIR, MOUNTAIN RESEARCH CENTRE FOR FIELD CROPS, KHUDWANI, ANANTNAG- 192102, J&K, INDIA

AMANDEEP HORA

SCHOOL OF LIFE SCIENCES, JAIPUR NATIONAL UNIVERSITY, JAIPUR, RAJASTHAN, INDIA

JYOTI USHAHRA

SCHOOL OF LIFE SCIENCES, JAIPUR NATIONAL UNIVERSITY, JAIPUR, RAJASTHAN, INDIA

C P MALIK

SCHOOL OF LIFE SCIENCES, JAIPUR NATIONAL UNIVERSITY, JAIPUR, RAJASTHAN, INDIA

EKTA KATHURIA

SCHOOL OF LIFE SCIENCES, JAIPUR NATIONAL UNIVERSITY, JAIPUR, RAJASTHAN, INDIA

DHEERA SANADHYA

SCHOOL OF LIFE SCIENCES, JAIPUR NATIONAL UNIVERSITY, JAIPUR, RAJASTHAN, INDIA

PUSHP SHARMA

DEPARTMENT OF PLANT BREEDING AND GENETICS, PUNJAB AGRICULTURAL UNIVERSITY, LUDHIANA-141 004, INDIA

HIMAKSHI BHATI-KUSHWAHA

SCHOOL OF LIFE SCIENCES, JAIPUR NATIONAL UNIVERSITY, JAIPUR, RAJASTHAN, INDIA

NISHA PAREEK

SCHOOL OF LIFE SCIENCES, JAIPUR NATIONAL UNIVERSITY, JAIPUR, RAJASTHAN, INDIA

M.L. JAKHAR

DEPARTMENT OF PLANT BREEDING AND GENETICS, SKN COLLEGE OF AGRICULTURE (SK RAJASTHAN AGRICULTURAL UNIVERSITY), JOBNER (JAIPUR), RAJASTHAN, INDIA

S.S. RAJPUT

DEPARTMENT OF PLANT BREEDING AND GENETICS, SKN COLLEGE OF AGRICULTURE (SK RAJASTHAN AGRICULTURAL UNIVERSITY), JOBNER (JAIPUR), RAJASTHAN, INDIA.

SUBHASH C KASHYAP

SKUAST-KASHMIR, MOUNTAIN RESEARCH CENTRE FOR FIELD CROPS, KHUDWANI, ANANTNAG, 192102, J&K, INDIA

G A PARRAY

SKUAST-KASHMIR, MOUNTAIN RESEARCH CENTRE FOR FIELD CROPS, KHUDWANI, ANANTNAG, 192102, J&K, INDIA

CONTENTS

1

Crop Improvement in Sustainable way through Genomic Interventions

Waseem Hussain, Gulzar S Sanghera, Navdeep Singh Jamwal and *Anila Badiyal*

The ultimate goal of agriculture science is to increase crop productivity coupled with the quality of the products, and maintain the environment. The increasing availability of DNA sequence information enables the discovery of genes and molecular markers associated with diverse agronomic traits creating new opportunities for crop improvement. Molecular approaches including modern genomics and genetic engineering technologies have emerged as powerful tools to assure rapid and precise selection for the trait(s) of interest. The genomic approaches viz. *structural genomics*—the study of the genome structure, *functional genomics*—the study of the genome function; and *comparative genomics*—the study of genome evolution. Functional genomics includes analyses of the *transcriptome*, the complete set of RNAs transcribed from a genome, the *proteome*, the complete set of proteins encoded by a genome, and *metabolome*, the complete set of small molecule metabolites and several others have been elaborated in literature during last decade. Attempt has been made in this chapter towards the utilization of aforementioned intervention that could increase the efficiency and precision of crop improvement in a systematic way.

Introduction

Plant breeding describes methods for the creation, selection, and fixation of superior plant phenotypes in the development of improved cultivars suited to needs of farmers and consumers. The ultimate goal of agriculture science is to increase crop productivity coupled with the quality of the products, and maintain the environment. There has been significant improvement in production and productivity of important cereal crops globally as a consequence of the "Green Revolution" and other initiatives (Briggs, 1998). However, today the stage has reached that the available traditional methods of crop improvement are not sufficient to provide enough and staple food grains to the constantly growing world population

(Barrett, 2010). This situation will be further more complicated because of climate change and may have adverse impacts on food production, food quality (Atkinson *et al.*, 2008), food security and thus difficult to maintain sustainability. In other words, the conventional plant breeding practices may not able to achieve the sustainability in today's agriculture. Under such circumstances molecular approaches including modern genomics and genetic engineering technologies have emerged as powerful tools to assure rapid and precise selection for the trait(s) of interest. Maintaining effective and environmentally friendly agricultural practices is a necessary prerequisite for maintaining sustainability. It is therefore expected that the genomics will be the integral part of the agricultural/ plant breeding practices in future for improving crop productivity leading to achieve food security and sustainable production.

Genome is the collection of genes contained within a haploid chromosome set and is derive root for new brand science 'genomics' that uses molecular characterization, transcript profiling and cloning of whole genomes to understand the structure, function and evolution of genes. The genomics subdiscipline is divided into *structural genomics*—the study of the genome structure, *functional genomics*—the study of the genome function; and **comparative genomics**—the study of genome evolution. Functional genomics includes analyses of the **transcriptome,** the complete set of RNAs transcribed from a genome, the **proteome,** the complete set of proteins encoded by a genome, and **metabolome**, the complete set of small molecule metabolites and several other relevant -omes. The main aim of genomics in relation to plant breeding is simply the prediction of phenotype from genotype by utilizing various novel genomic tools. The various genomic tools that help in crop improvement in sustainable way are being discussed under this chapter.

Genomics assisted breeding

The various genomic tools that are being utilized for crop improvement includes, molecular markers, QTL mapping, physical mapping, transcriptome/ genome sequencing and functional genomics. The ultimate aim of these novel tools is to find subset of candidate genes or markers that can then be utilized through MAS approaches for variety improvement (Fig.1). More interestingly, genomics tools and approaches are revolutionizing the breeding methodology, a procedure referred as 'genomics-assisted breeding' (Varshney *et al.*, 2005).

Structural genomics

Structural genomics concerns the organization and sequence of genetic information contained within a genome. Often, an early step in characterizing a genome is to prepare genetic and physical maps of its chromosomes. These maps provide information about the relative locations of genes, molecular markers, and chromosome segments, which are often essential for positioning chromosome segments.

Molecular markers

For successful utilization of genomics-assisted breeding approach in a crop, availability of basic molecular tools such as molecular markers, genetic maps, etc. is a pre-requisite (Gupta *et al.*, 2008). Molecular markers: a set of DNA-based markers that can detect DNA polymorphism at the level of specific loci and at the whole genome level. There are many types of molecular markers: the earliest to be developed were RFLPs (restriction fragment

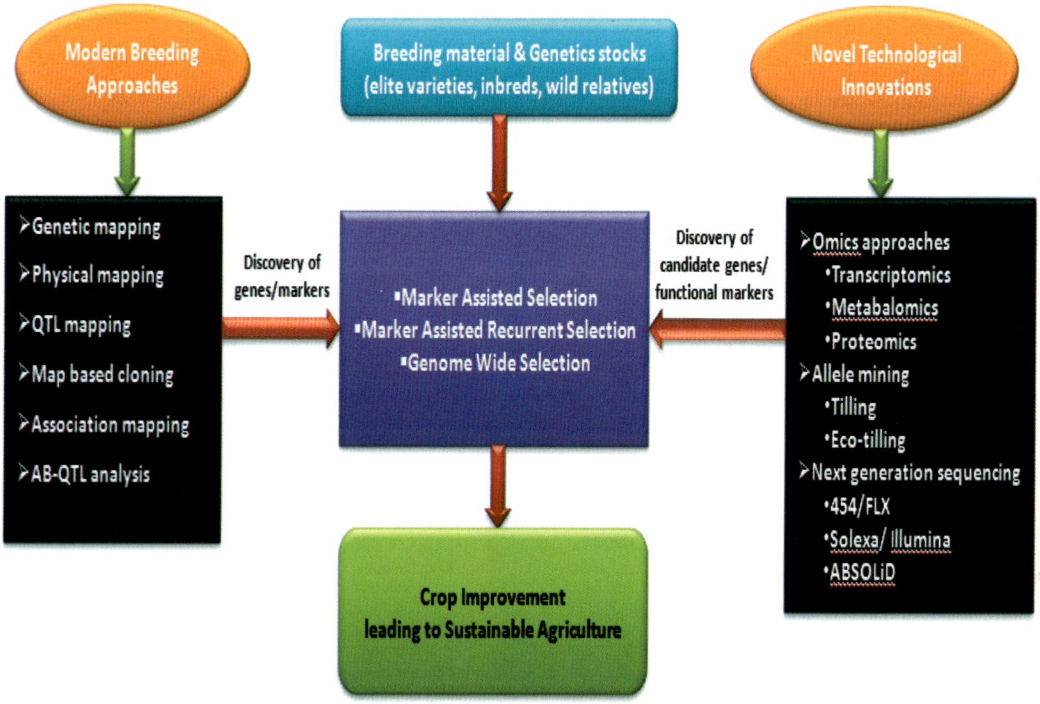

Fig-1: Modern breeding approaches in combination with novel technological tools helps in identification of markers (Genic/functional) or subset of candidate genes that may be then utilized in various selection methods like MAS, MARS or GWS for variety improvement in precise and sustainable way.

length polymorphisms) and others include RAPDs (random amplification of polymorphic DNAs), CAPS (cleaved amplified polymorphic sites), SSRs (simple sequence repeats) and AFLPs (amplified fragment length polymorphisms). The latest molecular markers developed include SNPs (single nucleotide polymorphisms) and SFPs (single feature polymorphisms) (Gupta and Varshney, 2000; Varshney *et al.*, 2007). Molecular markers have been utilized efficiently and effectively in mapping and tagging of genes, phylogeny and evolution, diversity analysis, markers assisted selection and gene pyramiding. Among the various molecular markers simple sequence repeats (SSRs), single nucleotide polymorphisms (SNPs) and diversity array technologies (DArTs) are widely used now. SSRs, or microsatellite markers, for various practical reasons, have been the leading marker types for breeding applications (Varshney, 2009). The DArT marker system has proven extremely effective in exposing genetic diversity in germplasm collections, and is well-suited for background selection in molecular breeding programmes (Varshney *et al.*, 2006). Finally, SNPs, which have been the prime marker platform in human and animal genetics for some time, are now rapidly gaining momentum and are overtaking SSRs as the marker type of choice in plants as well, mainly because of their amenability for high-throughput genotyping and cost effectiveness (Paterson *et al.*, 2009; Varshney *et al.*, 2009a).

QTL mapping

The detection of genes or QTLs controlling traits is possible due to genetic linkage analysis, which is based on the principle that genes and markers segregate via chromosome recombination (called crossing-over) during meiosis (i.e. sexual reproduction), thus allowing their analysis in the progeny (Paterson, 1996). QTL mapping is used to dissect the complex trait into simpler ones and QTL mapping has been enhanced in resolution with the availability of high density molecular maps. QTL mapping is used to determine the degree of association of specific region on the genome to the inheritance of trait of interest. It ends up letting us the degree of association of a stretch of DNA with a trait rather than pointing directly at the gene responsible for that trait. QTLs provide detailed information regarding the organization of the genes in the genome as well as the basis of interactions in genomes so that appropriate crop improvement strategies can be used.

Markers assisted selection

Once the markers associated with a trait of interest are identified through linkage mapping, the next step is to use these markers in the breeding programs either in MAS or MABC (Marker assisted backcrossing). Marker-assisted selection is a method that uses molecular markers associated with the traits of interest to select plants at the seedling stage, thus speeding up the process of conventional plant breeding and reducing the cost involved in maintaining fields. MAS facilitate improvement of traits that cannot easily be selected using conventional breeding methods. There are three major steps involved in MAS: (i) identification of molecular marker(s) associated with trait(s) of interest to breeders; (ii) validation of identified marker(s) in the genetic background of the targeted genotypes to be improved; and (iii) marker-assisted backcrossing (MABC) to transfer the QTL/gene from the donor genotype into the targeted genotype utilizing backcrossing technique. Important advantages of MAS are that, it can be effectively utilized for traits with low heritability, for gene pyramiding, selection can be made at seedling stage and above all there are no issues involving GE crops (Kuchel *et al.*, 2007).

Moreover in addition to routinely used marker-assisted Selection (MAS), marker-assisted recurrent selection (MARS) and genome wide selection (GWS) approaches are becoming popular now a day.

Marker-assisted recurrent selection (MARS)

In the majority of traits of interest quantitative variation is controlled by many QTLs each with minor effect. Moreover, minor QTLs show an inconsistent QTL effect in different environments and over different seasons. Even when the effect of these minor QTLs is consistent, their introgression into the desired genotype through MABC becomes extremely difficult as a larger number of progenies are required to select appropriate lines. In such cases, MARS has been proposed for pyramiding of superior alleles at different loci/QTLs in a single genotype (Edwards and Johnson, 1994; Bernardo and Charcosset, 2006).

Genome-wide or Genomic Selection (GS)

Although MAS has been practiced for the improvement of quantitative traits, it has its own limitations. Therefore in addition to MARS, Genomic Selection can be used to pyramid favourable alleles for minor effect QTLs at the whole genome level (Meuwissen *et al.*, 2001; Heffner *et al.*, 2009). A key to the success of GS is that, unlike MABC or MARS, it calculates the marker effects across the entire genome that explains the entire phenotypic variation. In simple terms genome-wide selection refers to marker based selection without significance testing and without identifying of a subset of markers associated with the trait (Meuwissen *et al.*, 2001).

Advanced-backcross (AB-QTL) analysis

To reduce the time and linkage drag in process of gene introgression through use MAS a new approach referred as advanced backcross QTL (AB-QTL) analysis was proposed by (Tanksley and Nelson, 1996). AB-QTL analysis is a method for simultaneous identification and transfer of favorable QTL alleles from un-adopted donor lines (e.g. land races and wild species) to elite lines for variety development. In this context, a superior cultivar / variety is crossed with a wild species leading to the production of a backcross population (BC_2, BC_3) and molecular markers are used to monitor the transfer of QTLs by conventional backcrossing. It is anticipated that the use of AB-QTL will be accelerated in a range of crops for improving important traits such as disease resistance as well as yield traits.

Map-based cloning (MBC)

The MBC approach involves the use of molecular markers for preparing a high-density genetic map around the region harbouring the gene of interest and ultimately, the local physical map to isolate the gene (Brem *et al.*, 2002).

Association mapping

Association mapping is also known as LD (linkage disequilibrium) mapping is a population based survey used to identify trait marker relationships based on LD. It is the 'non- random association of alleles at different loci' (Flint-Garcia *et al.*, 2003). It is the correlation between polymorphisms (e.g. SNPs) that is caused by their shared history of mutation and recombination. It relies on previous unrecorded sources of disequilibrium to create population wide marker phenotype associations (Tenaillon *et al.*, 2001; Buckler and Thornsberry, 2002).

In general, conventional linkage analysis is using bi-parental mapping population such as F_2 lines, BC and RILs is commonly used method for trait mapping. However such mapping populations are derived from few cycles of recombination events, hence limit the resolution of genetic maps and localize QTLs from 10 to 20 cm intervals and don't essentially use the germplasm that is being actively used in breeding programmes. In contrast association mapping based on LD measures the degree of non-random association between alleles of different loci. It doesn't require segregating population and in some cases more powerful than linkage analysis for identifying the gene responsible for variation in quantitative trait. Conventional mapping provides pertinent information about traits that tends to be specific to the some and genetically related populations, while results from association mapping more applicable to a much wider germplasm base. Thus association mapping offers three advantages over traditional linkage analysis- a) increase mapping resolution, b) Greater allele number c) Reduced research time. Association mapping employs one of the following two approaches -

1) Candidate meaning gene association mapping- which relates polymorphism in selected genes that have purported roles in controlling phenotypic variation for specific traits.
2) Genome wide association mapping- which surveys genetic variation in the whole genome to find the signals of association for various complex traits.

Omics Approaches

Omics is a collective, broad discipline largely referring to analysis of the interactions of biological information obtained from the profiling of the genome, transcriptome, proteome, metabolome, and several other relevant -omes. Essentially, the omics science is enabled by a host of diverse, high-throughput technologies and platforms (Briggs, 1998). The full range of omics technologies can now be applied to understand the same fundamental biological processes (Barrett, 2010). The basic aim of omics technologies is to provide a candidate set of genes that can be used for understanding the biology of a trait and for the development of perfect or diagnostic markers for used in MAS applied to crop breeding.

Transcriptomics

Measuring gene expression (particularly of mRNA) patterns simultaneously across all the genes in the genome, i.e. transcriptomics, is a uniquely powerful technology to explore the unknown players in a wide range of biological processes, diseases, traits and responses to stimuli. The technique is extremely powerful as a first step to implicate novel genes and pathways that may be involved or associated with a particular condition. Transcriptomics is an important first step to study traits that are under the control of several to many genes (i.e., polygenic traits) and responsive to external conditions and internal states (i.e., multifactorial traits). With this technology, the expression of thousands of genes is measured simultaneously. It provides a snapshot of all genes that are actively transcribed during a particular process. When we compare these measurements between conditions or treatments, those genes that are expressed at higher or lower level under a particular condition (stress and no stress; resistant and susceptible) can be identified. Transcriptomics technology is used to characterize the composition of the messenger RNA (mRNA) pools from each biological sample. The mRNAs are the transcripts of a gene that carry the information encoded in the gene to the site of protein synthesis. When a particular mRNA is present in a biological sample, it implies that the corresponding gene was expressed, and a template is available for the synthesis of the protein product of that gene. The abundance of each mRNA in the pool represents the level of expression of the corresponding gene. By comparing the relative proportional representation of each mRNA in the total mRNA pool among the samples, we can identify which genes differed in expression in response or relation to the compared conditions. The various common platform technologies used for genome analysis of gene expression are microarrays, serial analysis of gene expression (SAGE), massively parallel signature sequencing (MPSS), and next generation sequencing platforms (NGSPs) (Horak and Snyder, 2002; Rostoks *et al.*, 2005). Among these microarrays are widely used although the sequencing of the transcriptome is rapidly increasing in popularity. Microarrays are solid-based platforms (e.g., glass slides), containing millions of copies for thousands of 'reporter probes' that comprise part of the sequences of the genes in the genome. By binding (or 'hybridizing') fluorescent-labelled copies of the original mRNAs to the probes, measuring the label intensities for each position on the array, and associating these positions to their specific reporter probes, one can infer the presence and abundance of each transcript in the labelled RNA pool.

Proteomics

The term proteome refers to all proteins expressed by a genome in the targeted tissues at a defined time point. It encompasses a broad range of tools and techniques in determining the identity and quantity of the expressed proteins in cells/tissues, their 3D structure, and other interacting partners that help to disclose gene function. Change Proteomics deals with study of the structure and function of entire set of proteins, present in an organism and hence is essential for studying the whole metabolic pathway. This involves separation, identification, and determination of function and functional network of proteins allowing the integral study of many proteins at the same time (Agrawal and Rakaal, 2006). Proteomics not only enables the study of protein–protein interaction but also helps in identification of multisubunit complexes (Espagne *et al.*, 2007). Furthermore, proteomics can act as a powerful approach to organize and identify the proteome through development of 2DE gel protein reference maps of subproteomes in different plant species.

Metabolomics

Metabolome refers to a set of metabolites that are formed within a biological system and their types and levels can be regarded as the ultimate response of biological systems to genetic or environmental changes (Keurentjes *et al.*, 2006). Metabolomics is considered the ultimate level of postgenomic analysis as it can reveal changes in metabolite fluxes that are controlled by only minor changes within gene expression measured using transcriptomics and/or by analyzing the proteome that elucidates posttranslational control over enzyme activity (Keurentjes *et al.*, 2006; Kopka *et al.*, 2004). Metabolite profiling is a fast growing technology and is useful for phenotyping and diagnostic analysis of plants. It is also rapidly becoming a key tool in functional annotation of genes and in the comprehensive understanding of the cellular response to biological conditions. Metabolomic analysis consists of three distinct experimental parts:

(a) preparation of the sample, (b) acquisition of data using analytical chemical methods, and (c) data mining using appropriate chemometric methods (Keurentjes *et al.*, 2006). Essentially, all these steps are strongly interrelated and interdependent. Two main metabolite profiling strategies are (i) mass spectrometry (MS) and (ii) nuclear magnetic resonance (NMR). The gas chromatography–mass spectrometry (GC-MS), gas chromatography–time-of-flight mass spectrometry (GC-TOF-MS), and liquid chromatography–mass spectrometry (LC-MS) are extensively used MS-based techniques in metabolite analysis. The GC-MS technology enables the identification and quantification of over a few hundred primary metabolites within a single extract (Keurentjes *et al.*, 2006). The GC-TOF-MS offers fast scan times and higher sample throughput. On the other hand, LC-MS measures a far broader range of metabolites including primary and secondary metabolites (Meyer *et al.*, 2007). In addition to this, capillary electrophoresis–mass spectrometry (CE-MS) and Fourier-transformation cyclotron resonance–mass spectrometry (FTICR- MS) are also used. CE-MS is considered a highly sensitive methodology to detect low-abundance metabolites in plant samples (Keurentjes *et al.*, 2006). The FTICR-MS relies solely on very high-resolution mass analysis, which potentially enables the measurement of the empirical formula for thousands of metabolites, although it is somewhat limited by the lack of chromatographic separation. NMR approaches rely on the detection of magnetic nuclei of atoms after application of a constant magnetic field for metabolite profiling (Bino *et al.*, 2004). NMR can provide subcellular information and it is easier to derive atomic information for flux modeling from NMR than from MS-based approaches (Keurentjes *et al.*, 2006).

Allele mining

Unfortunately, efficient extraction and exploitation of the adaptive variation and valuable traits maintained in gene banks have yet to be fully achieved, though it remains a high priority of gene bank managers (Hoisington *et al.*, 1999; Richards, 2004). Traditional methods, which screen large, heterogeneous collections for phenotypic variation in agricultural traits, are not only logistically challenging but they may overlook valuable genotypic variation concealed by epistasis in non elite genetic backgrounds (Tanksley and McCouch, 1997). Furthermore, enormous sequence information is available in public databases as a result of sequencing of diverse crop genomes. It is important to use this genomic information for the identification and isolation of novel and superior alleles of agronomically important genes from crop gene pools to suitably deploy for the development of improved cultivars. Allele mining (dissection of naturally occurring variation at candidate genes/loci) is a promising approach to dissect naturally occurring allelic variation at candidate genes controlling key agronomic traits which has potential applications in crop improvement programs. Allele mining can be effectively used for discovery of superior alleles, through 'mining' the gene of interest from diverse genetic resources. It can also provide insight into molecular basis of novel trait variations and identify the nucleotide sequence changes associated with superior alleles. In addition, the rate of evolution of alleles; allelic similarity/dissimilarity at a candidate gene and allelic synteny with other members of the family can also be studied. Allele mining may also pave way for molecular discrimination among related species, development of allele-specific molecular markers, facilitating introgression of novel alleles through MAS or deployment through genetic engineering (Fig. 2)

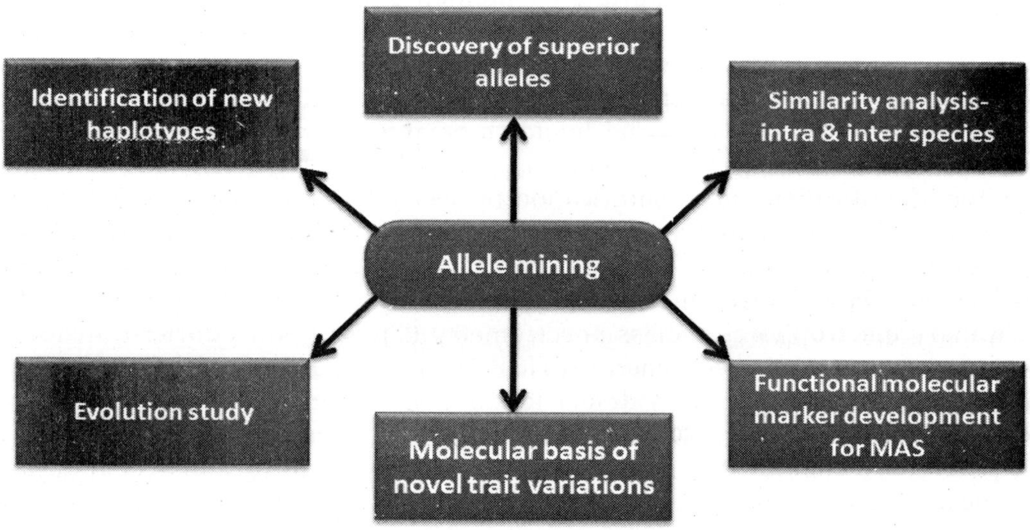

Fig-2: The possible role of allele mining for crop improvement is depicted above.

The major approach available for the identification of sequence polymorphisms for a given gene in the naturally occurring populations is a modified TILLING (Targeting Induced Local Lesions in Genomes) procedure called Eco-Tilling (Comai *et al.*, 2004). The TILLING strategy utilizes traditional mutagenesis followed by high throughput mutation discovery (Colbert *et al.*, 2001). The main steps in TILLING are mutagenesis, the development of a non-chimeric population, preparation of a germplasm stock, DNA extraction and sample pooling, screening the population for induced mutations, and the validation and evaluation of mutants (Fig.3). Eco-TILLING allows natural alleles at a locus to be characterized across many germplasms, enabling both SNP discovery and haplotyping.

Next generation sequencing (NGS) technologies

The term DNA sequencing refers to sequencing methods for determining the order of the nucleotide bases - adenine, guanine, cytosine, and thymine - in a molecule of DNA. It is

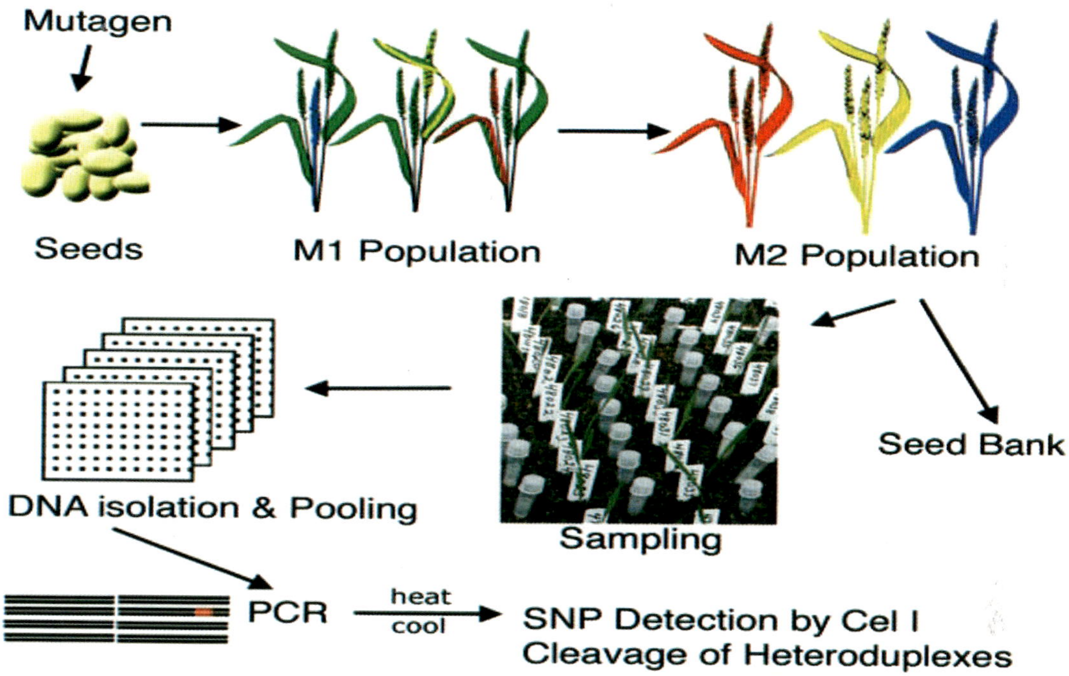

Fig-3: The TILLING Method. Seeds are treated with a chemical mutagen to induce genetic variation, and then planted. The resulting M1 population of plants is chimeric for mutations. Therefore, one seed from each M1 is planted to create the M2 population. M2 DNA is extracted from leaf tissue and M3 seeds from each plant are stored in a seed bank. DNA samples are pooled to increase throughput and PCR amplified with dye-labeled PCR primers specific to a target gene of interest. PCR products are denatured and allowed to reanneal to form heteroduplexes. Heteroduplex DNA is then cleaved by Cell and analyzed.

the most fundamental way of analyzing the structure of DNA, whether of recombinant plasmid, natural gene or whole genome, in terms of determination of sequences of bases of which DNA is composed. Sequencing methods have evolved from relatively laborious gel-based procedures to modern automated protocols based on dye labelling and detection in capillary electrophoresis that permit rapid large-scale sequencing of genomes and transcriptomes. The rapid speed of sequencing attained with modern DNA sequencing technology has been instrumental in the sequencing of the genomes of animals, plants, and microbial genomes.

The automated Sanger method is considered as a 'first-generation' technology, and newer methods are referred to as next-generation sequencing (NGS). These next-generation sequencing (NGS) technologies are able to generate DNA sequence data inexpensively and at a rate that is several orders of magnitude faster than that of traditional technologies. Advances in sequencing technologies are driving down sequencing costs and increasing sequence capacity at an unprecedented rate, making whole-genome resequencing possible. NGS technologies such as Roche/454, Solexa/Illumina and AB SOLiD have already demonstrated the potential to circumvent the limiting factors of Sanger sequencing. For example, sequencing can be multiplexed to a much greater extent by many parallel reactions at a greatly reduced cost (Hudson, 2008). Currently, Roche/454, Solexa and AB SOLiD are the technologies that are predominantly used in crop genetics and breeding applications. Although Roche/454 is superior to Solexa and AB SOLiD in terms of obtaining longer sequence reads, maximum data output is higher for both Solexa and AB SOLiD (Gupta, 2008, 2008a). In terms of costs per run or sequence data generation, Roche/454 is more expensive than either the Solexa or AB SOLiD technologies. The advantages of second-generation DNA sequencing are currently offset by several disadvantages. The most prominent of these include read-length (for all of the new platforms, read-lengths are currently much shorter than conventional sequencing) and raw accuracy (on average, base-calls generated by the new platforms are at least tenfold less accurate than base-calls generated by Sanger sequencing). Although these limitations create important algorithmic challenges for the immediate future, we should bear in mind that these technologies will continue to improve with respect to these parameters, much as conventional sequencing progressed gradually over three decades to reach its current level of technical performance. These advanced and rapid sequencing of many eukaryotic genomes has provided unprecedented opportunities to understand gene function, genome structure, and genome evolution.

Conclusion

Although the newly developed genetic and genomics tools will certainly enhance the crop improvement in sustainable way, but they will not entirely replace the conventional breeding and evaluation process. However these novel tools and approaches are available to plant breeders with ease to enhance the crop improvement in effective and precise manner. The application of DNA marker technologies in exploiting the vast and largely under-utilized pool of favorable alleles existing in the wild relatives of crops will provide a huge new resource of genetic variation to fuel the next phase of crop improvement. Nevertheless, marker-assisted breeding or marker-assisted selection will gradually evolve into 'genomics-assisted breeding' for crop improvement. Use of these genomic intervention approaches together with conventional breeding methodologies should prove very useful for enhancing the genetic gain leading to crop improvement in terms of increasing crop productivity; diversification of crops; enhancing nutritional value of food (biofortification); and reducing environmental impacts of agricultural production. Thus it may be concluded that only through judicious, rational, and science- and need-based exploitation of genetic resources through genomic technologies coupled with conventional plant breeding and genetic engineering will lead to sustainable agriculture.

REFERENCES

Agrawal GK and Rakwal R (2006) Rice proteomics: A cornerstone for cereal food crop proteomes. *Mass Spectrom Rev* 25: 1–53

Atkinson MD, Kettlewell PS, Poulton PR and Hollins PD (2008) Grain quality in the Broad balk wheat experiment and the winter North Atlantic oscillation. *Journal of Agricultural Science* 146: 541–549

Barrett CB (2010) Measuring food insecurity. *Science* 327: 8525-828.

Bernardo R and Charcosset A (2006) Usefulness of gene information in marker assisted recurrent selection: A simulation appraisal. *Crop Sci* 46: 614–621.

Bino RJ, Hall R D, Fiehn O, Kopka J, Saito K, Draper J, Nikolau B J, Mendes P, Roessner-Tunali U, Beale MH, Trethewey RN, Lange BM, Wurtele ES and Sumner LW (2004) Potential of metabolomics as a functional genomics tool. *Trends Plant Sci* 9: 418–425

Borevitz JO, Liang D, Plouffe D, Chang HS, Zhu T, Weigel D, Berry CC, Winzeler E, and Chory J (2003) Large-scale identification of single-feature polymorphisms in complex genomes. *Genome Res* 13: 513–523

Brem RB, Yvert G, Clinton R and Kruglyak L (2002) Genetic dissection of transcriptional regulation in budding yeast. *Science* 296: 752–755

Briggs SP (1998) Plant genomics: More than food for thought. *Proc Natl Acad Sci USA* 95: 1986–1988

Buckler ES and Thornsberry JM (2002) Plant molecular diversity and application to genomics. *Curr Opin Plant Biol* 5: 107–111

Colbert T, Till BJ, Tompa R, Reynolds S, Steine MN, Yeung AT, McCallum CM, Comai L, and Henikoff S (2001) High-throughput screening for induced point mutations. *Plant Physiol* 126: 480–484

Comai L, Young K, Till BJ, Reynolds SH, Greene EA, Codomo CA, Enns LC, Johnson JE, Burtner C, Odden AR and Henikoff S (2004) Efficient discovery of DNA polymorphisms in natural populations by Eco TILLING. *Plant J* 37: 778–786

Hoisington D, Khairallah M, Reeves T, Ribaut J-M, Skovmand B, Taba S, and Warburton M (1999) Plant genetic resources: what can they contribute toward increased crop productivity. *Proc Natl Acad Sci USA* 96: 5937–5943

Edwards M and Johnson L (1994) RFLPs for rapid recurrent selection. In: Analysis of Molecular Marker Data. Joint Plant Breeding Symposia Series, ASA, Madison, WI, USA, pp. 33–40

Espagne C, Martinez A, Valot B, Meinnel T and Giglione C (2007) Alternative and effective proteomic analysis in Arabidopsis. *Proteomics* 7: 3788–3799

Flint-Garcia SA, Thornsberry JM and Buckler ES (2003) Structure of linkage disequilibrium in plants. *Ann Rev Plant Biol* 54: 357–374

Gupta PK and Varshney RK (2000) The development and use of microsatellite markers for genetics and plant breeding with emphasis on bread wheat. *Euphytica* 113: 163-185

Gupta PK (2008a) Ultrafast and low-cost DNA sequencing methods for applied genomics research. *Proc Natl Acad Sci India* 78: 91–102

Gupta PK (2008b) Single-molecule DNA sequencing technologies for future genomics research. *Trends Biotechnol* 26: 602–611

Gupta PK, Rustgi S and Mir RR (2008) Array-based high-throughput DNA markers for crop improvement. *Heredity* 101: 5–18

Heffner EL, Sorrells ME and Jannink JL (2009) Genomic selection for crop improvement. *Crop Sci* 49: 1-12

Horak CE and Snyder M (2002) ChIP-chip: a genomic approach for identifying transcription factor binding sites. *Methods Enzymol* 350: 469–483

Hudson M (2008) Sequencing breakthroughs for genomic ecology and evolutionary biology. *Mol Ecol Resour* 8: 3–17

Keurentjes JJ, Fu J, de Vos CH, Lommen A, Hall RD, Bino RJ, van der Plas LH, Jansen RC, Vreugdenhil D and Koornneef M (2006) The genetics of plant metabolism. *Nat Genet* 38: 842–849

Kopka J, Fernie A, Weckwerth W, Gibon Y and Stitt M (2004) Metabolite profiling in plant biology: platforms and destinations. *Genome Biol* 5: 109

Kuchel H, Fox R, Reinheimer J, Mosionek L, Willey N, Bariana H and Jefferies S (2007) The successful application of a marker-assisted wheat breeding strategy. *Mol Breed* 20: 295–308

Meuwissen THE, Hayes BJ and Goddard ME (2001) Prediction of total genetic value using dense marker maps. *Genetics* 157: 1819–1829

Meyer RC, Steinfath M, Lisec J, Becher M, Witucka-Wall H, Törjék O, Fiehn O, Eckardt A, Wllmitzer L, Selbig J and Altmann T (2007) The metabolic signature related to high plant growth rate in *Arabidopsis thaliana. Proc Natl Acad Sci USA* 104: 4759–4764

Paterson AH, Bowers JE, Bruggmann R, Dubchak I, Grimwood J, Gundlach H, Haberer G, Hellsten U, Mitros T, Poliakov A and Schmutz J (2009) The *Sorghum bicolor* genome and the diversification of grasses. *Nature* 457: 551-556.

Richards CM (2004) Molecular technologies for managing and using gene bank collections. In: MC de Vicente (ed) The evolving role of genebanks in the fast developing field of molecular genetics. *Issues Genet Resour* 11: 13–18. IPGRI, Rome, Italy

Rostoks N, Borevitz JO, Hedley PE, Russell J, Mudie S, Morris J, Cardle L, Marshall DF

and Waugh R (2005) Single-feature polymorphism discovery in the barley transcriptome. *Genome Biol* 6: 54-58

Shendure J, Mitra RD, Varma C and Church GM (2004) Advanced sequencing technologies: Methods and goals. *Nat Genet* 5: 335–344

Tanksley SD (1993) Mapping polygenes. *Annu Rev Genet* 27: 205– 23

Tanksley SD and McCouch SR (1997) Seed banks and molecular maps: unlocking genetic potential from the wild. *Science* 277: 418–423

Tanksley SD and Nelson JC (1996) Advanced backcross QTL analysis: a method for simultaneous discovery and transfer of valuable QTL from unadapted germplasm into elite breeding lines. *Theor Appl Genet* 92: 191–203

Tenaillon MI, Sawkins MC, Long AD, Gaut RL, Doebley JF and Gaut BS (2001) Patterns of DNA sequence polymorphism along chromosome 1 of maize (*Zea mays* spp *mays* L.). *Proc Natl Acad Sci USA* 98: 9161–9166

Varshney RK (2009) Gene-based marker systems in plants: high throughput approaches for discovery and genotyping. In: Jain SM, Brar, DS (Eds.), Molecular Techniques in Crop Improvement, Vol II, Springer, The Netherlands, pp. 119-140

Varshney RK, Graner A and Sorrells ME (2005) Genomics-assisted breeding for crop Improvement. *Trends Plant Sci* 10(12): 621-630

Varshney RK, Hoisington DA and Tyagi AK (2006) Advances in cereal genomics and applications in crop breeding. *Trends Biotechnol* 24: 490-499

Varshney RK, Langridge P and Graner A (2007) Application of genomics to molecular breeding of wheat and barley. *Advances Genet* 58: 121-155

Varshney RK, Close TJ, Singh NK, Hoisington DA and Cook DR (2009) Orphan legume crops enter the genomics era. *Curr Opin Plant Biol* 12: 202–210

2

Translocation of Photoassimilates, Carbon Partitioning and Crop Productivity

Pushp Sharma

Plants are reliable source of food; therefore, there has been intense interest to improve crop productivity. Attempts to improve crop yields have increased largely through efforts to increase plant adaptability to environmental conditions, resistance to diseases and pests, improve agronomic practices and thus, improvement of genetic yield potential. Efforts are also made to increase agricultural productivity with the help of basic researches on biological determinants of yield i.e. photosynthetic efficiency and harvest index. Of all the biological yield determinants, photosynthesis continues to be considered a high priority area of research.

Carbon partitioning is influenced by both the supply and demand for photosynthate in plant. It is moderated by vascular connections and the storage capacity of the leaves and pathway tissues. The transport of photosynthate may be either for short distance (within the source or the sink) or for long distance transport (in the phloem) carbon partitioning of plants is influenced by a range of factors. If a particular organ or tissue is adequately supplied with photosynthates, then the organ can fully express its growth potential (a source or sink limitation) and depend on the direct supply and demand for carbon throughout development. The hormonal/nutritional factors indirectly control the growth of an organ. The vascular constraints may favour the growth of one organ over another. The buffering effect of carbon storage at different sites in the plant and in some instances photosynthetic rate (source activity) may also regulate the rate of carbon utilization (sink activity). The supply of photosynthate in various plant parts can also be modified either experimentally, or through action of environmental factors and disease. Thus, a series of factors, and not a single limiting event, controls the carbon partitioning in higher plants (Gifford *et al.*, 1981).

In thermodynamic terms the efficiency of crop production is defined as the ratio of energy output (carbohydrate) to energy input (solar radiation). Temperature and water supply are the main climatic constraints on efficiency. In India, the radiation and thermal climates are not uniform and rainfall is the main discriminate of yield between the regions. Total dry matter production by barley, potatoes, sugar beet and apple is strongly correlated with intercepted radiation and these crops form carbohydrate at about 1.4 g per M solar energy, equivalent to 2.5% efficiency. Crop growth may, therefore, be analyzed in terms of (a) the amount of light intercepted during the growing season and (b) the efficiency with which intercepted light is used. These variables are discussed in terms of the rate and duration of development phases.

Harvest Index: The two major determinants of high productivity in a crop area : the ability to produce high levels of photosynthates over a wide range of environmental conditions and the efficient transport and partition of the high proportion of those assimilates into economically important organs (Hay, 1995).

Yield increase has resulted mainly from empirical selection for high HI. Therefore, increasing the harvest indices of crops seems to be the most effective means for future yield improvements. If crop plant is not able to allocate a major portion of the fixed carbon into yields, high photosynthetic rates would not translate into increased food production. Therefore, bioregulation for more efficient allocation of fixed carbon to harvestable parts may hold the key to assure adequate world food supplies in the future. Since, agricultural yield (the economic product) is not necessarily synonymous with biological yield (total biomass), the concept of an efficiency index was first proposed by E.S. Beaven in 1914. In 1962, Donald proposed the most explicit term harvest index (HI) which is defined as the ratio of economic yield to total biomass production. Harvest index by definition, has a value less than unity and it ranges between 0 and 55% for most crops. Although HI, indicates how the products of photosynthesis are partitioned among different plant parts, a positive correlation between economic yield and HI, is not by itself unequivocal evidence of the value of the HI, as a selection criterion (Charles-Edwards, 1982). According to them HI integrates phenological, physiological and environmental factors, unless each factor is well defined, HI does not provide much aid in understanding crop performance. Therefore, direct analysis of the main parameters of economic yield is a more realistic means of evaluating crop yield than is the analysis of HI. This argument is consistent with the previous findings that HI values of many crops are sensitive to factors such as crop density, water status and N availability (Donald and Scheer, 1975, White and Wilson, 2006). They showed in cereal grains, that upto a point, as the population density was increased, a greater grain yield per unit area was achieved, but HI declined before maximum yield was attained (Sangoi and Salvodor, 1997). Likewise, under water stress conditions the harvest indices of several grain species (Poostchi *et al.*, 1972), mungbean(Sharma et al 2006) and oilseed crops (Sharma *et al.*, 2010, 2011) dropped. Nitrogen application usually increases biomass production but reduces HI of some crops (Donald and Hamblin, 1976; Li *et al.*, 2011).

Of the two alternatives to further improve crop yield – increasing photosynthetic rates and/or increasing harvest index – the potential for eventual success seems to be much greater if HI is manipulated by modifying carbon partitioning and allocation patterns. Although some description of how assimilates are partitioned in plants is emerging, no single aspect of it is fully understood (Robertson *et al.*, 2001). Crop plants are highly integrated systems consisting of multiple sources and sinks. Presumably, the whole system is so integrated that it maintains a balance of growth. Although, the balance determines the harvestable yield, its mechanism is poorly understood.

Partitioning of the photo assimilates: products of photosynthesis

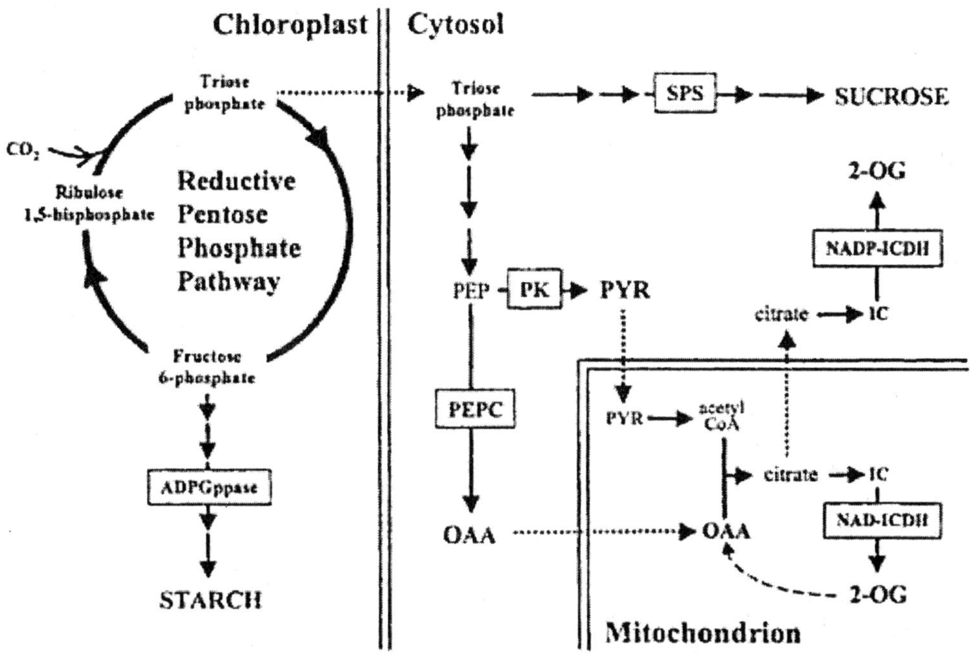

Fig 1 :Carbon partitioning between carbohydrate and respiratory generation of 2-oxoglutrate (2-OG) and other organic acids (OAA,PYR) .The scheme shows the shortest path for net 2-OG generation for glutamate synthesis .Likely key regulatory enzymes are boxed.ADPGppase,ADP-glucose pyrophosphorylase; IC(DH), isocitrate(dehydrogenase); OAA, oxaloacetate; PEP(C), phosphoenolpyruvate (carboxylase); PK, pyruvate kinase, PYR, pyruvate; SPS, sucrose phosphate synthase.

Most of the products of photosynthesis are exported out of the chloroplast to the cytosol as triose phosphate in exchange of inorganic phosphate (Pi). Triose phosphate is the substrate for the synthesis of sucrose in the cytosol and for the formation of cellular components in the source leaf (Fig 1). Sucrose is exported to other parts (sinks) of the plant via the phloem. Some species export oligosaccharides or sugar-alcohols rather than sucrose. Partitioning of the products of the calvin cycle within the cell is largely controlled by the concentration of Pi in the cytosol. If this concentration is high, then the rapid exchange for triose-phosphate allows export of most of the products of the calvin cycle. If the concentration of Pi drop, then the exchange rate will decline and the concentration of triose-phosphate in the chloroplast increases. Inside the chloroplast, the triose-phosphates are used for the synthesis of starch within the chloroplast. The partitioning, therefore of products of photosynthesis between export of the cytosol verse-storage in the chloroplast is largely determined by the availability in the cytosol. In intact plants the rate of photosynthesis is also reduced when the plant demand for carbohydrate is decreased i.e. reduced sink strength e.g. by the removal of part of the fruits or girdling of the petiole. Partitioning within the whole plant involves production of carbohydrates in photosynthetic organs (source), phloem loading and subsequent translocation and unloading at regions of growth or storage (sinks) (Geiger *et al.*, 1989). When the rate of incorporation/and/or export of the products of photosynthesis is low, the rate of photosynthesis may become restricted by feedback inhibition under such conditions phosphorylated intermediates of the pathway leading to sucrose accumulation, inexorably decreasing the cytosolic Pi in the chloroplast, the formation of ATP is reduced and the activity of calvin-cycle declines (Ranjan *et al.*, 2002). Therefore, less intermediates are available and less RUBP is regenerated so that the carboxylating activity of Rubisco and hence the rate of photosynthesis is dropped.

The role of feedback inhibition is important where sink has not been manipulated. For this oxygen sensitivity of photosynthesis must be determined. The rate of CO_2 assimilation normally increases when the CO_2 concentration is lowered than normal i.e. 21% to 1 or 2% because of the suppression of oxygenated reaction.

The Gap

Individual biochemical and physiological processes relating to photosynthesis and growth are well known but the factors primarily responsible for the control of carbon-partitioning in plants needs more understanding. Understanding of factors which control organ initiation and development i.e. source and sink formation and potential size, whether growth is sink or source (including possible sink-controlled photosynthesis) a detailed assessment of the role of storage in buffering developmental and environmental changes in sink and source activity. Indepth knowledge is needed on the role of hormonal and nutritional factors in regulating source and sink activity.

In unicellular green algae, each cell is capable of carbon fixation by photosynthesis in the light, the uptake of nutrients, growth in size and reproduction. However, in land plants, division of labour accompanies these processes between different cell types, tissues and organs, with an additional requirement for a suitable structural framework and an interconnecting transfer and transport system. The leaf chlorophyll is spatially separated from the nutrient absorbing root systems and the reproductive organs, yet each tissue and functional cell is a part of a closely integrated whole plant system. Considerable variation has developed in these systems that allow plants to grow and survive under a wide range of environmental conditions. This diversity of functions and the control of partitioning of dry matter to different plant organs is the platform on which crop yield is based and in many crops, improved yields have been associated not with the increase in total production but efficient partitioning of the available carbon to the organs being harvested (reviewed by Gifford and Evans, 1981).

Photosynthetic rate given under environmental conditions are constant must be initiated by some message/signal that crosses the chloroplast membrane. It appears that an increase in photosynthetic rate as sink demand is increased is not a typical response in short-term studies. However, since, there is need for adjustment within the plant, this response is generally observed in long term experiments. Whether the adjustment is a direct result of sink demand or an incidental effect of associated change in hormone synthesis or availability of crucial nutrients for the photosynthetic apparatus is difficult to establish.

Local transport

Studies of source-sink relations have been largely empirical in nature, change is either the supply or demand for photosynthate in plants. Phloem transport can occur over large distances, but a sink is generally supplied with photosynthate from the nearby source, with the association between specific sources and sinks continuously changing as growth proceeds; new sources develop and the photosynthetic rate of the older leaves declines.

Hierarchy of sinks

In higher plants, there is a hierarchy of sinks. The fruit and seed growth dominate the growth of vegetative tissue, although flowers in contrast to fruits appear to be poor competitors under source limiting conditions. Shoot growth generally dominates that of the roots, but many underground storage organs have the same ability as fruits. This concept is supported by experiments in which the supply of photosynthate or size and number of growing organs were manipulated.

Source -sink relationship in crop productivity

Sink demand and photosynthetic capacity

Alterations in the source sink ratio influences the rate of photosynthesis i.e. partial or complete removal of sinks generally leads to reduction in net photosynthesis. This results in the decline of assimilates in the leaves, which alters photosynthetic rates, however, evidence for this is equivocal.

Photosynthetic rate response to altered sink activity is not necessarily a universal phenomenon. Certain conditions must be present for such feed back to be effective. Herald (1980) described some requirements for demonstrating when sink demand is communicated to the source. To demonstrate a positive correlation between photosynthetic rate and sink demand, three conditions are required.

1. The activity of sink components, other than manipulated sinks, should remain constant.
2. The direct, vascular link between the supplying organ and the receiving sink should be maintained.
3. Potential should exist and must be maintained for enhanced photosynthetic rate.

Two kinds of sinks may be defined for carbon partitioning: primary sinks, which are the cytoplasm of photosynthetic cells and secondary sinks, which refer to distant, new photosynthetic cells such as roots or storage tissues. In this view, the chloroplast is considered to be the source, the ultimate barrier between source and sink is chloroplast envelope and changes in photosynthate.

SUCROSE CLEAVING ENZYMES

One of the crucial function of source leaves in the synthesis of energy-rich molecules for the transport of carbon, whereas heterotrophic sink organs such as developing fruits, seeds, roots and tubers depends on the import and utilization of these compounds. In most plant species, assimilated carbon is transported as sucrose, cleavage of sucrose initiates utilization and in plants, this reaction is catalyzed by two enzymes with entirely different properties, invertase and sucrose synthase. Invertase is a hydrolase which cleaves sucrose into two monosaccharides. In contrast sucrose synthase is a glycosyl transferase, which in the presence of UDP - converts sucrose into UDP-glucose and fructose. Cleavage of sucrose by invertase is generally correlated with growth and cell expansion. Sucrose synthase is located in the cytoplasm (Zeeman *et al.*, 1995). In general, cleavage of sucrose by sucrose synthase is associated with anabolic process in which UDP-glucose is the precursor of numerous compounds. It is also located in the sieve tube – companion cell complex (STCC-complex) of source leaves where it is involved in the catabolism of sucrose to generate ATP for phloem loading.

Sucrose transport from source leaves into sink organs are controlled by sink strength. Experiments to unravel the molecular nature of this force have been identified as one of the key steps. By contrast, sucrose synthase is responsible for feeding assimilated carbon into sink metabolism. Studies on maize (Sugiharto *et al.*, 1990, Nagyen *et al.*, 1990) and faba bean seeds (Heather and Howard, 1992)) and carrot tap roots (Sekova *et al.*, 1995) support this model.

It is evident that a turgor pressure gradient based on sucrose concentration is the driving force for assimilate transport from source leaves into sink organs. The sucrose concentration gradient depends on the rate of sucrose utilization in sink tissues. Thus, cleavage of sucrose by all sink-located sucrose cleaving enzymes might contribute to sink strength. In fababean seeds, sink strength is generated by sucrose synthase activity rather than by cell wall invertase. The conversion of an osmotically active sugar into an insoluble polymer such as starch creates a strong sucrose-sink, making cell wall invertase activity unnecessary. The hypothesis is further supported by developing potato tubers and tomato fruits which have little acid invertase activity but high sucrose synthase levels. A reduction in sucrose synthase activity in potato tubers by anti-sense technology leads to marked reduction in their starch content and to a marked reduction in tuber yield (Zrenner *et al.*, 1995). During tomato fruit development sucrose synthase and not acid invertase is positively correlated with fruit relative growth rate and with starch content of the pericarp tissue.Leaf photosynthesis is normally down-regulated by the sink and the presence of sinks viz fruit prohibits or retards leaf senescence; several sinks can compete for photo-assimilates; after fruit removal, i.e. harvest, the roots become the dominant sink. In apple, fruit removal led to starch accumulation in the leaf chloroplasts, while the soluble sugars remained largely unaffected. In tomato, fruit harvest lead to an instant decline in photosynthesis, carbohydrate surplus, release of vacuolar nitrate into the cytoplasm with nitrate assimilation by cytoplasmic nitrate reductase as a consequence (Blanke, 2009).

In seeds of maize and faba-bean, expression of cell wall invertase is restricted to only a few cells at the transition zone, at the point where sucrose exist the phloem and enters the storage tissue. In faba-bean and cells of *Chenopodium rubrum*, expression of cell wall invertase is coupled to the expression of a plasmamembrane located hexose transporter facilitating the entry of hexoses into the storage tissue (Ehness and Roitsch, 1997). These examples, emphasizes that

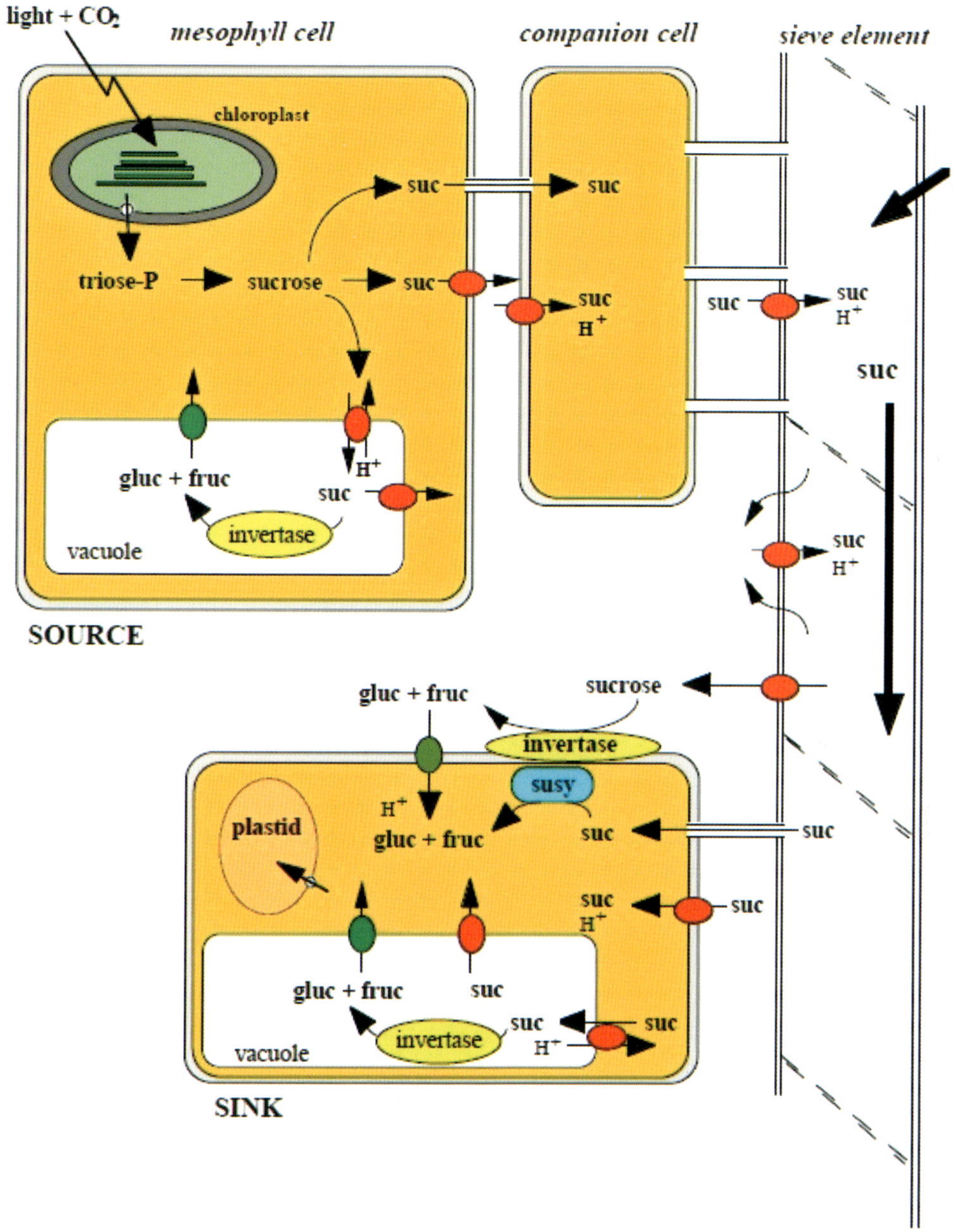

Fig 2: Sugar fluxes in the higher plants. Assimilates derived from photosynthesis are transported via the phloem, mainly in the form of sucrose, to the sink tissues. Loading and unloading of the phloem can occur either symplastically through the plasmodesmata, or apoplasmically mediated by sucrose transporters (red spheres). Sucrose export or leakage and reuptake also occur along the phloem. In sink tissues, invertase and susy activity together with hexose transporters (green spheres) help to establish a sucrose gradient from source to sink. (modified Lalonde *et al.,* 1999)

both cell wall invertase and a sugar transporter appear to contribute to sink strength and also to control assimilate partitioning by sink-located transfer and transport processes (Weber *et al.*, 1997). After hexose import into the storage tissue, sucrose is partially resynthesized. During maturation, starch deposited during early fruit development is reconverted into sucrose and depending on the activity of acid invertase in the vacuole, is stored mainly as sucrose or as a mixture of glucose and fructose. Antisense repression of the fruit vacuolar invertase converts hexose storing fruit into sucrose-storing fruit without affecting the allocation of assimilated carbon to the sink organs (Klann *et al.*, 1996). Post-harvested cold treatment of potato tubers leads to enhanced starch degradation and resynthesis of sucrose. Subsequently, a cold-inducible vacuolar invertase hydrolyzes the disaccharide to glucose and fructose. Antisense repression of this cold inducible invertase in transgenic tubers does not change the total quantity of soluble sugars that are released in response to low temperature. However, it does alter the ratio of hexose to sucrose in favour of disaccharide. Confirming the role of vacuolar invertase in the control of sugar composition. Role of these hydrolytic enzymes in the elongation of cotton fibre development has been vividly illustrated by Sharma and Malik (2009).

Source – sink regulation by sugar and stress

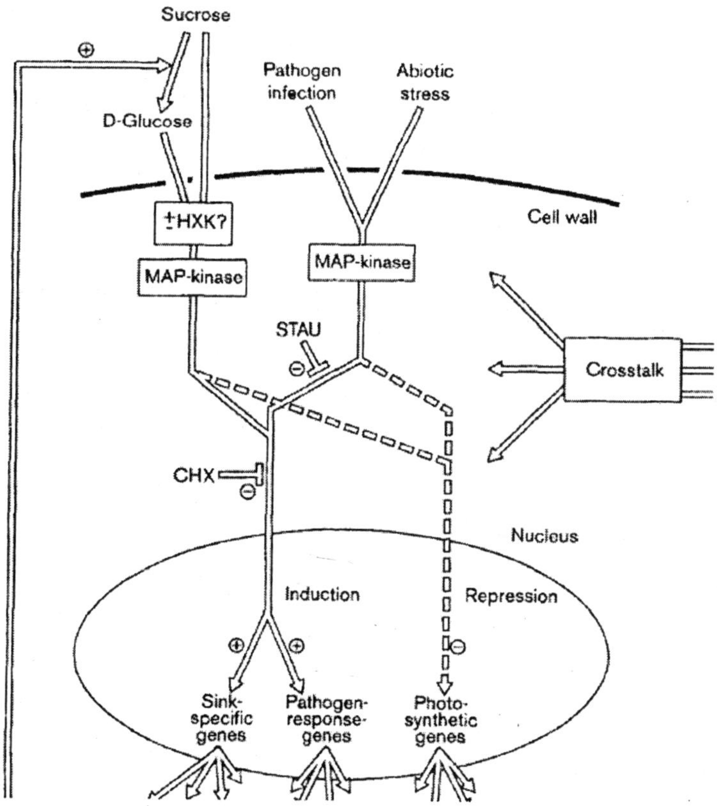

Fig 3 Model for the regulation of sink metabolism, photosynthesis and defence responses by sugars and stress related stimuli. (Source Roitsch, 1999)

Sugars and stress related stimuli activate different signal transduction pathways that are ultimately integrated to co-ordinately regulate gene expression.Intracellular signaling may involve hexokinase and MAP kinases and is modulated by other signal transduction pathways. Any signal that unregulated extracellular invertase will be amplified and maintained via the sugar induced expression (Fig 3, Roitsch, 1999). Sucrose is unloaded from the sieve elements of the phloem into the apoplast by sucrose transporters(Suc TP) as depicted in Fig 4, the diasaccharide (Suc) is cleaved by an extra cellular invertase and the hexose monomers (Fru and Glc) are taken up from the sink cell by monosaccharide transporters. Extracellular invertase is regulated by glucose, as well as by phytohormones and stress related stimuli. Sugars are substrate for heterotrophic growth and function as signals for gene regulation.

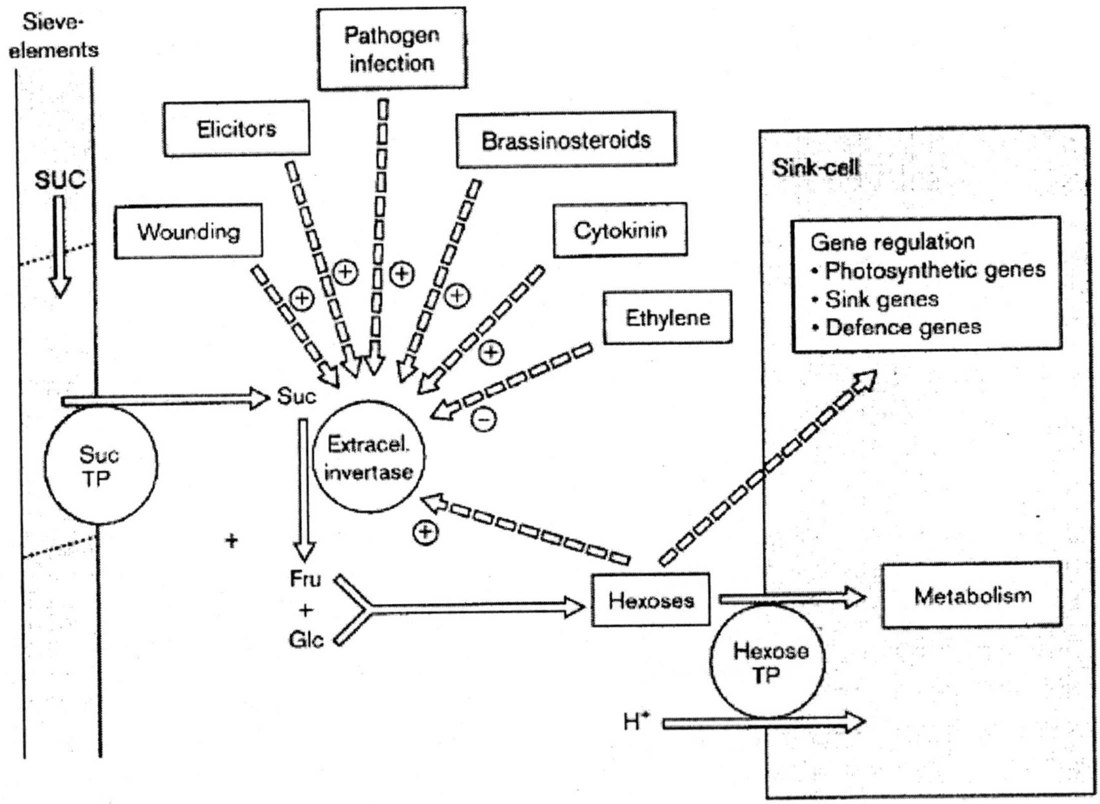

Fig 4 Model for apoplastic phloem unloading in sink tissues and regulation of extracellular invertase (Roitsch, 1999)

IMPROVED CANOPY STRUCTURE

Shoot density and leaf geometry improves the photosynthetic input in a plant without changing the photosynthetic potential of individual leaves. Leaf angle is important for improving plant productivity as shown in cereal crops. Upright leaves in rice and maize isogenic lines improved yields. However, the dry matter production of soybean was a function of interception of solar radiation regardless of the planting pattern. In order to adapt their shape to the constantly changing environment, and in response to competition from nearby plants, individual plants gradually develop their architecture according to source-sink activities and endogenous signals. This adaptation can result in complex structures, great efforts have been made from the mid-1990s onwards to develop functional-structural (FS) plant models by combining physical and/or physiological processes (photosynthesis, assimilate partitioning, etc.) with explicit description of plant structure (Perttunen *et al.*, 1996; Room *et al.*, 1996; LeRoux *et al.*, 2001; Godin and Sinoquet, 2005; Fourcaud *et al.*, 2008; Hanan and Prusinkiewicz, 2008; Vos *et al.*, 2010).

Elevated CO_2 concentration

Many ecological processes affect photosynthesis through their impact on plant demand for carbohydrate. The processes which increased carbohydrate demand cause an increase in photosynthetic rate and vice-versa. Although defoliation leads to reduction of photosynthetic tissue, causes a compensatory increase in photosynthetic rate of remaining leaves.

Increasing the CO_2 concentration of air is one way to increase net CO_2 exchange (NCE). Accumulation of carbohydrates has been observed in many studies under CO_2- enrichment because photosynthetic rate exceeds the sink capacity to utilize the photosynthate for growth. Therefore, attention has been paid to work out the causal relationship(s) between carbohydrate accumulation and suppression of photosynthesis. Feed back inhibition due to high level of carbohydrate, seems to be the possible mechanism for suppression. Decrease in the amount of key components of photosynthetic apparatus, due to accumulation of carbohydrates directly or indirectly has been suggested another feed back mechanism(s) by Drake *et al.*, (1997).

In many species the increase of starch content by CO_2 enrichment is greater than soluble sugars. Extreme enlargement of starch grains may lead to the physical damage of chloroplast, but accumulation of starch hinders CO_2 diffusion in chloroplast (Jones *et al.*, 1997; Sionit *et al.*, 1984). The extent to which starch and soluble sugars accumulate in response to CO_2 depends on species. For example bean, cotton, soybean and clover preferentially accumulate starch whereas in wheat, sunflower and rice, soluble sugars accumulated greater than the starch. The down-regulation of photosynthesis by CO_2- enrichment observed when photosynthate exceeds the carbohydrate utilization for growth. Some species e.g. potato and radish do not show any down regulation of photosynthesis whereas they genetically belong to starch accumulating species. The biomass of the storage roots is markedly enhanced under CO_2-enrichment and consequently no-over accumulation of carbohydrates is observed in leaves (Tukiello *et al.*, 1999; Reddy *et al.*, 1999). Similarly in rice, the leaf sheaths act as a temporary sink for carbohydrate and the absolute amounts of carbohydrates in the leaf blades are considerably small. So, the down regulation of photosynthesis in rice may be small as compared with that in bean, cotton or soybean. These species accumulate a great deal of carbohydrates in the leaves especially as the starch in chloroplasts.

Decrease in leaf N content by CO_2-enrichment is commonly found in many species. This decline could be attributed to change in N allocation at the morphogenetic level of the whole plant. During long-term growth under elevated CO_2, plants reallocate N away from leaf blades are considerably small. Therefore, the down regulation of photosynthesis in rice may be small compared with that in bean, cotton or soybean.

Long term response to CO_2-enrichment is related to differences in sink-source status of the whole plant depending on the developmental stages CO_2 enrichment leads to a greater stimulation of biomass production in young seedlings than in mature plants such as soybean, cotton, alfalfa, tobacco and rice. It promotes the development of tillers but lowers the carbohydrates levels in tobacco (Geiger *et al.*, 1998, Poorter and Nagel, 2000). At seedling stage growth is source-limited and the carbohydrates increased by CO_2-enrichment can be efficiently utilized for additional sink for the development of new tillers/secondary shoots.

Elevated CO_2 levels accelerate development of whole plant like radish and tobacco, however, no effect was recorded in rice (Allen *et al.*, 1995). Down regulation of photosynthesis by CO_2-enrichment in tobacco appears due to shift in the timing of natural ontogenetic decline of photosynthesis to an earlier onset associated with leaf senescence. An acceleration of leaf senescence has been observed in other species like wheat, sunflower and rice. Leaf senescence is accelerated by CO_2-enrichment in rice and wheat, however, no shift in the timing of photosynthetic stage of leaf ontogeny. Thus, different responses of photosynthesis to CO_2 observed at the level of a single leaf may result from different growth strategies at the plant level (Bunce, 1998; Portis, 1995).

RESPONSE TO ABIOTIC STRESSES

C-partitioning is sensitive to environmental changes that may differentially influence the relation between growing organs and between growth and photosynthesis.

N-nutrition results in a greater partitioning of dry matter to the roots of soybean (Nakamura *et al.*, 1997), thus, changing the recognized pattern away from shoot dominance. Wheat grown under low phosphorus conditions the grain is less dependent on the supply of photosynthate from the flag leaf (Fageria and Baliger, 2005). Low nutrition often results in the accumulation of reserve carbohydrates in the stems and leaves, which indicates that growth is retarded more than photosynthesis. However, a further complication may be carbohydrate accumulation under low nutrient levels may in turn result in reduced photosynthetic rates (Shinano *et al.*, 1994).

Low temperature

Low temperature may result in a change in the partitioning of dry matter between growing organs and also an accumulation of starch and soluble carbohydrates (Farrar, 1988) e.g. in Peanut, reduction in leaf photosynthesis with the fall in temperature (20 to 30°C) is related to a rise in storage of photosynthate resulting from slow growth and a low demand for photosynthate at lower temperature, rather than a direct effect of temperature on leaf function.

Accumulation of sucrose is a common response to low temperature in many species. For example, cold induced sweetening of potato tubers is a well known phenomenon of major economic importance. Higher sucrose concentrations lower the freezing point of the cells content, helping to prevent ice-crystal formation at moderate sub-zero temperature. In addition to a cryoprotective function. Sucrose reserves allow the cell metabolism to recover quickly when more favorable conditions return.

Some C_3 but a few C_4 plants develop freezing tolerance after previous exposure to low but not freezing temperatures. This accumulation distinguishes winter from spring varieties of crop species. The activities of the cytosolic F1, 6 BPase and SPS increased in winter varieties but not spring varieties of wheat, rape and rye, when plants were exposed to chilling temperature. According to these findings, they are part of the acclimation response, allowing continued photosynthesis and growth in the cold SPS activity also increased in potato tubers in the cold and is brought about by an increase in the activation state of SPS and by the appearance of new form of the enzyme with a slightly higher molecular mass (Deitging *et al.*, 1998). Protein phosphatase 2A activity was almost completely inhibited when alfa-alfa cells were exposed to low temperature by Monroy *et al.* (1998). This suggests that the cold-induced activation of SPS is unlikely to occur by a type 2A protein phosphatase and, therefore, differs from the activation of SPS by light.

Moisture stress

Under water stress, the level of water-soluble carbohydrates in the leaves of species such as sunflower (Bunce, 1982) cotton (Mauney *et al.*, 1978) and Brassicas (Kaur, 2012) increased. This accumulation in the leaves of plants under moisture stress also probably results from a reduction in growth and a slowing in carbohydrates utilization prior to a reduction in photosynthesis (Munns, 1988). However, the response to drought is dependent on the stage of plant development and in cereals the developing grains continue to grow under severe water stress, with the result there is a relatively greater use of both the limited supply of current photosynthate and stored carbohydrates.

The imposition of water or salt stress led to sucrose accumulation in the leaves of several C_3 plants. A stress-induced change in photosynthetic partitioning in the favour of sucrose was observed in several plants and was correlated with an increase in SPS activity and the rate of sucrose synthesis in *Phaseolus vulgaris* but decreased the rate of starch synthesis even more. So, a change in partitioning in favour of sucrose was also seen in this species. Decreased export or turnover of sucrose, might also contribute to sucrose accumulation. In resurrection plant, *Craterostigma plantagineum* which can withstand extreme desiccation, dehydration induces the transcription of one of the two SPS genes, leading to increased levels of SPS protein and activity (Ingram *et al.*, 1997). In contrast to most C_3 plants, the sucrose content of C_4 plants viz. maize and sugarcane leaves, rises very little/even decreased when the plants were subjected to water stress (Foyer *et al.*, 1998). At the same time, there was an increase in hexoses associated with higher activity of soluble acid invertase, whereas the activities of cytosolic F, 1, 6 BPase in sugarcane and of SPS in both species decreased. The increase in free hexose levels are too low to make a significant contribution to osmotic adjustment within the cell but could be important in signaling as part of the adaptive response to the plants. Productivity of legumes in response to abiotic stresses has been critically reviewed by Sharma and Sardana (2008).

CONTROL AT SOURCE

Types of sources

Leaf photosynthesis is generally the main source of carbon for growth and storage in other parts of the plant. However, other chlorophyll containing organs such as stems, inflorescence parts and fruits also contribute significantly to the overall supply of photosynthate and thus, influence the pattern of carbon partitioning. The formation of new leaves and their longevity is the major factor in the supply of photo assimilates for growth that can influence the source-sink relationships.

Regulation in leaf

Carbon incorporation into the leaf during photosynthesis faces a range of alternative metabolic pathways, transfer through several organe!!e and cell membranes, temporary storage as starch, oligosaccharides, sugar and possible recycling back to the atmospheric following respiration all of which are factors either directly or indirectly linked to the availability of carbon for translocation out of leaf via phloem known as 'phloem loading'. Carbon partitioning between starch and sucrose in source leaves are depicted in Fig 5 and the different phloem unloading routes in Fig 6.

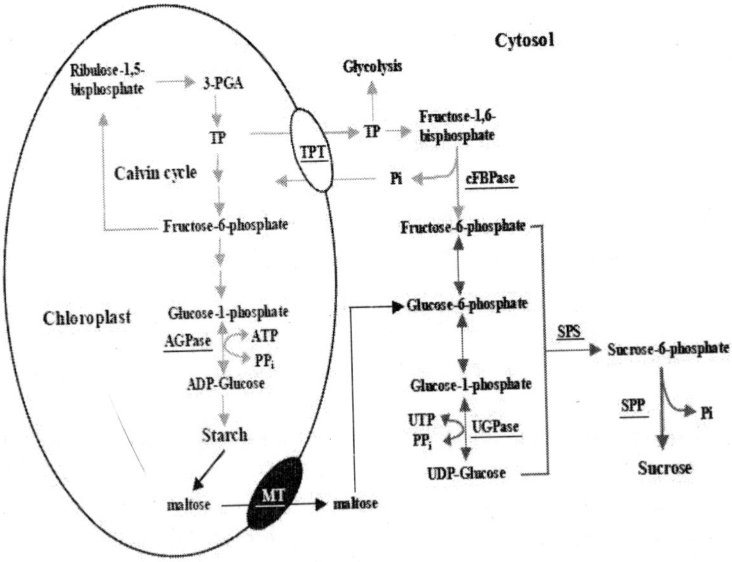

Fig 5 Schematic representation of sucrose and starch synthesis in a photosynthetically active mesophyll cells.3PGA : 3-Phosphoglycerate; AGPase : ADP-Glucose pyrophosphorylase; cFBPase : cytosolic fructose-bisphosphatase; MT : maltose transporter; Pi : Inorganic pyrophosphate; PPi : inorganic pyrophosphate SPP : sucrose phosphate phosphatase; SPS : sucrose phosphate synthase; TP; triose phosphate; TPT : triose – phosphate translocator;UGPase –UDP- glucose pyro phosphorylase . The grey arrows represent the reactions occurrening in light, the black arrows represent the reactions in the dark and the red arrows represent the reactions in both light and dark.

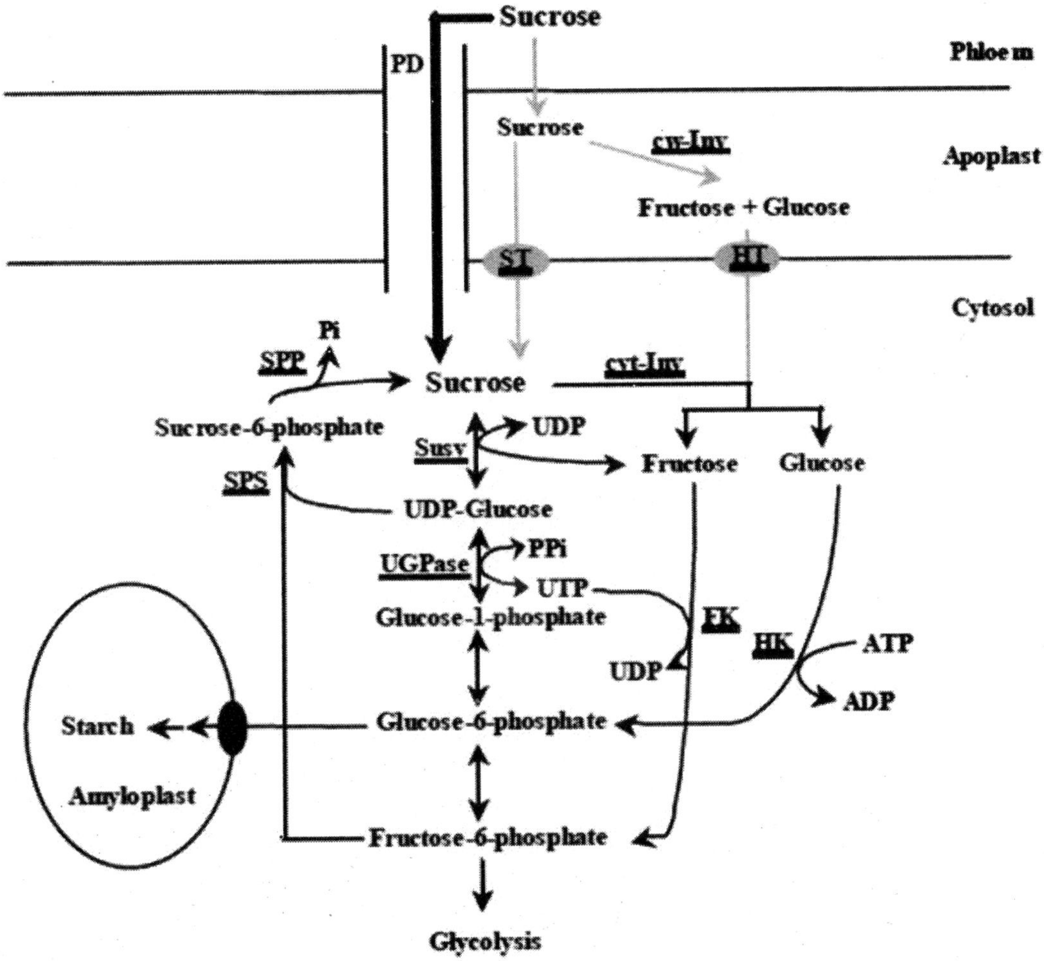

Fig 6 Possible route for sucrose unloading and pathways of sucrose metabolism in sink tissues .cw –Inv; cell wall invertase; cyt-Inv : cytosolic invertase; FK fructokinase; HK : hexokinase; HT : hexose transporter; Pi: inorganic phosphate; PPi : inorganic pyrophosphate; SPP : sucrose phosphate phosphatase; SPS : sucrose phosphate synthase; ST : sucrose transporter; Susy : sucrose synthase; UGPase – glucose pyro phosphorylase . The black arrows represent the major flux of phloem unloading in the sink tissues and the grey arrow represents minor flux

Spatial and biochemical partitioning

C_3-species

In C_3 species, the first step in the carbon fixation in the light is the carboxylation of ribulose 1,5 bisphosphate to form two molecules of P-glycerate. However, ribulose bisphosphate carboxylase/oxygenase (Rubisco) the enzyme responsible for carbon fixation also converts RUBP to P-glycolate in the presence of oxygen, resulting in the photorespiratory pathway and loss of CO_2. Net CO_2 exchange by the leaf is, therefore, largely a balance between the carboxylase and oxygenase i.e. photorespiration activity. P-glycerate in turn forms triose-P which may continue as part of the photosynthetic carbon reduction cycle (PCR) to produce further substrate (RUBP) for CO_2 fixation or diverted into starch in the chloroplast, or transferred from the chloroplast into the cytosol. Under normal conditions a large part of the carbon would be incorporated into sucrose. In many plants, carbon is translocated through the phloem of vascular system from the leaf to other parts of the plants, however, sucrose and other sugars are stored in vacuoles of mesophyll cells. Based on this pattern, it should be evident that there are many processes that may be involved in regulating the movement of carbon from the chloroplast to the vascular tissue. The enzyme sucrose-phosphate synthase (SPS) is also involved in the synthesis of sucrose from uridine 5^1 diphosphate glucose (UDPG) and fructose-6-phosphate (F-6-P). It also regulate F-6-P formation from fructose 1, 6 bisphosphate (F, 1-6P_2) by fructose 2, 6 bisphosphate (F,1-6P_2). The role of latter may be important in the regulation of photosynthesis when the demand for carbon in growth is slow. Slowing the transfer of triose-P out through the chloroplast membrane to the cytosol and hence lowering the reverse exchange of P_i into the chloroplast (Fig7). This favours starch synthesis in the chloroplast.

C_4 species

In C_4 species, where the first products of carbon fixation in the mesophyll are the C_4 acids, the system of carbon transfer is perhaps even more complex than C_3 species (Hatch, 1987). In maize, malate is the main C_4 acid transferred from the mesophyll to the bundle sheath cells where it is decarboxylated and reassimilated in the PCR cycle. The carbon from P-glycerate may continue in the PCR cycle to regenerate RUBP, or be siphoned off into starch. However, a part of the P-glycerate returns to the mesophyll cells where it is converted to triose phosphate which in turn is either converted to sucrose or returned to the bundle sheath cells to re-enter the PCR cycle in the chloroplast. Several processes could operate to regulate photosynthesis and carbon transfer in this system. The photorespiratory pathway is not in C_4 species, but other controls could be similar for both C_3 and C_4 plants. It is well known that C_4 acids control the activity of PEP carboxylase and malate accumulation could be expected to reduce photosynthesis in maize.

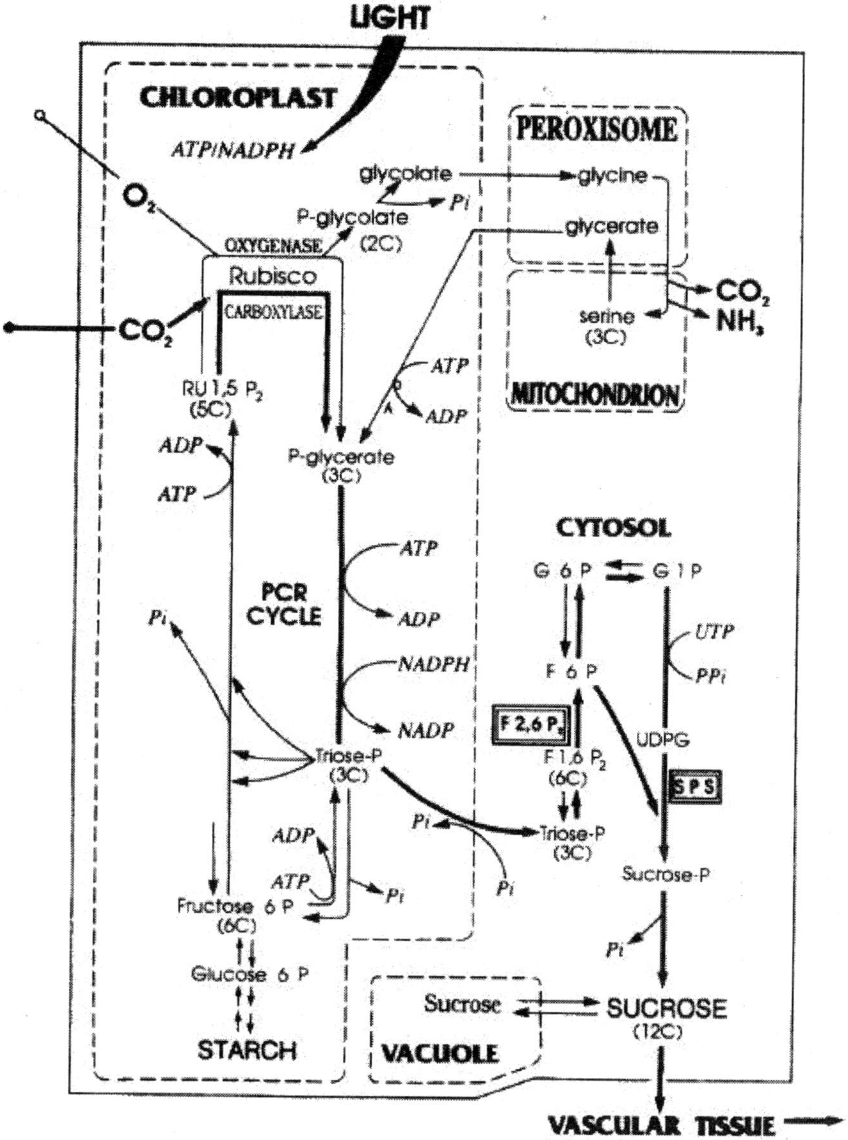

Fig 7 Diagram illustrating the physical and metabolic pathway of carbon transfer from the photosynthetic fixation of COg to the synthesis of sucrose in the mesophyll of C_3 species. Values in parentheses indicate the length of the carbon chains. Rubisco, ribulose bisphosphate carboxylase oxygenase; Rul,5P2, ribulose 1, 5-bisphosphate; PCR cycle, photosynthetic carbon reduction cycle; F1, 6P2, fructose 1, 6-bisphosphate; F6P, fructose 6-phosphate; G6P, glucose 6-phosphate; GIP, glucose 1-phosphate; UDPG, uridine-5'-diphosphate glucose; F2, 6P2, fructose 2, 6-bisphosphate; SPS, sucrose phosphate synthase.

VEIN LOADING AND VASCULAR CONSTRAINTS

Path of transfer

In sugarbeet, amaranthus, vicia and pea (C_3), the movement of sugars from the mesophyll to vascular tissue in the leaves is thought to involve an apoplast step prior to loading into the companion cell-sieve element complex, because of selective transfer of leaf metabolities to the sieve elements; the low number of plasmodesmatal connections between the companion cells or sieve tubes and adjacent parenchyma cells, the high osmotic potential of the sieve tube-companion cell complex in comparison with either the mesophyll or adjacent phloem parenchyma cells i.e. maintenance of a concentration gradient, is energy dependent.

The apoplast step could provide another important control in the export of carbon from the leaf, governed by the rate of uptake of sugars across the companion cell-sieve tube membrane and the area of membrane available for uptake. Concentration gradient between the photosynthetic tissues and the vascular system are smaller, discrimination between metabolities transferred is not so apparent, there are numerous plasmodesmatal connections between the conducting elements and adjacent cells and free space exudates collected from $^{14}CO_2$ fed leaves are low in ^{14}C-labelled photosynthate.

In C_4 species like maize the path of movement of carbon is further complicated by rather complex pattern of transfer of carbon between the mesophyll and bundle sheath cells. The enzyme necessary for sucrose synthesis occur mainly in the mesophyll and not in the bundle sheath cells as a result of the carbon fixed by the mesophyll must be transferred first to the bundle sheath cells for incorporation into 3 PGA and then returned to the mesophyll for reduction to triose phosphate and conversion to sucrose before being transferred back through the bundle sheath cells to the vascular tissue.

The final transfer is not fully understood but presumably must involve movement through the plasmodematal connections between the mesophyll and bundle sheath, because of the suberized wall between the two cell types. These additional transfer steps would appear to negate, at least in part. It claims that C_4 species have an advantage over C_3 species in relation to photosynthate export because of the shorter distance of transfer and fewer cells between the mesophyll and the vascular system.

The role of sucrose synthase (SUS) in sucrose metabolism during Arabidopsis seed development has been illustrated by Fallahi *et al.,* (2008). During the early stages of seed development sucrose enters the seed coat via the vascular bundle of the funiculus. Sucrose is cleaved by either sucrose synthase (SUS) or invertase (Inv). The UDP-glucose produced by SUS activity is used in biosynthesis of starch. The starch is temporarily stored in the seed coat, whereas the resultant hexoses, produced by Inv activity, are used by developing embryo and endosperm (Fig 8A). During later stages of seed development, sucrose is delivered directly to the embryo from the maternal tissues. SUS activity in the embryo uses sucrose to provide precursors for the biosynthesis of storage protein and lipids. In addition, the temporary starch reserves are remobilized from the seed coat to the embryo via the aleurone layer. SUS activity in the aleurone layer may be involved in this remobilization process. (Fig 8 B)

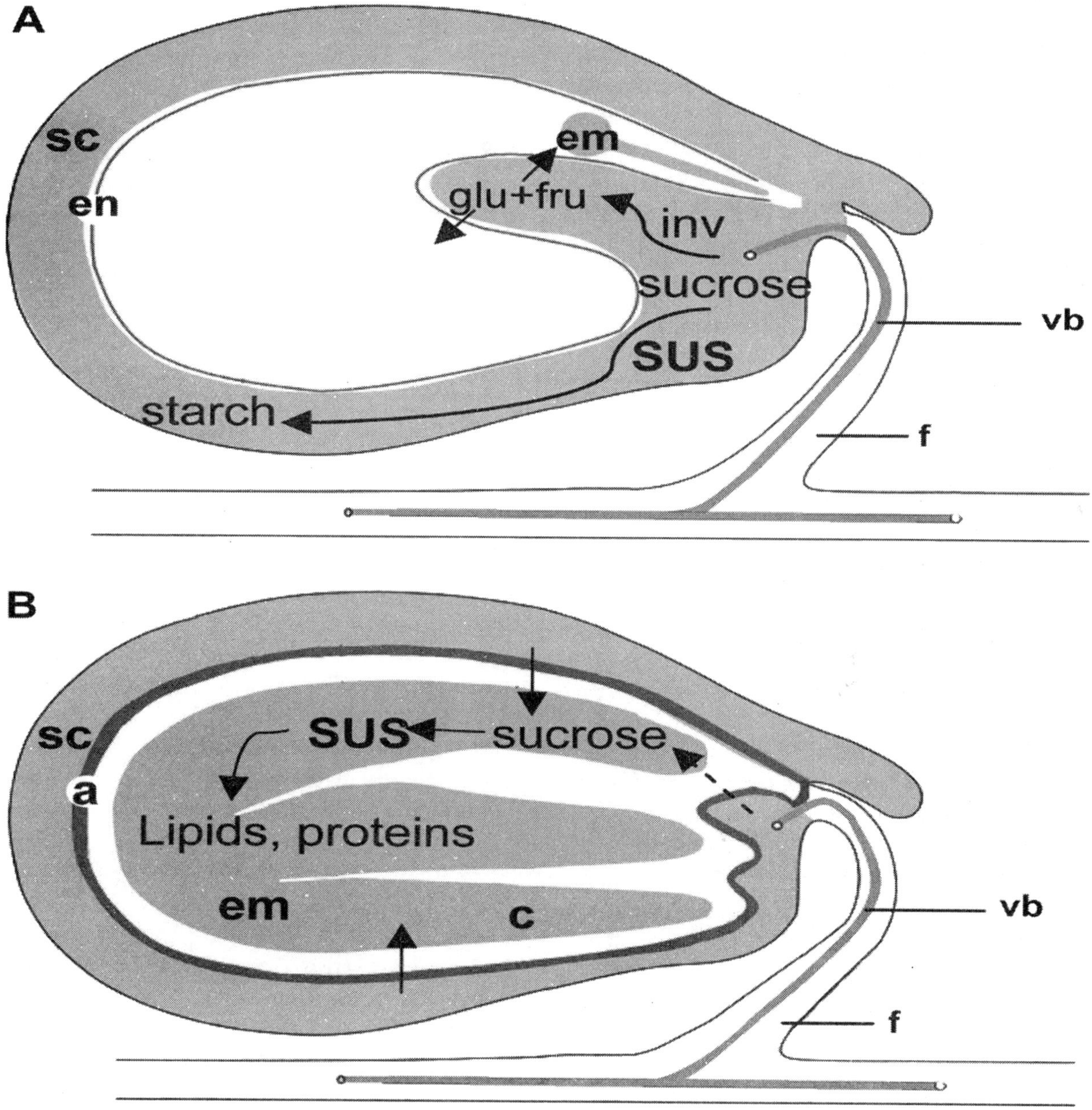

Fig 8 A model illustrates the diverse role of SUS in sucrose metabolism during Arabidopsis seed development. al, aleurone layer; c, cotyledon; em, embryo; en, endosperm; fru, fructose; f, funiculus; glu, glucose; Inv, invertase; SUS, sucrose synthase; sc, seed coat; vb, vascular bundle.(Fallahi *et al.*, 2008)

The main carbohydrate transported in higher plants is sucrose. The phloem provides a means for long distance movement. However, cereal grains have no direct vascular connections with the parent plant and so a short distance transport mechanism operates to move sucrose from vascular tissues to the endosperm. Wheat grains possess a furrow running along the length of the kernel with a vascular bundle embedded at the bottom. Nutrient unloading occurs along the length of the bundle, and has to pass through three distinct layers before reaching the inside of the grain. Solutes are unloaded from the phloem symplastically through the chalaza and then into the apoplast via the specialized transfer cells in the nucellus, before it reaches the outer layer of the endosperm where transfer cells in the aleurone layer redirect solutes back into the symplasm and into endosperm cells (Thorne,1985). The pathway of starch synthesis in non photosynthetic storage tissue involves the conversion of sucrose into ADPglucose (ADPG), and the subsequent conversion of this soluble precursor into insoluble polyglucan (Fig 9). The enzymes involved in the synthesis of starch from ADPG are the starch synthases, branching enzymes and debranching enzymes, and are located exclusively within plastids. (Smith, 1999; 2004, 2005)

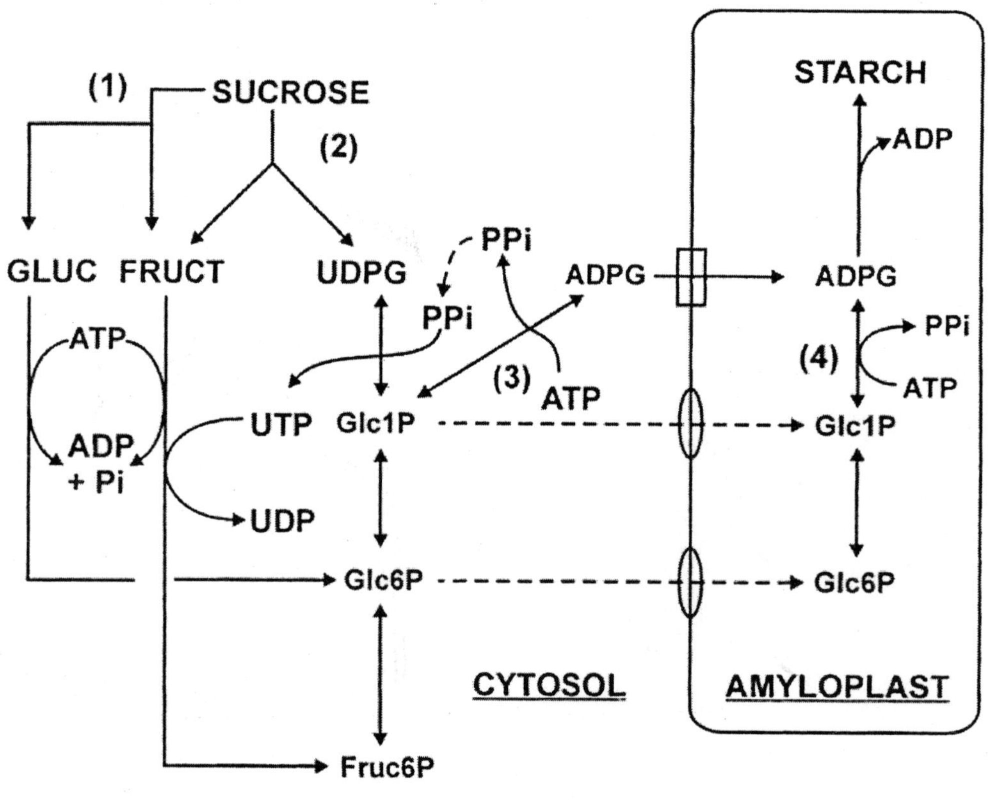

Fig 9 Interconversion of sucrose to starch in developing endosperm. Enzymes shown are (1) invertase, (2) sucrose synthase, (3) cytosolic AGPase, and (4) plastidic AGPase.

Uptake by transport tissue

The minor veins of sugarbeet are 13 times as extensive as the major veins and a 10 µm length of minor vein services approximately 8 to 9 mesophyll cells; the minor veins are considered to play a major part in the initial uptake of sugars from the mesophyll. The sieve elements of these veins are very narrow in both sugarbeet and sunflower. There is evidence that the initial site of uptake and short distance transfer to a better developed vascular pathway is by the relatively large intermediately cells.

In gramineae, the smallest veins in the grasses are the transverse veins cross-linking the larger parallel veins running the length of the leaf. These cross-veins could act as a direct loading point for photosynthate from the mesophyll, a transfer that in rice and wheat may involve an apoplast step. Therefore, division of labour exists with the smaller veins acting as a collecting system for photosynthate, which is then transferred to the larger veins for export.

VENATION

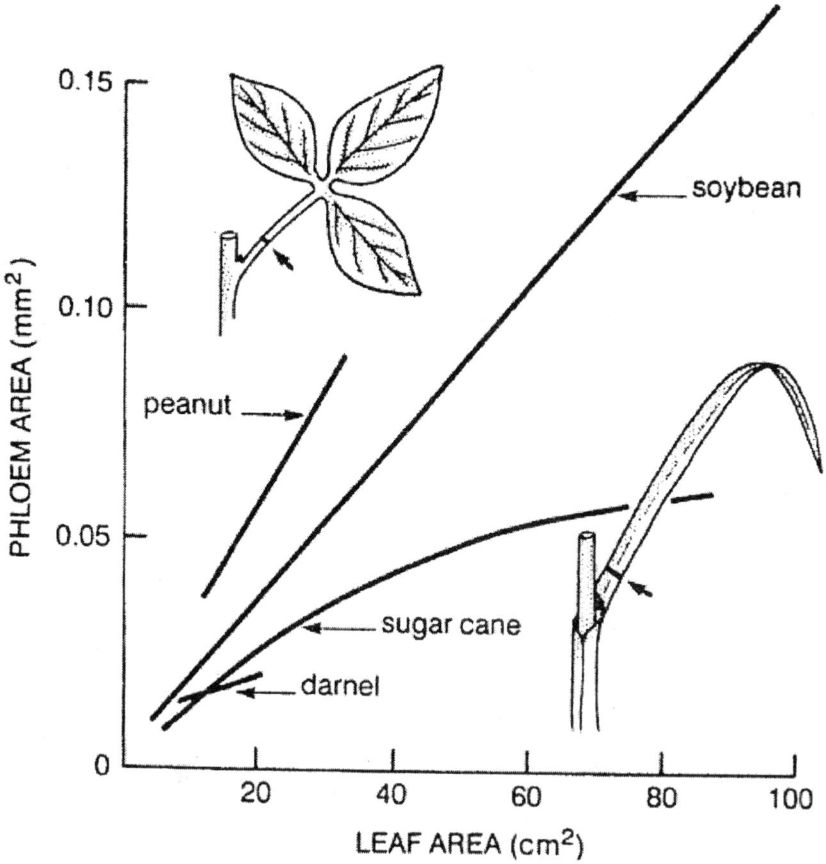

Fig 10 The relation between leaf area and the cross-sectional area of phloem in the petiole, or at the base of the leaf blade (see arrows) for a range of species (Source: Wardlaw,1990)

Both the vein patterns and the dimensions of the vascular system (sieve tube cross sectional area-leaf area) as demonstrated by Wardlaw1990 (Fig 10) may be important in providing a transport system capable of matching the potential photosynthetic rates of a leaf. There is considerable variation in the interveinal distance of leaves between grass species ranging from around 90 µm in the C_4 species maize and sorghum to more than 300 µm in some C_3 species. C_4 species have closely spaced veins, a high rate of assimilate transfer and a high rate of net CO_2 exchange.

Responses to carbon transport

Carbon inputs

Carbon assimilation and export by leaves have been analyzed in detail, resulting in models that include spatially and metabolically separated carbon pools both in series (direct export path) and in parallel (storage), which vary in size and exchange capacity(Rocher and Prioul, 1987). Systems are so complexed that it is often impossible to establish precise relations between individual factors and the rate of carbon transfer into the vasculature, although this has been attempted in relation to the rates of net CO_2 exchange and amounts of sugar and starch in the leaf.

Suc Membrane Transport in Different Parts of the Plant. In mesophyll cells, vacuolar influx/efflux is driven by concentration gradients characteristic of facilitated diffusion, and the presence of Suc/H+ symporters implies energized efflux. Efflux across the plasma membrane occurs by an unknown mechanism and retrieval by SUTs may be a general phenomenon; efflux may be greater in the vicinity of the sieve-element/companion-cell complex. Uptake of Suc from the apoplasm into the phloem (i.e. phloem loading) occurs in the collection phloem and is catalyzed by SUTs.(Fig 11: Top) Suc is released from the transport phloem to enter transient storage reserves and nurture flanking tissues. Suc released from transient storage reserves are most likely retrieved into the phloem by the same SUTs involved in phloem loading (Fig 11: Middle). Suc and other nutrients are distributed from the release phloem symplasmically through plasmodesmata or across membranes to the apoplasm. Suc in the apoplasm is recovered into recipient cells directly by SUTs or is hydrolyzed by cell wall invertases and recovered by monosaccharide transporters. Suc accumulation in vacuoles appears to be catalyzed by Suc/H+ antiport and Suc/H+ symporters on the tonoplast imply energized efflux from vacuoles. (Fig 11)

Integration of Carbohydrate Demand in Sink Organ with Photosynthesis in Source Leaves. White Path: (Fig 12) Suc starvation in developing cotyledons activates expression of SUTs that remove Suc from the seed apoplasm to increase apoplasmic solute potential (Ψp). Higher Ψp in the apoplasm promotes osmosis into seed-coat cells to increase hydrostatic pressure and activate nutrient release from the post-phloem symplasmic domain of the seed coat. (Fig 12: Top) Reduced Suc in the phloem relieves inhibition on SUT expression in source-leaf phloem to increase activity. Phloem loading increases apoplasmic Ψp, and may increase nutrient release from the mesophyll symplasm (possibly from phloem parenchyma cells) in a fashion analogous to release from the seed coats; may operate through a Suc-specific signaling pathway; or release may be controlled solely by the enhanced Suc-concentration gradient between the phloem apoplasm and the mesophyll symplasm. Regulation at this step is speculative, as indicated by (Ψp (?)). Efficient Suc transport out of the mesophyll stimulates photosynthesis. Grey Path: (Fig 12: Bottom) High levels of Suc in the developing seeds, indicating low sink demand, repress SUT expression. Apoplasmic Ψp is reduced because nutrients are not absorbed and osmosis to the apoplasm relieves hydrostatic pressure in the cells of the seed coats, leading to reduced nutrient release. (Fig 10:Top) Increased Suc in the phloem represses SUT expression and reduces the rate of phloem loading. Apoplasmic Suc increases and, through an unknown mechanism (Ψp (?)), reduces Suc efflux from the mesophyll symplasm to promote feedback inhibition on photosynthesis.

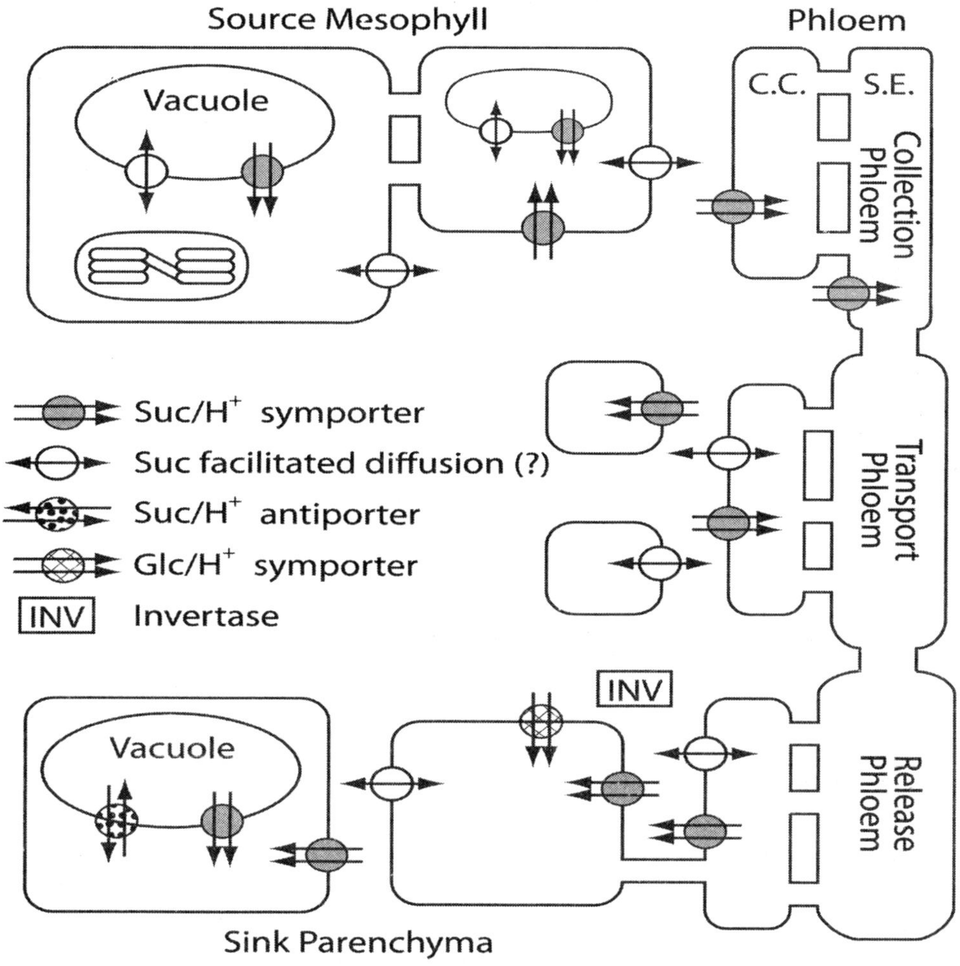

Fig 11 Diagram of Suc Membrane Transport in Different Parts of the Plant. Top: In mesophyll cells, vacuolar influx/efflux is driven by concentration gradients characteristic of facilitated diffusion, and the presence of Suc/H+ symporters implies energized efflux. (Ayre, 2011)

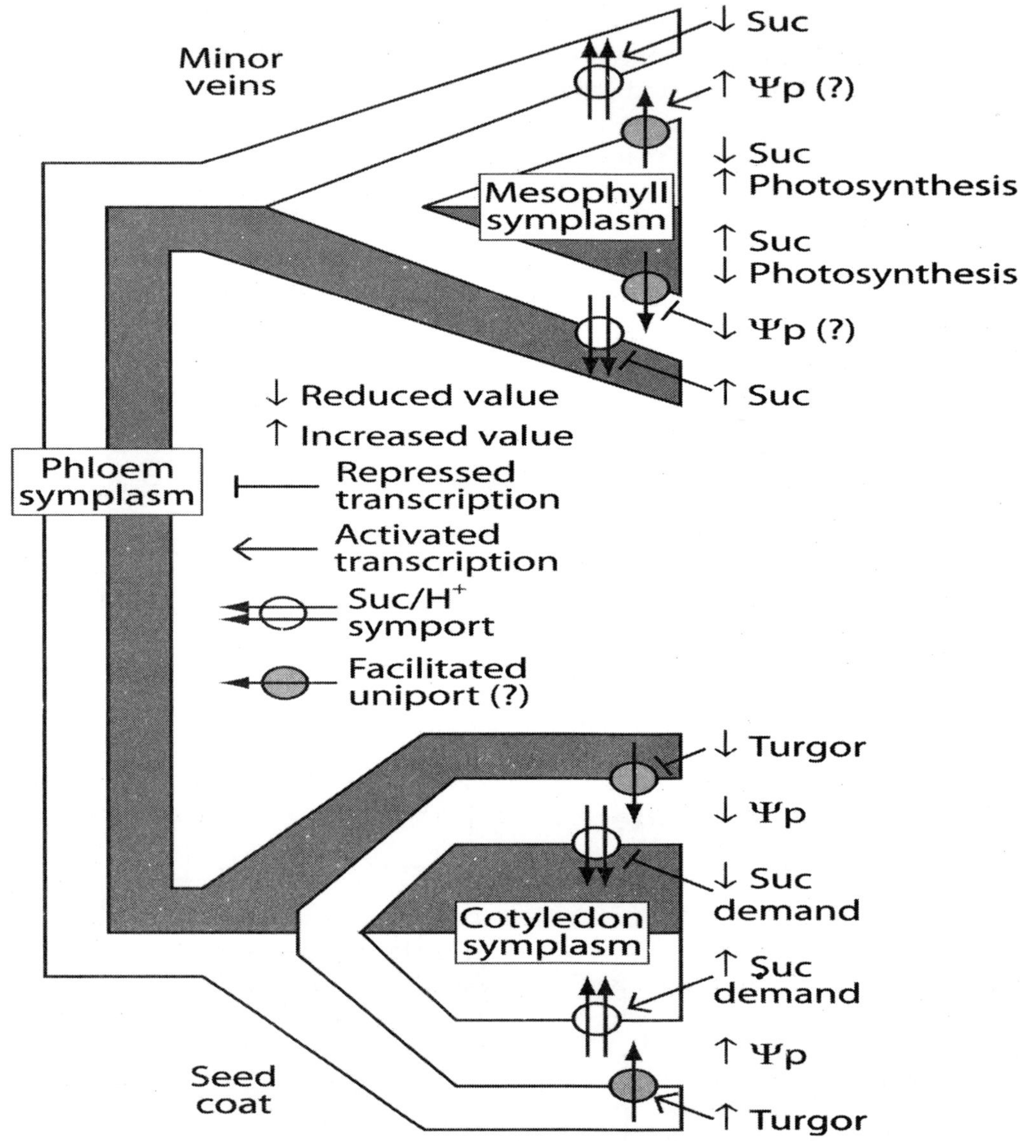

Fig 12 Integration of Carbohydrate Demand in Sink Organs with Photosynthesis in Source Leaves.White Path: (Bottom) Suc starvation in developing cotyledons activates expression of SUTs that remove Suc from the seed apoplasm to increase apoplasmic solute potential (Ψp) (Source: Ayre B G 2011)

Photosynthetic rate

The export of carbon from the leaves has been positively correlated with the rate of day time photosynthesis, but seldom with the amount of sucrose in the leaf; even though the latter is the main form of carbon exported. There is a strong positive relation between photosynthesis and carbon export in tomato (Ho, 1978) above a threshold rate of 2 mg C $dm^{-2}h^{-1}$. Similar correlation between photosynthesis and the export of carbon have been reported for a range of species including cotton (Hendrix and Huber, 1986), sugarbeet (Servaites *et al.*, 1989), sorghum (Wardlaw, 1989), etc. However, the correlations within a species do not necessarily hold between species where there can be big differences in the proportion of carbon that is stored in the leaf or translocated.

In leaves, light immediately follows darkness and there may be an initial period during which translocation lags behind the rapid increase in photosynthesis (Eschrich and Burchardt, 1987). During the day, there can be changes in storage and translocation within the leaf without changes in net CO_2 exchange (Terry and Mortima, 1972). The failure to correlate sucrose concentrations in photosynthesizing leaves with the rate of carbon export is not difficult to understand if the pool of sucrose associated directly with transport is small, mobile and continuously balanced by storage in the vacuole. So, any correlation between carbon export and sucrose in a leaf during photosynthesis might be expected only at a very early stage of carbon fixation following a dark period or under low light conditions. Indirect evidences support the importance of sucrose synthesis in transport with a correlation between sucrose export in the light and the sucrose phosphate synthase activity in the leaves of a number of species including soybean (Rogers and Israel, 1984) and cotton (Hendrix and Huber, 1986).

Translocation of carbon from a leaf is frequently determined from the loss of ^{14}C photosynthate following a pulse labeling with $^{14}CO_2$, under specific conditions and with caution. The rate of loss of ^{14}C from the leaf depends on pool size and respiratory losses as well as the rate of carbon export through the phloem in Lolium (C_3) and Sorghum ($C\text{-}_4$) species. Leaf photosynthesis is normally down-regulated by the sink and the presence of sinks viz fruit prohibits or retards leaf senescence; several sinks can compete for photo-assimilates; after fruit removal, i.e. harvest, the roots become the dominant sink. In apple, fruit removal led to starch accumulation in the leaf chloroplasts, while the soluble sugars remained largely unaffected. In tomato, fruit harvest lead to an instant decline in photosynthesis, carbohydrate surplus, release of vacuolar nitrate into the cytoplasm with nitrate assimilation by cytoplasmic nitrate reductase as a consequence. (Blanke, 2009)

Mobilization of Stored carbon

When the light intensity falls at the end of the day, photosynthesis declines, there is a point of time when the supply of current photosynthate is inadequate to maintain translocation, and carbon reserves built up during the day may be mobilized and exported. It is well postulated that a base rate of export of carbon in the leaf, which if not maintained by current photosynthesis will require the mobilization of leaf reserves. If starch reserves are still available in the morning as light gradually increases, starch can continue to be mobilized until photosynthesis is adequate to maintain translocation with sinusoidal change in light during the day, net carbon export can, therefore, remain steady despite changes in photosynthesis. There are differing opinions on the amount of turnover of starch in illuminated leaves. Simultaneous synthesis and degradation of starch in the leaves of tobacco and isolated chloroplasts from spinach has been reported. However, in pea and pepper variance with this view has been observed.

When plants are transferred abruptly from light to dark it was shown for a range of species that there was an initial export of sucrose prior to the mobilization of starch. In spinach leaves, with a gradual decrease in light towards the end of the day (Servaites *et al.*, 1989). There was an increase in export of accumulated sucrose prior to the mobilization of starch. The rate of export of carbon during the night is much lower than during the day (Kalt-Torres, 1987)and may(Geriger and Bate, 1967) or may not (Muller and Koller, 1988) be related to sucrose concentration in the leaf. The maximum export of carbon during the night can be related to the rate of starch mobilization and sucrose synthesis in a leaf in soybean (Mullen and Koller, 1988). A balance between carbon metabolism and transport has been observed by all the workers.

Day length

Same continuity in the supply of carbon from the leaf in growing organs over a diurnal cycle is maintained by the build up and mobilization of reserves. However, this may be more than just a simple response to a changing photosynthetic input, as it has been shown that starch accumulation in the leaf of a range of species can be increased upto five fold when they are shifted from long to short days (Baysdofer and Robinson, 1985, Greiger, 1985). The mobilization of these reserves during darkness may be slower under short days (Hewitt *et al.*, 1985). Soybean is one of the species in which starch storage in the leaf is inversely related to day length (Chatterson and Silvius, 1979) but this response is less evident during seed filling when the demand for photosynthate is high (Carlson and Brun, 1984).

The duration of high light (photosynthetic) period rather than the photoperiod per se, regulates the rate of starch synthesis in soybean leaf. Under long days, in comparison to short days, the lower rate of starch accumulation in the leaf was found to be associated with greater sucrose-phosphate synthase activity, higher leaf sucrose and higher rate of photosynthate export (Huber *et al.*, 1984b). The endogenous rhythms differ between sugars (approx 26 hrs) and starch (approx 23 hrs) which may also be a factor in the photoperiodic regulation of starch and sugar metabolism in leaves.

Demand

Individual leaves do not necessarily respond to a change in the demand for photosynthetic inputs as storage outside the leaf and the development of alternative sinks may moderate the response. However, there are many reports where a change in assimilate demand has resulted in either a change in the partitioning of carbon between storage and translocation in the leaf or a change in the rate of photosynthesis or both. Artificially induced changes in the demand for photosynthate can clearly influence the rate of leaf photosynthesis (Ferrar and

Osmond, 1989). However, the operation of this control under natural conditions is somewhat doubtful (Geiger, 1976). Barnett and Pearce (1983) found that problem lies in the proper identification of source–sink relationship. The defoliation of maize increased the net CO_2 exchange (NCE) of the residual source leaf in genotypes with a large source-sink ratio, while ear removal decreased NCE only in genotypes with a small source sink ratio. In rice, a high NCE in cultivars with the low source-sink ratio and a low NCE in cultivars with high source-sink ratio and concluded that it should be possible to improve the yield of many rice cultivars by improving the sink (Lafitte and Travis, 1984). These two examples refer to reproductive growth and probably the strongest case of sink limited photosynthesis and provide the relation between high photosynthetic rates and fruiting.

In Potatoes (Nosberger and Humphries, 1965) and tuberous roots (Hozyo and Park, 1971) a similar response is associated with other rapidly growing storage organs. In soybeans, rather anomalous results have been obtained when pod set and the growth rate were increased by an early thinning treatment and were accompanied by an increase in the photosynthetic rate of young leaves and no change in mature leaves (Lauer and Shibles, 1987). However, decreasing the demand by pod removal did not result in a decrease in photosynthesis.

Based on a simulation model for the vegetative growth of tobacco (Hackett, 1973) the photosynthetic rate of leaves was commonly suppressed due to shortage of demand. The interpretation of these source-sink interactions is often clouded by the possibility that the changes in demand may influence photosynthesis not directly but indirectly through factors other than the accumulation of photosynthate possibly hormones(Herold, 1980). Mechanisms are available to explain the end product inhibition of photosynthesis based on the accumulation of sucrose in the mesophyll cytosol (Sitt *et al.*, 1987).

Genetic variation in net CO_2 exchange

Yield is usually limited by photosynthetic rates under field conditions, genetic improvements in photosynthetic potential have not been yet explored and achieved till-date.

The rates of photosynthesis relate directly to the photosynthetic potential of a leaf, or are governed by other factors like genetic variation in photosynthetic rates, dark respiration in lines/related species. These factors are directly concerned with the consequences of the differences in relation to the partitioning of carbon.

In wheat, with increased ploidy and the change from primitive to modern lines there has been a decrease in the rate of photosynthesis per unit leaf area, although there has been a compensation with larger and thinner leaves in modern cultivars. Plant form has changed total biomass has not greatly altered and improved yields are associated with a greater harvest index (Siddique *et al.*, 1989). No relation between peak photosynthetic rate of leaves at different positions (third and flag leaf) and between photosynthetic rate and yield has been deciphered from the study conducted on 6 cultivars and 120 progeny of wheat (Rawson *et al.*, 1983). Therefore, it was concluded that the major determinants of photosynthetic inputs were leaf area and maintenance of photosynthesis. Increased rate of grain growth was observed with CO_2 by Wheeler *et al.*, (1996) which could be attributed partly to a change in partitioning of assimilate the grain. In contrast, the primary effect of warmer temperature was to shorten the duration of grain filling.

In legumes a better relationship has been observed between leaf net CO_2 exchange and yield. The high yielding varieties of soybean tended to have high photosynthetic rates and seed yield and canopy photosynthesis were significantly related (Harrison *et al.*, 1981). However, in 36 cultivars of soybean, where photosynthesis ranged from 12 to 24 mg CO_2 $dm^{-2}h^{-1}$, there was no correlation between net CO_2 exchange after anthesis and total dry matter, pod yield and harvest index (Curtis *et al.*, 1969). Thus, there is correlations evidence for an improved biomass and harvest index in response to increase in photosynthesis (Babu *et al.*, 1985) but experimental evidence are required to confirm these differences which are source and not sink driven. Some difficulties in extrapolating from single leaf and canopy photosynthesis to yield are outlined in review by Gifford (1987). However, despite the practical difficulties there does not appear to be any established reason why yield should not be improved through an increased photosynthetic input if the appropriate partitioning of carbon is maintained.

UTILIZATION OF PHOTOSYNTHATES

Sink characteristics and limitations

Factor which controls carbon partitioning and also ensures balanced development is the timing of organ initiation and growth. Once a potential sink is established for example floral initiation or the formation of axillary bud, the success or competitive ability of that organ is dependent on the development of an adequate vascular link for the C supply and possibly more important the characteristics of the growing organ. The less direct controls e.g. amount of mineral nutrients delivered to the growing tissue through phloem and xylem and often quoted, but not well understood controls imposed by growth regulators.

Number of features within a growing organ that gives a competitive advantage in terms of controlling carbon partitioning are: the size of surface area across which metabolites are transferred from the vascular system to the zone of utilization (unloading area), the efficiency of transfer of carbon from vascular system to the sink (Fig 13), the spatial/biochemical isolation of assimilates in growing/storage organs once they leave the vascular system.

The first attribute would make size a major factor in sink dominance. However, it is suggested that organs with a high relative growth rate (RGR), rather than size per se, will be competitive for a limited supply of photosynthate (Starck and Ubysz, 1974). This suggests that the efficiency of photosynthate transfer and the spatial and/or biochemical isolation of photosynthate in sinks must also be important in sink dominance. Basis of sink strength (assuming a pressure flow mechanism of translocation) is the ability to effectively lower the concentration of photosynthate in the sieve elements serving the sinks and thus, establish a favourable concentration gradient between the sink and the source (Marcelis, 1996).It is not possible to include a complete survey of all the data relating to the features of major and minor sinks in the plants which influences their demand for the utilization of photosynthates. Observations have been restricted to the controls that may operate in expanding vegetative tissues, growth of the fruit and seeds and the critical period of seed set, where there appears to be some anamalies in the source-sink relationship.

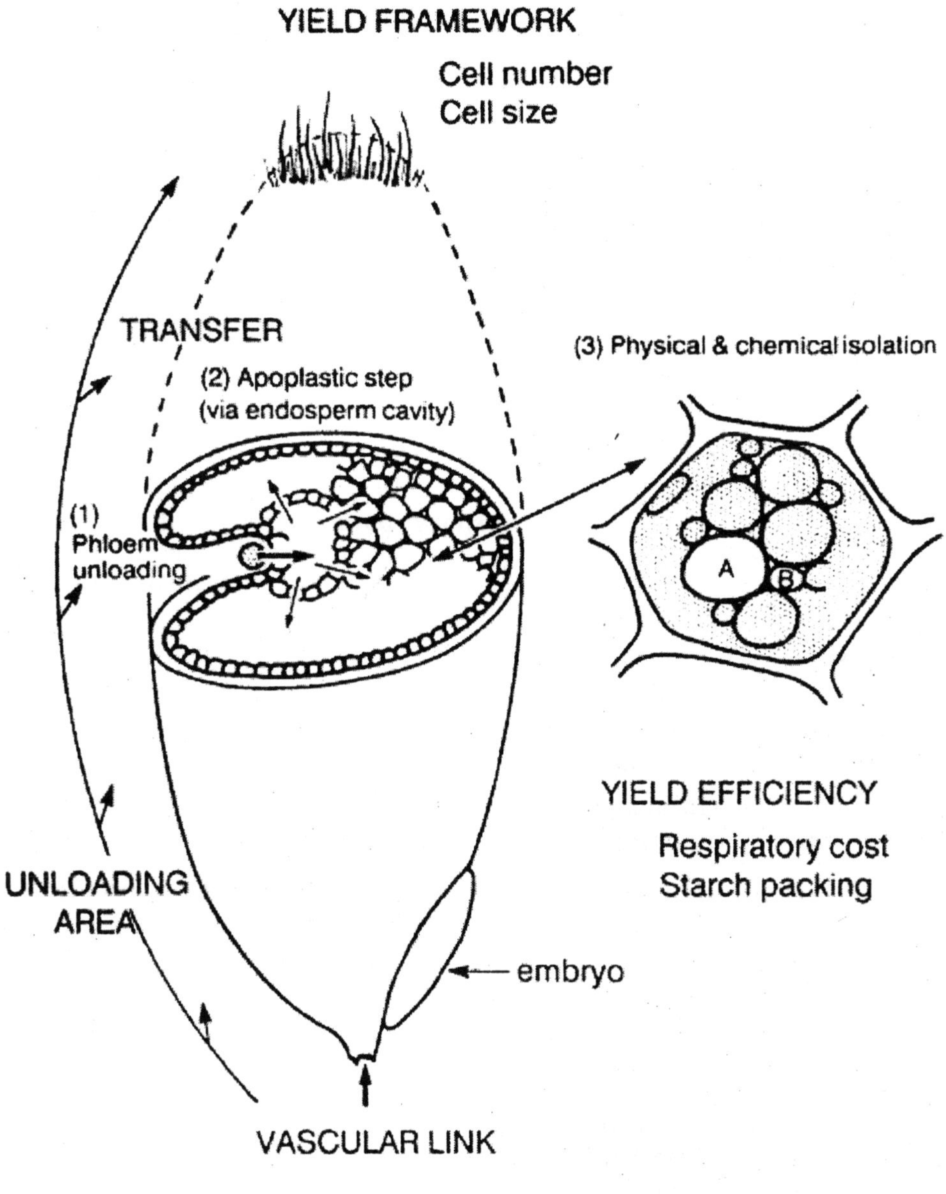

Fig 13 The wheat grain as sink model. Featuring the importance of vascularization; the transfer of sugars from the phloem to the endosperm cavity (apoplasm); the uptake of sugars from the endosperm cavity by the endosperm cells and the incorporation of carbon into starch (physical and chemical isolation); the physical constraints of endosperm cell number, cell size and starch packing; the cost of transfer and synthesis in terms of carbon lost in respiration (Wardlaw, 1990)

VEGETATIVE TISSUES

Transport and transfer limitations

The growth of the vegetative shoot apex is dependent on the movement of photosynthate over considerable distances, presumably through the still undifferentiated procambial strands linked to the mature vascular system below. In young leaf primordial of Coleus (Jacobs and Morrow, 1958), the sieve elements are not differentiated until the leaves are longer than 0.4 mm and it is not until the leaf reaches a length of 1.4 mm that the sieve tubes are closer to the leaf tip than 142 μm.

The young leaf primodial may be limited by the supply of photoassimilates. The first differentiated sieve tubes in the protophloem of developing leaf may be formed in the absence of recognizable companion cells and appear to have plasmodesmatal connections to the surrounding cells, which is a cellular organization that may favour vein loading and the supply of photosynthate to the apex via the symplast (Gunning, 1976).

In dicot leaves, which are capable of their own photosynthesis as they expand, the maximum import of photosynthate occurs at about 15-20% of full size (Ho and Shaw, 1977). However, this will vary with conditions and in the presence of pathogens, which acts as an additional sink for photosynthate, import continues for a much longer period (Thrower and Thrower, 1966). As growth proceeds the terminal part of the leaf matures first and the apparent anomally can arise where photosynthate is being imported into the growing base of the leaf and simultaneously exported from the apex in the leaf of eastern cottonwood (Larson *et al.*, 1972). Such apparent bi-directional movement of photosynthate through the petiole of a developing leaf is likely to result from movement in different bundles in the petiole and not within a single sieve element (Turgeon, 1989). In eastern cottonwood differentiation of the photosynthetic tissue and stomata proceeds basipetally in the leaf lamina, while simultaneous differentiation of major vein network proceeds acropetally. Maturation of the minor veins (generally associated with vein loading) proceeds basipetally and laterally in line with their export role. The mature leaves are not parasitic on the rest of the plant and will not import carbon even under very low light conditions at the base of the canopy, they are induced to became minor sinks under extreme conditions (Fischer and Eschrich, 1987). This transition occurs without any apparent anatomical changes.

In grasses, the stem and leaves growth occurs at an intercalary meristems, the growing tissue is enclosed by leaf sheaths and unlike many dicotyledonous plants growth and the source of photosynthate are spatially separated. Actively growing tissue, therefore, occurs between regions of mature tissue and the question is not only one of photosynthate supply to the region of growth, but also transport through the growing region. Rate of differentiation of sieve elements may be sufficient to maintain phloem continuing across the intercalary growth region of the stem/lamina, the tenuous nature of this connection would be expected to favour retention of assimilates in the tissue (Wardlaw, 1980).

Roots are of special interest, as poor relation exists when it comes to the allocation of limited supply of photo assimilates (Strand, 1997). In millets, the number of both phloem and pericycle cells is proportional to the diameter of the stele at the base of primary root, thus, vascular capacity matches with root size. So, the growth of roots is not limited by long-distance phloem transport, rate of translocation and growth of root system can be increased experimentally if the internal competition between roots is reduced (Passioura and Ashford, 1974). These responses are probably the basis of adaptation to drought, where part of the root system is inhibited by lack of water but remainder will grow to a greater depth in the moist soil (Sharp and Davies, 1985). The root associated e.g. mycorrhiza (Koch and Johnson, 1984), nematodes (Dropkins and Nelson, 1986) and nodulated legume roots capable of N-fixation (Pate and Her ridge, 1978, Caldwell *et al.*, 1984) all stimulate the transport of carbon through the root system, apparently by creating a greater demand for photosynthate. If transport processes involved in limiting the root growth involves transfer from young developing vascular tissue to the actively expanding and meristematic tissue at the root apex rather than long distance transport. The transport of photosynthate in the root from phloem to the pith/cortex behind the apex may follow symplastic path. Sucrose does not hydrolyse during unloading in roots (Giaquinta *et al.*, 1983), although in corn, there is an effective cell wall invertase in the apoplast. Autoradiographic studies reveals an apoplastic step in the movement of carbon from the end of vascular channels in the root to the root apex, a symplastic transfer of photoassimilates from vascular system to expanding apical tissues of both root and shoot appears to be the favoured mechanism.

Metabolic inhibitions

Acid invertase activity is positively correlated with growth rates as evident in sugarcane. This is an important enzyme for sucrose metabolism both for growth and storage (Schaffer *et al.*, 1987). However, it is difficult to confirm, the activity of acid invertase is more than just an expression of growth, and the rate of cell elongation can be correlated with β-glucosidase (Sharma and Malik, 2005).The activity of sucrose synthase may be more pronounced than invertase, as a measure of sink strength in roots of several species and is in parallel to the rate of leaf expansion in maize (Huber *et al.*, 1989).

Turgor and cell wall relaxation

Turgor pressure and cell wall relaxation play important role in controlling growth. Cosgrove (1986) stated that cell growth starts with a reduction in wall stress because of irreversible yielding of the wall. Cell turgor is an important link that coordinates water uptake with cell wall yielding. Turgor threshold is essential for growth and controls cell wall extensibility (Boyer, 1988). An effect on cell wall extensibility may explain in light of the stimulation of stem elongation (Cosgrove *et al.*, 1989) by gibberellins and auxins (Fry, 1989). However, still proper identification is needed for the control and the differences in growth between and within species.

Physical constraints

Given the concept of turgor and cell wall extensibility perhaps more consideration should also be given to the possible role of physical constraints in regulating growth, as evident from the growth of rice kernels(Matsushima, 1970) and expansion of tillers in wheat (Williams and Metcalf, 1975). Dale (1988) had commented on the possible mechanical control of leaf expansion by the epidermis. It is not difficult to implicate a physical limitation on growth as an important control in the pattern of plant development, but in a finely tuned system this concept needs further experimental support.

TRANSITION FROM FLOWER TO FRUIT

In many species, the potential for seed production is very high, but this potential is not realized because of flower drop, failure of fertilization or abortion of developing seeds. Experiments in which light, leaf area or number of developing fruits varied indicate the supply of photosynthates to be critical at these stages. The dry bean yield showed a positive correlation with the rate of photosynthesis at the time of pod set. However, the actual requirement for photosynthate by flowers and small fruits at this critical stage is relatively low.

A threshold level of C is required to stimulate both fertilization and initiation of pod development. Once pod starts growing, it lowers the availability of C to a critical level in parts of the shoot and prevents at least some of the remaining flowers from developing into fruits. This interaction involves vegetative tissue defoliation, removal of apical and axillary shoots at flowering increases the seed number as it curtails the competition for assimilates.

Possibility of less direct effects of photosynthate supply on flower and fruit development is examplified through a change in the movement of nutrients/hormones to the shoot from the roots, the roots being the organs most likely to respond to source limitation or a form of apical dominance where a more advanced fruit could regulate the growth and development of later. This critical stage of development in many species and knowledge of the contents that operate over this period could provide the key to HI and yield.

Fruit growth

The dominance of fruit and seed growth over vegetative organs has been well recognized for a long time (Wardlaw, 1990). Many basic processes involved in the fruits and seeds growth have been identified, but the basis of this dominance still needs clarification. There is a definite contrast between kernels of cereals and the seed of legumes. In kernels of cereals, the growth consists largely of storage in an endosperm where cell development is complete by the time. The kernels are about a third of their maximum dry weight and only the outer layers of cells of the endosperm i.e. aleurone layers are still viable at maturity. In legume seeds, storage largely occurs in the cotyledons of the embryo with cells that remain viable throughout development.

Seed composition also varies between cereals and legumes and the cost of carbon in producing 1 g of seed inherently greater in legumes, which have high protein content than in the cereals where starch is the main form of storage (Penning de vries, 1975).

VASCULAR CONSTRAINTS

Vascular links

The role of vascular system in regulating the carbon partitioning is still not clear, although evidences suggest that the carrying capacity of phloem is not a major factor determining yield (Wardlaw, 1980). The vascular links do appear to govern specific source-sink relationship. Temporary storage of photosynthate along the path of movement acts to moderate the effect of a change in the activity of either a source or a sink.

Retention of photoassimilates and later transfer

During the long distance phloem transport between a source and sink the loss of assimilates from the STCC-complex is small compared with the amount transported (Chirsty and Fisher, 1978), a characteristic feature of vascular system in carbon partitioning. In *Phaseolus vulgaris* (Minchin and Thorpe, 1987) and *Vicia faba* (Aloni *et al.*, 1986), the phloem unloading in the stem operates by passive leakage with the net transfer depending on the balance between membrane permeability and active uptake. In *Riccinus communis* (Malek and Baker, 1977) the sugar uptake by phloem involves a sugar-proton co-transport system. Based on structure and the frequency of plasmodesmatal connections, the transfer of photosynthate from ST-CC complex to other tissues in the stem of *Phaseolus vulgaris* could be either by symplast or apoplast (Hayes *et al.*, 1985) while in sugarcane apoplast is clearly involved (Glasziou and Gayler, 1972). In dicots, the tangential movement of photosynthate can occur around the stem of the sink demand in a longitudinal direction is low. In monocots (Patrick and Wardlaw, 1984), the vascular bundles of the stem running between the nodes are clearly isolated and there are no cross connecting veins there appears to be very little tangential movement. However, in both dicot and monocots, the exchange of carbon between stem and leaf traces entering the stem is largely governed by linkages in the nodal regions.

Proximity of source-sink and vascular connections

Studies between source and sink have led to the conclusion that distance over which the transfer of assimilates has to occur is not a major limiting factor when the leaves and fruits are in close proximity, an appropriate defoliation showed photosynthate could move readily from a leaf to fruit as in grapes (Menhard and Milon, 1963, Peelo and Ho, 1970)). The increasing distance between the source leaves and apple (Hansen, 1977) fruits had no significant effect on fruit growth. A large sinks dominate the overall supply of current photosynthate with the smaller sinks having to rely on local supplies and storage. Although, the proximity of source-sink is an important factor in regulating the carbon partitioning, vascular links and phyllotaxis also have a part to play in this transfer. The movement of photosynthate in sunflower follows the phyllotaxis link between leaves and growing organs (Wardlaw, 1990). This link between leaves and potato tubers becomes less noticeable if the source is restricted to a single leaf (Oparka and Davied, 1985).

Carrying capacity

Evidences exists for translocation where measurements and estimates were made of the specific mass transfer of dry matter via phloem there appeared an upper limit for transport of about 3-5 g cm^{-2} phloem h^{-1} or 15-25 g cm^{-2} sieve tube area h^{-1} (Evans *et al.*, 1970). Suggesting the carrying capacity of phloem may be limiting growth. However, recent data has shown that specific mass transfer can be much greater than the estimates suggested. Confirmation for an excess carrying capacity in phloem of cereals can be visualized in the failure to observe a reduction in grain growth of sorghum (Fischer and Wilson, 1975) or Wheat (Wardlaw and Moncur, 1976).

Xylem: An alternative transport system

Phloem continuity in the peduncle of wheat is destroyed by a heat treatment, there is inhibition in photosynthate's movement from the flag leaf to the ear; although phloem mobile mineral nutrients may bypass this block by transferring to the xylem (Martin, 1982). The distinction in the role of phloem and xylem in transporting sugars and mineral nutrients is less evident in deciduous trees where there is considerable storage of sugar in the xylem (Essiamah, 1980). However, the subsequent movement of these stored sugars is often uncertain (Sauter and Ambrosiua, 1986).

Storage path

Photosynthate storage in the leaf and in tissues along the path of transport between the source and sink provides an important control in relation to carbon partitioning of storage, a passive response to an excess supply of photosynthate, with remobilization occurring when current photosynthesis is low or demand is high. Control of storage is more complex than the suggested model as its regulation, cost in terms of respiration and C turnover, as well as influence on carbon partitioning needs careful consideration. Pathway storage like in the leaf, can have a buffering action on the diurnal fluctuations in the supply of photosynthate available for growth. Little is known about the control of pathway storage, the capacity for storage or the extent to which storage tissues are competitive sinks for carbohydrates. Most detailed analysis of pathway storage in stems is that of sugarcane, where the stored sugar can reach upto 50% of the total dry matter of the stalk (Glaszion and Gayler, 1972). The form of storage in trees can also depend on temperature, e.g. *Populus* (Sauter *et al.*, 1973), where an increasing degradation of starch and accumulation of sugars, in the rate autumn to mid-winter is reversed when temperature fall to zero, the resynthesis of starch begins in spring as temperature rises above 5°C. However in contrast, the ray cells of other genera, *Quercus, Fraxinus* and *Larix*, there is little starch degradation at low temperature.

Remobilization

Carbohydrates stored in tissues adjacent to the translocation pathway are used for growth when remobilized. Maintenance of storage tissue, transfer of metabolites across the membranes, and elaboration of new materials require the use of energy and hence involve a cost to the system in terms of C lost in respiration (Hayashi *et al.*, 1967) calculated the economic ratio i.e. the efficiency of conversion of reserve substances into growing organs, for the formation of a new shoot in potato tubers and germinating rice was 0.4-0.5. It is speculated that less than half of any reserve carbohydrate may be effective in new growth (Satoh and Ohyama, 1976). Extensive literature on utilization of reserves carbohydrates during plant growth which covers regrowth in pastures to seed germination has been reviewed by Wardlaw (1990).

Till date, there is no firm indication that either short distance transfer out of leaves or into growing organs or long distance transport between source and sink are major factors regulating the overall growth although actual patterns are influenced by vascular linkages, which determines the growth of individual organs under source limited conditions.It is debated whether dry matter production is source or sink limited, which vary with genotype, stage of development and environmental conditions. There still exists scope for improving yields, both through increased efficiency of canopy or single leaf photosynthesis and through improved partitioning of dry matter to the harvestable organ.

Source-Sink relationship in annual crops

Physiological aspects

Limiting factors to dry matter production/photosynthetic rates per unit leaf area can be related to source or sink capacities when source capacity is beyond particular sink capacity, sinks control rates of dry matter production. Conversely, when sink capacity is beyond source capacity, dry matter production is controlled by source capacity. Several sinks can compete for assimilate from similar sources. Three fundamental growth phases occur with cereal crops: vegetative, reproductive and grain filling. Major sinks during vegetative growth are leaves, roots and tillers. During the reproductive phase, major sinks include developing panicles, internodes and several leaves at the top of each culm. Grains constitute the major sink during grain filling. Based on source-sink concept of grain yield formation, rice grains are the major sink during plant maturation, while potato has both leaves and tubers as major sinks throughout growth (Tanaka, 1972). Improvement in cereal grain yields may be obtained by improving either sinks or sources during the ripening phase of growth which is most practical depends upon several conditions.

Potential capacities of sinks during grain filling may be expressed in terms of yield components: number of panicles per unit area, number of spikelets per panicle, size of hull, number of filled grains. These yield components are determined at panicle initiation/flowering except number of filled grains. The potential capacities of sources may be expressed by LAI, leaf longevity, leaf extinction co-efficient to light and potential photosynthetic rates of leaves. Of these leaves longevity and potential photosynthetic rates of leaves may be altered considerably by crop management practices after flowering (Tanaka, 1980). Thus, the more active sources during vegetative and reproductive phases, the larger sinks will be during reproductive and grain filling stages. In this manner, source-sink relationship show sequential development with crop growth.

Harvest organs of root crops start growing during early growth stages. After commencement of tuber growth, both tubers and new leaves of upper stems/ branches become sinks for photosynthesizing leaves (source) for long periods. They may compete with each other if source capacity cannot fulfill demands that sinks develop. New leaves may compete with harvested organs as sinks during same growth stages and will become source for harvested organs during later growth stages. For this reason, analysis of yield based on source-sink relationship is more complicated for root crops than for cereals.

Vegetative, reproductive and pod filling growth phases overlap each other for grain legumes. For example, after commencement of flowering in common beans, pods start to grow and both new leaves and flower primodia continue to differentiate on growing stems for fairly long periods. Young pods constitute the major sink during flowering, pods (sink) and leaves (source) grow simultaneously and these organs may compete with each other when source capacity is insufficient to meet demands of sinks. Sink capacities during ripening (number of pods) are decided by competitive conditions during short periods at or after flowering. The nature of source-sink relationships and their effects on yields of grain legumes (Sharma and Sardana, 2009, 2010) are intermediate between cereals and root crops.

Yield manipulation relative to nitrogen

Nitrogen is very important determinant of source-sink capacities of crop plants during most of their growth cycle. N determines both source (leaf canopy) and sink (inflorescence) size. The effects of N on sink sizes during the growth cycles of 45 lowland rice cultivars were grouped by days to maturity. Very short (< 110 days), short (111 to 120 days), medium (121 to 130 days) and long (>130 days) grown with N at 0 and 90 kg N/ha (60 kg basal + 30 kg top dressing at panicle initiation) were determined at IRRI, 1987. Cultivar differences for N absorption ability were observed upto maximum tillering but not thereafter. Concentration of N in plants at flowering for both 0 and 90 kg N/ha rates were highly correlated with growth duration, sink size and yield. Optimum growth duration for sink size and yield was observed at 125 days, highest concentrations of plant N to sinks at flowering, to grow yield were also recorded at same duration. Fate of carbon and nitrogen in developing storage organs has been reviewed by Sharma *et al.* (2011)

Yield decreased when growth duration was shorter than optimum because of small sinks, due to low amount of plant N during late stages of spikelet initiation and did not affect number of spikelets, although N contributed to grain size. On the other hand, when growth duration was longer than optimum yields decreased because of sink shortages due to high number of degenerated/non-viable spikelets. Percentages of degenerated spikelets were correlated with growth duration, inspite of large amounts of plant N at flowering.

Drought was unfavorable to sink sizes. When water stress occurred during flowering, rice with shorter growth cycles yielded more than those with longer growth cycles. Short-growth cycle cultivars had lower potential sink sizes and had less damage from stress than long-growth cycle cultivars (Fageria, 1992). At critical growth stages, the N uptake was correlated with growth duration in dry and wet seasons under similar cultural practices. N uptake at flowering was highly correlated with the amount of sink and yield in both the season. However, yield was governed by sink strength inspite of large differences in percentage of ripened grains during the two seasons. Sink contribution was more to yield in the dry seasons.

N absorption during early growth stages was more important for short duration cultivars than the medium or long duration cultivars. Lower yields of long duration cultivars were attributed to degenerated sinks due to longer vegetative lags. Hence, optimum growth duration can be explained primarily by duration of vegetative lags. Nitrogen was primarily responsible for increase in maize grain yields during the past 50 years (Olson and Sander, 1999). N is essential for C flow and protein synthesis of higher plants (Sugiharto *et al.*, 1990). Bruns and Abel (2003) reported by increasing N concentration in maize plant tissue was positively associated with grain yield. Lu *et al.* (2010) demonstrated that the varieties tolerant to low nitrogen, had longer active grain-filling stage, higher maximum filling rate, longer duration of maximum LAI, and more harmonious sink-source relation, while less tolerant species, had shorter active grain-filling hours, lower maximum filling rate, lower mass increment and LAI under maximum grain-filling rate after silking, and significantly decreased source supply capacity. Low nitrogen stress increased the yield difference among the test varieties significantly. Further to improve maize yield, Xue *et al.* (2010) indicated to that strengthen artificial selection for adapting to natural selection, improve source-sink trait of maize population under stress, enhance leaf photosynthetic efficiency from silking stage to maturity, strengthen source and promote sink, and improve production capacity and adaptability under stresses.

Grain filling rate, grain filling duration, grain weight and grain yield also increased with the increasing levels of nitrogen. Harvest index, however, did not increase with increasing levels of nitrogen (Warraich *et al.*, 2002). Modern wheat cultivars were frequently characterized as sink-limited (Matthew and Foyer, 2001, Acreche and Slafer, 2009) and to attain maximum grain set and yields; these cultivars require an ample supply of N fertilizer (Kichey *et al.*, 2007). Post-anthesis nitrogen supplies could increase grain yield by decreasing the sink limitation (Madani *et al.*, 2010). Urea fertilizer caused much more shoot N accumulation than did symbiotic N2 fixation, but this shoot N enrichment did not result in an increase in seed yield or seed N content. These findings indicate that there is a limit to grain yield and N harvest index in lupins. Although exogenous application of cytokinin can increase sink strength and cause a redistribution of yield components between the main stem and lateral branches, it did not increase the grain yield of the whole plant (Ma *et al.*, 1998)

Modification for Cultivar Improvement

Crop production has been significantly improved by modification of source-sink relationships during past few decades. Major improvements have been made for rice, wheat, maize and potato and to some extent for legume crops. Source strength has improved partially through modification of plant architecture to intercept more solar radiations. In addition sink size increased by increasing HI and grain size. HI of traditional tall cultivars increased about 0.3 and that of semi-dwarf cultivars 0.5. In principle, HI could be increased further (about 0.65) which would improve sink capacity. Recently Xie *et al.* (2011) cited that HI and residue factor (RE)were essentially significant for crop production .In the last 20 years HI of cereals was significantly improved and in near future grain yield will be improved mainly through a substantial increase in biomass instead of HI. Crop yield results from the product of seed weight and number of seeds. These components may vary with genotype and with availability of resources. Achieving maximal yield depends on the plant's capacity to tune resources (offer) to its needs (demand) at each phase of seed development and growth. Regulation of each phase by resource availability results in a trade-off between the different yield components (Peltonen-Sainio and Jauhiainen, 2008; Gambin and Borras, 2010). Characterization of the interaction between architecture and source-sink relationship has been worked out by Jullien *et al.* (2010) in winter oilseed rape (*Brassica napus*), rice (Venkateswarlu and Visperas, 1987) and soybean (Soheil and Keyvan, 2011))

Management strategies for maximizing source-sink relationship

Yield potentials of cereals and legume crops have increased through plant breeding, primarily by improving sink capacity (Gifford *et al.*, 1984). Modern cotton cultivars partition greater proportions of dry matter into fiber and seeds than obsolete older cultivars (Pace *et al.*, 1999). Modern cultivars are early maturing than the older ones.

Source/sink ratios depend on the interaction between genotype and environment, which can be influenced by crop management factors such as planting date, population density, nutrient supply, adequate soil moisture and control of biotic factors such as diseases, insects and weeds. Timing and amount of nitrogen application can particularly improve source-sink relationship. Studies have reported that adequate N-in plant leaves at critical growth stages can improve both source and sink capacities when proper environmental conditions prevail i.e. soil moisture availability, N top dressing during reproductive phases, increase in source and sink capacities usually occur during ripening from increases of spikelet numbers per panicle. So, grain yield can be increased if source capacity during ripening is sufficiently large enough to support increased sink capacity (Tanaka, 1980). In soybean the high change in seed weight may not always be indicated of source limitation (Kobraea and Shami, 2011). However this relationship has been reviewed by Venkateswarlu and Visperas (1987) in crop plants. Breeding crop cultivars possessing optimum source-sink balances for responses to N and other management factors, offers another strategy to maximize yields.Manipulation of source/sink ratios by artificial reduction in grain number per inflorescence has been used to estimate potential kernel weight and study grain – filling processes of several cereals (Bruckner and Frohberg, 1991). Actual kernel weight is usually less than potential kernel weight because of plant competition for light, water and nutrients (Peterson, 1983). Lu *et al.* (2010) demonstrated that the varieties tolerant to low nitrogen, e.g., Xianyu 335 and Zhengdan 958, had longer active grain-filling stage, higher maximum filling rate, longer duration of maximum LAI, and more harmonious sink-source relation; while less tolerant species, e.g., Shaandan 902 and Yuyu 22, had shorter active grain-filling hours, lower maximum filling rate, lower mass increment and LAI under maximum grain-filling rate after silking, and significantly decreased source supply capacity. Low nitrogen stress increased the yield difference among the test varieties significantly. Further to improve maize yield, Xue *et al.* (2010) indicated to that strengthen artificial selection for adapting to natural selection, improve source-sink trait of maize population under stress, enhance leaf photosynthetic efficiency from silking stage to maturity, strengthen source and promote sink, and improve production capacity and adaptability under stresses

HORMONE DIRECTED TRANSLOCATION

Growth regulators have long been implicated in assisting translocation in established source-path sink systems. Most discussion relates to growth regulators released from sinks, hormones from sources have also been considered. When IAA and other regulators like cytokinins, ethylene and gibberellic acid are applied to a cut stem surface or ABA is dissolved in the rooting solution of *Phaseolus* plants (Karmoker *et al.*, 1979), assimilates accumulate in the region of application. Combinations of growth regulators can have additive, synergistic or inhibitory effects. The idea that such IAA directed transport is due solely to stimulation of growth rate of tissue at the site of application argued from the outset and there is evidence for direct effects on transport (Patrick, 1979). However, the systems studied usually involve very low translocation rates and may not reflect the most important controls for intact plants. In bean, seedlings, which were regenerating root and shoot apices following their excision, the main control over the distribution of sucrose between root and shoot sinks was attributed to hormonal influences (IAA and cytokinin) on sink activity per se, although some superimposed influence on relative sucrose availability to the respective sinks could not be excluded. By contrast, the enhanced accumulation of ^{14}C assimilates in pea pods caused by pod warming was due both to an effect on ovule growth directly and to an influence on transport outside the warmed zone (William and William, 1978); this is suggestive of a hormonal effect emanating from the sink. Assimilate transport into developing seeds may be controlled at least partly by phytohormones especially ABA. However, no clear evidence supporting an essential role of ABA in phloem unloading or control of sink strength. ABA may stimulate unloading into the seed coat apoplast but the last few years have seen a consensus that seed is not limited by endogenous concentrations of ABA. Reproductive sink may have a high content of ABA, however, since ABA is intensively translocated from source leaves to sink regions together with assimilates.

Reports highlights that IAA or growth rate in the fungal toxin fusicoccin enhance phloem loading. Treatment with ABA inhibits sucrose loading. In additional benzyladenine (BA) or kinetin also enhances sucrose loading in the stem of *Riccinus*. This is based on experiments where petioles were perfused with cytokinins and exudates from the petioles were sampled for sucrose and K^+. The loading of sucrose into isolated phloem tissue of *celery* occurs at greater rates when the tissue is placed in a medium adjusted to 200 to 300 m osmolarity (with non-penetrating solute, PEG 3350). IAA @ 0.1-100 µM or GA_3 @ 10 µM promote greater rates of uptake in 200 m osmolal medium but not in medium with 50 or 400 m osmolal. Application of GA_3 to excised mature leaves of broad beans enhances sucrose export from source leaves. This promotion of phloem loading by GA_3 occurs after duration of only 10 minutes. The accumulated data indicate that both the synthetic cytokinin, 6-benzlamine purine (BAP) or ABA added to excised seeds stimulate photoassimilate unloading from excised bean seed coats. The cytokinin effect is almost immediate while ABA imparts its effect within 12-minutes of application. IAA, NAA, GA_3 and ACC (1-aminocyclopropane-1-carboxylic acid) are inactive in the unloading system. The action of ABA may be through restricting the ATPase driven proton carrier, thereby allowing the sucrose/proton sympost carrier activity to be enhanced. Recently, it has been reported that cytokinin effect in membrane transport is believed to be unrelated to their hormonal action and more likely it reflects changes in purine metabolism through changes in levels of cytokinins in the cytoplasm.

Although exogenous application of cytokinin can increase sink strength and cause a redistribution of yield components between the main stem and lateral branches, it did not increase the grain yield of the whole plant in lupins (Ma *et al.*, 1998).

Application of ABA to filling grains of wheat and barley has been found to enhance the mobilization of recently fixed photo assimilates to the grain filling. Promotive action of ABA is upon unloading process and similar effect reported in barley grains appears to be inversely related to endogenous ABA content. Two weeks after anthesis i.e. young ears, showed promotory effect of ABA. However, higher concentration of ABA @ 10^{-3}M inhibited assimilates import when applied 3 weeks after anthesis i.e. older ears, when their endogenous ABA content increased five fold. This might indicate that while ABA can promote mobilization in cereals, high concentrations might be inhibitory. Borkovec and Prochazaka (1992) did interaction study between cytokinins and ABA and its efforts on transport of ^{14}C-sucrose into the kernels. Application of ABA @ 10^{-4}M during pre or post anthesis reduce transport of ^{14}C-sucrose into developing kernels. This ABA inhibitory effect is reserved by addition of cytokinins, but only applied at anthesis. Application of cytokinins @ 10^{-6}M to developing wheat kernel increased transport of ^{14}C-sucrose into the grain. However, this increase is observed when ABA was added to pre-anthesis (Ranjan *et al.*, 2004). Crop responses to interaction between growth regulators and nutrients have been vividly discussed by Mir *et al.* (2010)

Phytohormones may be involved in the regulation of sink-pontential by regulating cell division and differentiation of developing sinks. This process may involve not only a particular hormone, but possibly a balance between cytokinins, ABA, IAA and GA_3. The balance between cytokinins and ABA may be critical for establishing the tolerance of maize kernel to heat stress and determines sink potential by regulating cell division and seed set. The balance between cytokinin and IAA in developing endosperm of maize at the time of cell differentiation particularly during DNA endoreduplication may also be critical. At this critical stage of kernel development, there is an increase in IAA and decrease in zeatin and zeatin riboside.

Finally, partitioning may be directly altered by regulating the duration of seed fill. This can occur by either delaying leaf senescence or extending the seed filling period by delaying the onset of seed maturation. Though minimal information is available, it is reasonable to postulate that hormones may play role in regulating partitioning of photosynthate.

CONCLUSION

Improvement in potential crop yield by breeding has largely been by selecting plants for high yield of the sink of economic interest. Clearly, there must be a limit to how far this can go. We know from CO_2 and light enrichment studies that crop yield is frequently, perhaps usually, photosynthetically limited. Yet genetic improvement in growth rate/ photosynthetic rate potential has not occurred so far. Continuation of improvement in the ratio of economic sink to total plant still seems to be the most effective route, until; it is at the expense of light-intercepting leaf surface or robustness of the crop. Improvement in the maximum rate of leaf photosynthesis may then become essential to further increase in yield potential. In the meantime, the key to understand the distribution of photoassimilate to particular organs lies not so much in the leaf mesophyll or the phloem loading system, nor in the translocation system, but more in the determination of the properties of the sinks themselves. It lies in understanding what determines the establishment and premature abortion of sinks, what determines the duration of sink growth, what is about some sinks which enables them to stimulate leaf photosynthesis, how a sink controls unloading of the phloem, and what determines the response of sink growth to the sucrose concentration in its free space.

REFERENCES

Acreche MM, Slafer GA (2009) Grain weight, radiation interception and use efficiency as affected by sink-strength in Mediterranean wheat released from 1940 to 2005. *Field Crops Research* 110: 98–105

Alkio MA, Schubert W, Diepenbrock, Grimn E (2003) Effect of source sink ratio on seed set and filling in sunflower (*Helianthus annus* L). *Plant Cell Envir* 26: 1609-1619

AllenLH, Baker IT, Albrecht SL, Boote KJ, Pan D, Vu JVC (1995) Carbon dioxide and temperature effects on rice. pp.258-277. In S. Peng, K.T. Ingram, H.V. Nene and L.H. Ziska (eds.) climate Change and Rice, New York, Sprinder-Verlag, Int., Las Banas, Phillippines, *Rice Res. Inst.* 4 (2): 300-302

Asthir B, Rai PK, Bains NS, Sohu VS (2012) Genotypic variation for high temperature tolerance in relation to carbon partitioning and grain activity in wheat. *Am. J of Pl Sci* 3: 381-390

Ayre BG (2011) Membrane Transport systems for sucrose in relationto whole plant carbon partitioning. *Molecular Plant* 4(3): 377-394

Barnett KH, Pearce RB (1983) Source-sink ratio alteration and its effect on physiological parameters in maize. *Crop Science* 23: 294 – 299

Baysdorfer C, Robinson JM (1985). Sucrose and starch synthesis in spinach plants grown under long and short photosynthetic periods. *Plant Physiology* 79: 838-842

Blanke MM (2009)Regulating mechanism s in source –sinl relationships in Plants: a review. *Acta Hort* 835: 13-20

Bruckner PL, Frohberg RC (1991) Source-sink manipulation as a post anthesis stress tolerance screening technique in wheat. *Crop Sci.* 32: 326-328

Bruns HA, Abel CA (2003) Nitrogen fertility effects on Bt. 8-endotoxin and nitrogen concentrations of maize during early growth. *Agron J.* 95: 207-211

Bunce JA (1988) The temperature dependence of the stimulation of photosynthesis by elevated carbon dioxide in wheat and barley. *J Exp Bot.* 49: 1555-1561

Caldwell CD, Fensom DS, Bordeleau L, Thompson RG, Drouin R. didsbury R (1984). Translocation of ^{13}N and ^{11}C between nodulated roots and leaves in alfalfa seedlings. *Journal of Experimental Botany* 35: 431-443

Carlson, DR, Brun, WA (1984). Effect of shortened photosynthetic period on ^{14}C-assimilatetranslocation and partitioning in reproductive soybeans. *Plant Physiology* 75: 881-886

Charles-Edwards DA (1982) Physiological determinants of crop growth. Academic Press, New York, p. 106-112

Chatterson NJ, Silvius JE (1979). Photosynthate partitioning into starch in soybean leaves I Effects of photoperiod versus photosynthetic period duration. *Plant Physiology* **64:** 749-753

Christy AL, Fisher DB (1978) Kinetics of ^{14}C-photosynthate translocation in morning glory vines. *Plant Physiology* 61: 283-290

Cosgrove DJ, Sovonick-Dunford SA (1989) Mechanisms of gibberellin-dependent stem elongation in peas. *Plant Physiology* 89: 184-191

Cosgrove PJ (1998) Cell wall loosening by expansions. *Plant Physiol.* 118: 333-339

cultivation history at high latitudes. *Field Crops Research* 108: 101–108

Curtis PE, Ogren WL, Hageman RH (1969) Varietal effects on soybean photosynthesis and photorespiration. *Crop Science* 9: 323-327

Deiting U, Zrenner R, Stitt M (1998) Similar temperature requirement for sugar accumulation and the induction of new forms of sucrose phosphate synthase and amylase in cold stored potato tubers. *Plant Cell and Environ.* 21: 127-138

Donald CM, Scheer WPA (1975) Growth regulators increase yield and reduce length of harvest of high bush blueberries. *Hort Science* 10: 260-261

Donald CM (1962) In search of yield. *J Austrial Inst Agri Sci.* 28: 171-178

Drake BG, Gonzalez-Meler MA, Lang SP (1997) More efficient plants: A consequence of rising atmospheric CO_2. *Ann Rev. Plant Physiol Plant Mol Boil.* 48: 609-639

Echarte MM, Alberdi I, Luis A (2012) Post -flowering assimilate availability regulates oil fatty acid composition in Sunflower grains. *Crop Sci.* 52(2): 818-829

Ehness R, Roitsch T(1997) Coordinated induction of mRNAs for extracellular invertase and a glucose transporter in *Chenopodium rubrum* by cytokinins. *Plant Journal* 11: 539-548

Fageria NK, Baligur VC (2005) Enhancing nitrogen use efficiency in crop plants. *Adv Agron.* 80: 97-185

Fageria NK (1992) Maximizing crop yields. New York, Marcel Dekker.

Farrar JF (1985) Luxes of carbon in roots of barley plants. *New Phytol* 99: 57-69

Fourcaud T, Zhang XP, Stokes A, Lambers H, Korner C (2008) Plant growth modeling and applications: the increasing importance of plant architecture in growth models. *Annals of Botany* 101: 1053–1063

Foyer CH, Valadier MH, MiggeA, Becker TW (1998) Drought-induced effects on nitrate reductase activity and mRNA and on the coordination of nitrogen and carbon metabolism in maize leaves. *Plant Physiol.* 117: 283-292.

Geiger DR, Shieth WJ, S aluke RM (1989) Carbon partitioning among leaves, fruits, and seeds during development of *Phaseolus vulgaris* L. *Plant Physiology* 91: 291-297

Gunning BES (1976) The role of plasmodesmata in short distance transport to and from the phloem. In: *Intercellular Communication in Plants: Studies on Plasmodesmata* (Ed. by B ES Gunning &A. W Robards), pp. 203-227. Springer-Verlag, Berlin.

Gambin BL, Borras L (2010) Resource distribution and the trade-off between seed number and seed weight: a comparison across crop species. *Annals of Applied Biology* 156: 91–102

Geiger DR, Shieh WJ, Salube RM (1989) Carbon partitioning among leaves, fruits and seeds during development *of Phaseolus vulgaris* L. *Plant Physiol* (1) 291-297

Gifford RM, Yhorne JH, Hitz WD, Giaquinta RT (1984) Crop productivity and photoassimilate partitioning. *Science* 225: 801-808

Godin C, Sinoquet H (2005) Functional-structural plant modelling. *New Phytologist* 166: 705–708

Guo Y, Fourcaud T, MarcJ, Zhang X, Baoguo Li (2011) Plant growth and architectural modelling and its applications. *Annals of Botany* 107: 723–727

Hanan J, Prusinkiewicz P (2008) Foreword. Studying plants with functional- structural models. *Functional Plant Biology* 35: i–iii

Hatch MD (1987) C-4 photosynthesis - a unique blend of modified biochemistry, anatomy and ultrastructure. *Biochemica et Biophysica Acta* 895: 81-106

Hay RKM (1995) Harvest index its use in plant breeding and crop productivity. *Annals of biology* 126: 197-216

Heather AR, Howard VD (1992) Purification and characterization of sucrose synthase from cotyldons of vicia faba L. *Plant Physiol* 10: 1008-1013

Herald A (1980) Regulation of photosynthesis by sink activity - The missing link. *New Physiol* 86: 131-144

Ingram J, Chandler JN, Gallagher L, Salamini F, Bartels D (1997) Analysis of cDNA clones encoding sucrose-phosphate synthase in relation to sugar interconversions associated with dehydration in the resurrection plant *Crateroshgme plantaginium. Plant Physil.* 115: 113-121

Jeuffory MH, Bertrad N (1997) Crop physiology and productivity. *Field crop Res* 53: 3-16

Jones P, Allen Jr, LH, Jones JW, Boote K J, Campbell WJ (1984) Soybean canopygrowth, photosynthesis and transpiration responses to whole season carbon dioxide enrichment. *Agron J.* 76: 633-637

Jullien A, Mathieu A, Allirand JM, Pinet A, Philippe de R, Cournede PH, Ney B (2010) Characterization of the interaction between architecture and source sink relationship in winter oilseed rape (*Brassica napus*) using Green Lab model .*Annals of Botany* 105: 1-15

Kalt Torres W, Kerr P S, Usuda H, H uber S C (1987). Diurnal changes in maize leaf photosynthesis. I. Carbon exchange rate, assimilate export rate, and enzyme activities *Plant Physiology* 83**:** 283-288

Koch K E, J ohnson C R (1984). Photosynthate partitioning in split-root citrus seedlings with mycorrhizal and non-mycorrhizal root systems. *Plant Physiology* 75**:** 26-30

Karmoker JL, Van Steveninck RFM (1979) The effect of abscissic acid on sugar levels in seedlings of *Phaseolus vulgaris* L. cv. Red land Pioneer - *Planta* 146: 25-30

Kichey T, Hirel B, Heumez E, Dubios F, Le Gouis J (2007) In winter wheat (*Triticum aestivum* L.), post-anthesis nitrogen uptake and remobilisation to the grain correlates with agronomic traits and nitrogen physiological markers. *Field Crops Research* 102: 22–32

Klan EM., Hall B, BennettAB (1996) Antisense acid invertase (TIVI) gene alters soluble sugar composition and size in transgenic tomato fruits. *Plant Physiol.* 112: 1321-1330

Lafitte H R, Travis R L (1984) Photosynthesis and assimilate partitioning in closely related lines of rice exhibiting different sink: source relationships. *Crop Science* 24**:** 447-452

Larson P R, I sebfands JG, D ickson (1972) Fixation patterns of ^{14}C within developing leaves of eastern cottonwood. *Planta* 107: 301 314

Lauer JS, Simmonis SR (2012) Photoassimilate partitioning of main shoot leaves in field grown spring barley (*Hordeum vulgare L.) Crop Sci* **25**(5): 851-855

Lauer M J, S hibles R (1987). Soybean leaf photosynthetic response to changing sink demand. *Crop Science* 27: 1197-1201

Le Roux X, Lacointe A, Escobar-Gutierrez A, Le Dizes S (2001) Carbon-based models of individual tree growth: a critical appraisal. *Annals of Forest Science* 58: 469- 506

Li LH, Luo Y, Ma JH (2011) Radiation use efficiency and the harvest index of winter wheat at different nitrogen levels and there relationship to canopy spectral reflectance . *Crop and Pasture Sci* 62 (3): 208-217

Lu HD, Xue JQ, Ma GS, Zhang RH, Zhang XH (2010) Effects of low nitrogen stress on source-sink characters and grain-filling traits of different genotypes summer *Ying Yong Sheng Tai Xue Bao.* 21(**5**): 1277-82

M eynhardt J T, Malan A H (1963) Translocation of sugars in double-stem grape vines. *South African Journal of Agricultural Science* 6: 337-338

M inchin P E H, T horpe M R (1989) Carbon partitioning to whole versus surgically modified ovules of pea: an application of the *in vivo* measurement of carbon flows over many hours using the short-lived isotope carbon-11. *Journal of Experimental Botany* 40: 781-787

Ma Q, Longnecker N, Atkins C (1998) Exogenous cytokinin and nitrogendo not increase grain yield in narrowleafed lupins .*Crop Sci.* 38(3): 717-721

Madani A, Shirani Rad A H, Pazoki A, Nourmohammadi Gh, Zarghami R (2010) Grain filling and dry matter partitioning responses to source: sink modifications under post-anthesis water and nitrogen deficiencies in winter wheat (*Triticum aestivum* L.). *Acta Scientiarum Agronomy* 32: 145–151

Madani1A, Shirani-RadA, Pazoki A, Nourmohammadi1 G, Zarghami G, Malek F, Baker D A (1977) Proton co-transport of sugars in phloem loading. *Planta* 135: 297-299

Marcelis LFM (1996) Sink strength as a determinant of dry matter partitioning in the whole plant. *J of Expt Bot* 47: 1281-1291

Marini R P, Barden, JA (1981) Seasonal correlations of specific leaf weight to net photosynthesis and dark respiration of apple. *Photosynth. Res.* 2: 251-258

Matsushima S (1970) *Crop Science in Rice. Theory of Yield Determination and its Application.* Fuji Publishing Co., Tokyo

Matthew J P, Foyer CH (2001) Sink regulation of photosynthesis. *Journal of Experimental Botany,* 52: 1383–1400

Mauney JR., Fry KE, Guinn G (1978) Relationship of photosynthetic rate to growth and fruiting of cotton, soybean, sorghum and sunflower. *Crop Sci* 18: 259-263

Mir MR, Mobin M, Khan NA, Bhat MA, Lone NA, Bhat KA, Razvi SM, Mokhtassi-Bidgoli A (2010) The impact of source or sink limitations on yield formation of winter wheat

(*Triticum aestivum* L.) due to post-anthesis water and nitrogen deficiencies. *Plant Soil Environ* 56(5): 218-227

Monroy AF, Sangwen V, Dhindsa RS (1998) Low temperature signal transduction during cold acclimation protein phosphatase 2A as an early target for cold- inactivation. *Plant J* 13: 653-660

Munns R (1988) Why measure osmotic adjustment? *Australian Journal of Plant Physiology* **15:** 717-726

Nguyen QB, Michelue K, Huber SC, Leahary Y (1998) Sucrose synthase in developing maize leaves.*Plant Physiol* 94(**2**): 516-520

Nosberger JH, Umphries EC (1965). The influence of removing tubers on dry-matter production and net assimilation rate of potato plants. *Annals of Botany* 29: 579-588

Nakamura TO, Osaki M, Shinano T, Tadano T (1997) Different mechanisms of carbon-nitrogen interaction in cereal and legume crops. pp. 913-914. In: T. Tndo, K. Fujita, T. Mae, H. Tsumoto, S. Mori and J. Sekiya (eds.). Plant nutrition for sustainable food production and enrichment Dordrecht. The Netherland Kluwer Acad. Publ.

Olson RA, Sander DH (1999) Corn production. pp. 639-686. In: G.F. Sprangne and J.W. Dudley (eds.) Corn and Corn Improvement. *Agron. Monogr.* 18, Third Edition Madison, WT: *Am. Soc. Am. Crop Sci Soc Am* and *Soil Sci. Soc. Am.*

Oparka KJ, Davies HV (1985) Translocation of assimilates within and between potato stems. *Annals of Botany* 56**:** 45-54

Pate J S, H erridge DF (1978) Partitioning and utilization of net photosynthate in a nodulated annual legume. *Journal of Experimental Botany* 29**:** 401 -412

Pace PF, Cralle HT, Cothren JT, Senseman SA (1999) Photosynthate and dry matter partitioning in short and long season cotton cultivars. *Crop Sci* 39: 1065-1069

Passioura JB, Ashford AE (1974). Rapid translocation in the phloem of wheat roots. *Australian Journal Plant Physiology* 1: 521-527

Patrick JW (1972) Distribution of assimilates during stem elongation in wheat. *Aust J Biol. Sci.* 25: 455-67

Patrick JW, Wardlaw IF (1984) Vascular control of photosynthate transfer from the flag leaf to the ear of wheat. *Australian Journal of Plant Physiology* 11: 235-241

Patrick JW (1979) An assessment of auxin-promoted transport in decapitated stems and whole shoots of *Phaseolus vulgaris* L. *Plant* 146: 107-112

Peel AJ, Ho LC (1970) Colony size of *Tuberolachnus salignus* (Gmelin) in relation to mass transport of ^{14}C-labelled assimilates from the leaves in willow. *Physiologia Plantarum* 23: 1033-1038

Peltonen-Sainio P, Jauhiainen L (2008) Association of growth dynamics, yield components and seed quality in long-term trials covering rapeseed cultivation history at high latitudes. *Field Crops Research* 108: 101–108

Penning De Vries FWT (1975) Use of assimilates in higher plants. In: *Photosynthesis and Productivity in Different Environments* (Ed. by J PCooper) pp 459-480 Cambridge

University Press

Perttunen J, Sieva¨nen R, Nikinmaa E, Salminen H, Saarenmaa H, Va¨keva¨ J (1996) LIGNUM: a tree model based on simple structural units. *Annals of Botany* 77: 87–98

Peterson D M (1983) Effects of spikelet removal and post-heading thinning ondistribution of dry matter and N in oats. *Field Crops Res.* 7: 41-50

Poorter H, Nagel O (2000) The role of biomass allocation in the growth response of plants to different levels of light, CO_2, nutrients and water: a quantitative review 27: 595-607

Poostchi I, Rouhani I, Razmi K (1972) Influence of levels of spring irrigation and fertility on yield of winter wheat *Triticum aestivum* L. under semi-arid conditions. *Agron J.* 64: 438-442

Portis AR (1995) The regulation of Rubisco by Rubisco activase. *J. Exp. Bot.* 46: 1285-1291

Rocher J P, P rioul J L (1987). Compartmental analysis of assimilate export in a mature maize leaf. *Plant Physiology and Biochemistry* 25: 531-540

Ranjan R, Purohit S S, Prasad V(2004) Hormones in photosynthate partitioning and grain filling. In Plant Hormones Action and Application Studies in Plant Physiology Series o.5 (eds.) Ranjan, R., Purohit, S.S. and Prasad V. *Agrobio India.* pp. 215-219

Ranjan R (2002) Photosynthesis: Physiological, Biochemical and molecular aspects: In Studies in Plant Physiology Series No.2 (eds SS Purohit, Ranjan R) *Agrobios, India* pp 216-246

Reddy KR, Davidonis GH, JohnsonAS, Vinyard BJ (1999) Temperature regime and carbon dioxide enrichment alter cotton boll development and fiber properties. *Agron J.* 91: 851-858

Robertson MJ, SlimS, ChauhanYS, Ranganathan R (2001) Predict ting growth and development of pigeon pea: biomass accumulation and partitioning .*Field crops Res* **70**: 89-100

Roistsch T (1999) Source sink regulation by sugar and stress. *Current opinion in Plant Biology* 2: 198-206

Room P, Hanan J, Prusinkiewicz P (1996) Virtual plants: new perspectives for ecologists, pathologists and agricultural scientists. *Trends in Plant Science* 1: 33–38

Stitt M, H uber S, K err P (1987) Control of photosynthetic sucrose formation. In: *Photosynthesis* (Ed. by M D Hatch and N K Boardman), *The Biochemistry of plants,* vol. 10, pp 327-409. Academic Press

Sangoi L, Salvador R (1997) Dry matter production and partitioning of maize hybrids and drawf unes at four plant populations. *Ciencia Rural Santa Maria* 27: 1-6

Saute J J, Iten W, Zimmermann MH (1973) Studies on the release of sugar into the vessels of sugar maple *{Acer saccharum). Canadian Journal of Botany* 51: 1-8

Sauter J J, Ambrosius T (1986) Changes in the partitioning of carbohydrates in the wood during bud break in *Betula pendula* Roth *Journal of Plant Physiology* 124: 31-43

Schaffer A A, Sagee O, G old Schmidt T E E, Goren R (1987) Invertase and sucrose synthase activity, carbohydrate status and endogenous IAA levels during Citrus leaf development. *Physiologia Plantarum* 69: 151-155

Sebkov, Unger C, HardeggaM, Sturn A (1995) Biochemical, Physiological and molecular characterization of sucrose synthase from Dacus carota. *Plant Physiol* 108(**1**): 75-83

Servattes JC, F ondy B R, Li B, Geiger D R (1989). Sources of carbon for export from spinach through out the day.*Plant Physiology* 90: 1168-1174

Sharma P, Sardana V (2009) Increasing yield potential of grain legumes: Physiological deteriments in relation to cereals. In: Biotechnology: Cracking New Pastures (ed. Malik C P and Verma Aman) MD Publishers Pvt. Ltd New Delhi PP 173-193

Sardana, V., Sharma P, Sheoran P (2010) Growth and production of pulses. EOLSS Publication. In Soils, Plant growth and Crop Production (Ed. Willy H. Verheye), in Encyclopedia of Life Support Systems (EOLSS), Developed under the Auspices of the UNESCO, EOLSS Publishers, Oxford, UK (http: //www.eolss.net)

Sharma P, Sardana V, Sheoran P (2011) Fate of carbon and nitrogen in developing storage organs. *Journal of Plant Science Research* 27(**1**) 1-20

Sharma P, Malik CP (2009) Hormonal regulation of fibre elongation and the enzymic hydrolysis of absicic acid conjugate in developing cotton (*Gossypium arboreum* L) fibres. *Ind J Plant Physiol.* Vol 14 No 4 (NS): 360- 363

Sharma P, Sardana V (2008) Physiological processes, nitrogen fixation, growth and productivity of legumes in response to abiotic stresses. IN Advanced Topics in Biotechnology and Plant Biology (eds Malik CP, Bhavneet K and Chitra W) pp 323-398

Sharma P, Sardana V (2011) Carbon and nitrogen assimilation: Mechanisms for productivity .In Plant Environment and Sustainability (P.C.Trivedi Ed) *Agrobios, India* pp 39-61

Sharma P, Sardana V (2011) Nitrogen dynamics: Growth, uptake and distribution in crops: An ecophysiological Perspective .In Plant Environment and Sustainability (P.C. Trivedi Ed) *Agrobios, India* pp 63-82

Sharp RE, Davies WJ (1985) Root growth and water uptake by maize plants in drying soil. *Journal of Experimental Botany* 36: 1441-1456

Shinano T, Osaki M, Tadano T(1994) ^{14}C-Allocation of ^{14}C-compounds introduced to a leaf to carbon and nitrogen components in rice and soybean during ripening. *Soil Sci Plant Nutr.* 40: 199-209

Siddiquae KHM, Belford RK, Perry MW, Tennent D (1989) Growth, development and light interception of old and modern wheat cultivars in a Mediterranean type environment. *Australian Journal of Agricultural Research* 40: 473-87

Smith AM, Zeeman SC, Smith SM (2005) Starch degradation. *Annu Rev Plant Biol.* 56: 73–98

Smith SM, Fulton DC, ChiaT, Thorneycroft D, Chapple A, DunstanH, Hylton C, Zeeman SC Smith AM (2004) Diurnal changes in the transcriptome encoding enzymes of starch metabolism provide evidence for both transcriptional and posttranscriptional regulation of starch metabolism in Arabidopsis leaves. *Plant Physiol.* 136**:** 2687–2699

Sionit N, Rogers HH, Bingham GE, StrainBR. (1984) Photosynthesis and stomatal conductance with CO_2 enrichment of container and field grown soybeans. *Agron J.* **76:** 447-451

Soheil K, Keyvan S (2011) Source –sink relationship in soybean. *Annals of Bilogical Research* 2(4): 334-342

Starck Z (1971) Pattern of ^{14}C-assimilate distribution in relation to their supply and demand in sunflower. *Acta Societatis Botanicorum Poloniae* 40: 653-667

Starck Z, Ubysz L (1974) Effect of limited supply of assimilates on the relationships between their sources and acceptors. *Acta Societatis Botanicorum Poloniae* 43: 427-445

Strand Å, Hurry V, Gustafsson P, Gardeström P (1997) Development of *Arabidopsis thaliana* leaves at low temperature releases the suppression of photosynthesis and photosynthetic gene expression despite the accumulation of soluble carbohydrates. *Plant J*: 605 – 614

Stitt M., Quick W P, Schurr U, Schulze ED, Rodermel SR, Bogorad L (1991) Decreased ribulose 1, 5-bisphosphate carboxylase oxygenase in transgenic tobacco transferred with antisense rbc SII. Flux control coefficients for photosynthesis in varying light, Co_{-2} and air humidity. *Planta* 183: 555-566

Sugiharto BK, Miyata H, Nakamoto H, Sasakawa Sugiyama T (1990) Regulation of expression of carbon assimilation enzymes by nitrogen in maize leaf. *Plant Physiol* 92: 936-969

Suhmaninder K (2012) Identification and characterization Morpho- Physiological traits associated with response response to water stress in *Brassicas*. MSc. Thesis submitted to Punjab Agricultural University, Ludhiana, India

Tanaka A (1972) The relative importance of the source and the sink as the yield limiting factors of rice. Tech. Bull. No.6 Taipei City, Taiwan Food Fertil. Tech. Center.

Tanka A, Yamaguchi J (1972) Dry matter production, yield components and grain yield of the maize plant. *J Fac Agri Hokkaido Uni., Japan* 77: 71-132

Thorne JH, Koller HR (1985) Influence of assimilate demand on photosynthesis, diffusive resistance, translocation and carbohydrate levels of soybean leaves. *Plant Physiology* 54: 201–217

Tubiello F N, Rosenzwerg C, Kimball BA, Pinter Jr, P J, Wall GW, Hunsaker DH, LaMorte R L, Garcia RL (1999) Testing CERES – wheat with free air carbon dioxide enrichment (FACE) experiment data CO_2 and water interactions. *Agron. J.* 91: 247-255

Venkateswarlu B, Visperas RM (1987) Source sink relationships in crop plants. *IRRI Research Paper Series* No. 125

Vos J, Evers JB, Buck-Sorlin GH, Andrieu B, Chelle M, de Visser PHB (2010) Functional–structural plant modelling: a new versatile tool in crop science. *Journal of Experimental Botany* 61: *2101–2115*

W arraich EA, N azir A, S hahzad MA Basra I rfan A (2002) Effect of Nitrogen on Source-Sink Relationship in Wheat International Journal of Agriculture and 4 (2): 300-302

W illiams RF, Metcalf R A (1975) Physical constraints and tiller growth in wheat. *Australian Journal of Botany* 23: 213-223

Wani SA, Nowsheeba W, Sabina A, Shazia Rashid, Nasir HM, Payne WA (2010) Crop responses to interaction between Plant growth regulators and nutrients. *Journal of Phytology* 2(10): 9-19

Wardlaw I F (1980) Translocation and source-sink relationships. In: *The Biology of Crop Productivity* (Ed. by P. S. Carlson), pp. 297-339. Academic Press, New York, London, Toronto, Sydney, San Francisco

Wardlaw I F, Moncur L (1976) Source, sink and hormonal control of translocation in wheat. *Planta* 128: 93-100

Weber H, Boresjrik L, Wobus U (1997) Sugar import and metabolism during seed development. *Trends Plant Sci* 2: 169-174

White E M, Wilson FEA (2006) Responses of grain yield, biomass and harvest index and their rates of genetic progress to nitrogen availability in the winter wheat varieties. *Irish J of Agri. And Food Res.* 45: 85-101

Williams A M, Williams KR (1978) Regulation of movement of assimilates into ovules of *Pisum sativum. Austral. J. Plant Physiol.* 5: 295-300

Xie Guang-hui, Han Dong-qian, Wang Xiao-yu, Lü Run-hai (2011) Harvest index and residue factor of cereal crops in China. *Journal of China Agricultural University 2011-01*

Xue JQ, Z hang RH, Guo-Sheng MA, Hai-Dong LU, Z hang XH, Li FY, Hao YC, Tai SJ (2010) Effects of Plant Density, Nitrogen Application, and Water Stress on Yield Formation of Maize. *Acta A gronomica Sinica* 36 (6): 1022-1029

Zrenner R *et al* (1995) Evidence for the critical role of sucrose synthase for sink strength using transgenic plants. *Plant J.* 7: 97-107

3

Advances in Hybrid Rice Technology through Applications of Novel Technologies

Sajad Hussain Dar, Waseem Hussain and *Gulzar S Sanghera*

Rice is the world's most important food crop and in order to increase its productivity and production heterosis breeding is one of the most feasible methods. But the major problems in hybrid rice breeding are the limited number of parental lines with specific desirable traits, a lower frequency of maintainers and restorers among elite breeding lines, the narrow genetic base, lack of resistance to biotic stresses, and poor grain quality of some parental lines. These problems can be resolved to a great extent by the application of novel molecular breeding methods. Advanced biotechnological tools with traditional breeding approaches will critically help in identification of resistance genes, incorporate, and pyramid multi resistance to parental lines. Marker-assisted identification of fertility restorer genes and their introgression into parental lines can be done and this transgenic restorer line can be directly used for hybrid production. Genes leading to different races or biotypes being resistant to a disease or insect pest can be pyramided together to make a line with multi-race or multi-biotype resistances. SSR and STS markers were used to confirm purity, which was considerably simpler than conventional methods. Marker Assisted screening of genotypes for the presence of wide compatibility (WCG) genes can be done. Further, the future hybrids are expected to have wider adaptability to different environments, better yield capability, increased insect and disease resistance, and value-added grain with high nutritional content such as iron, vitamin A, and lysine.

Introduction

Rice is a global food crop and billions of people around the world depend on rice for their energy, protein and vitamins. In India, rice is grown on about 44.5 Mha and provides food for more than 70% of the population and serves as the principal energy source for most of the people. The demand for rice will continue to increase with increase in population and decrease in land. In order to keep pace with the growing population, the estimated rice requirement by 2025 is about 130 Mt. Plateuing trend in the yield of HYV's, declining and degrading natural resources like land and water and acute shortage of labor make the task of increasing rice production quite challenging. The current situation necessitates looking for some innovative technologies to boost rice production.

The lesson learned from the development and adoption of hybrid rice in China shows that more rice could be produced even on less land with hybrid rice. Hybrid rice is a proven and successful technology for rice production, having contributed significantly toward improving food security, raising rice productivity and farmers' income, and providing more employment opportunities over the past three decades. Hybrid rice technology had already played a pivotal role in increasing rice production and productivity in India and is one of the components of the national food security mission launched in 2007 to boost rice production. Some popular released hybrids grown in the country are PA 6444, PHB-71, KRH-2, Pusa RH-10, PA 6201, Suruchi, JKRH-2000, PSD-3, Sahyadri, and DRRH-2 Hybrid rice based on cytoplasmic male sterility (CMS), increases grain yield by more than 20% relative to improved inbred rice varieties.

The major problems in hybrid rice breeding are the limited number of parental lines with specific desirable traits, a lower frequency of maintainers and restorers among elite breeding lines, the narrow genetic base, lack of resistance to biotic stresses, and poor grain quality of some parental lines. Rice hybrids are affected by many pests and diseases such as blast, bacterial blight, sheath blight, yellow stem borer, brown plant hopper, white backed plant hopper, leaf folder, and gall midge. For the large scale adoption of this technology, hybrids need not to be resistant to the major pests and diseases prevailing in the target areas but also should have desirable quality traits. To breed improved heterotic rice hybrids, we need to adopt new strategies to enhance the frequency of availability of maintainers and restorers, ensure a constant supply of genetically diverse parental lines and resistant to biotic and abiotic stresses. Major advances in biotechnology offer us a variety of new tools such as direct gene transfer, development of molecular markers to assist in marker assisted selection, to enable scientists to develop hybrid rice varieties with increased yield potential, improved grain quality and multiple resistance to or tolerance of various biological and environmental stresses. The application of novel molecular breeding methods is playing a strong role in accelerating hybrid rice breeding. Transgenic technology has great prospects in breeding for resistance to herbicide, pests, and stress; improving quality; and increasing yield potential. Further development of biotechnology would certainly bring a revolutionary change to hybrid rice breeding particularly in developing countries. Thus modern molecular breeding technologies combined with conventional breeding methods to breed new varieties of super high-yielding rice more quickly and efficiently. The possible role of new biotechnological tools for hybrid rice improvement is briefly given in this chapter.

Improving parental lines against diseases and insect pests via genetic engineering

Hybrid rice breeding involves the three-line system in which a cytoplasmic male sterile line (CMS) used as a female parent is crossed with a fertility restoration (R) line to produce a hybrid. The third line, a maintainer (B) line, is used for maintaining/ producing male sterile plants. A hybrid is resistant to a disease/insect when one of the above parents is resistant because of a dominant gene(s) (Virmani, 2001). Resistance to biotic stresses is one of most important subjects for hybrid rice development. Although no evidence showed a significant effect on resistance to/tolerance of biotic and abiotic stress with hybrids derived from wild abortive (WA) cytoplasm (Faiz, 2000), in the past few years, bacterial blight, blast, false smut, and kernel smut have been reported in many Asian countries for serious damage to hybrid rice grain and seed production, especially the new epidemics of kernel smut, false smut, and brown plant hopper. There is a lack of knowledge whether these threats are specifically associated with hybrid rice germplasm. Advanced biotechnological tools with traditional breeding approaches will critically help in identification of resistance genes, incorporate, and pyramid multi resistance to parental lines. Incorporation of a resistance gene(s) in a CMS or maintainer or restorer line is expected to make the hybrid resistant to the target disease or pest. While it is difficult to produce a transgenic CMS plant every time, one-time genetic transformation of a maintainer line with a resistance gene(s) will automatically result in a resistant CMS line by backcrossing (Alam *et al.*, 1999). The transgenic restorer line, in contrast, can be directly used for hybrid production (Tu *et al.*, 2000). They successfully transformed two transgenic maintainer lines (IR68899B, IR68897B) and two restorer lines (MH63 and BR-827-35R) with the truncated chimeric *Bt* gene, *cryIAb* (driven by 35S and PEPC), and/or the hybrid *Bt* gene *cryIAb/Ac* driven by the actin 1 promoter. These lines showed a wide range of expression (low to high) of *Bt* proteins and the protein content was stably inherited (Datta *et al.*, 1998; Alam *et al.*, 1999; Tu *et al.*, 2000). A selected homozygous MH63 *Bt* was hybridized with CMS line Zhenshan 97A to produce the first-ever hybrid *Bt* rice (Shanyou 63). Further, in an effort to breed resistance to both disease and insect pests, a wild-rice-derived dominant gene Xa21 conferring multi-race resistance against BB and a fused *Bt gene cry1Ab/ cry1Ac* conferring resistance to lepidopteran insects have been individually introduced into the same genetic background of an elite indica cytoplasm male sterile (CMS) restorer line Minghui63 by means of marker-assisted selection (MAS) (Chen *et al.*, 2000) and by genetic engineering (Tu *et al.*, 2000), respectively.

To develop durable resistance in parental lines, gene pyramiding is a very useful approach and feasible option. Genes leading to different races or biotypes being resistant to a disease or insect pest can be pyramided together to make a line with multi-race or multi-biotype resistances. Gene pyramiding was used to combine the fused Bt insecticidal gene and the dominant BB resistance gene Xa21 into the same target plant of an elite indica CMS restorer line Minghui 63 using MAS by (Tu *et al.*, 2000).

Grain quality of future rice hybrids can be improved to the desired level by using parental lines with the desired grain quality and critically evaluating the derived heterotic rice hybrids (in comparison with the check varieties) before their release for commercialization. Hybridity per se is not the cause of the poor grain quality of some of the hybrids commercialized so far.

Marker-assisted identification of fertility restorer genes and their introgression into parental lines

Restorer lines are difficult to be identified using conventional procedures as they are much labour intensive. Employing molecular markers can effectively and efficiently may help in identification and tagging of restorer genes among the elite lines, which latter can be utilized in MAS (Marker assisted selection) for their introgression into desirable and adapted lines. For better hybrid production these putative restorers can then be test crossed with appropriate CMS lines to confirm their fertility restoration and check the magnitude of heterosis. For example, SSR markers have been used for mapping and tagging of fertility restorer genes in rice (Ahmadikhah *et al.*, (2006) and (2009); Bazrkar *et al.*, (2008), Sattari *et al.*, (2009). Furthermore, at Indian Agricultural Research Institute (IARI), New Delhi, studies of fertility restoration in two rice hybrids showed a dominant monogenic inheritance in both F_2 crosses viz., IR58025A/IR40750 and IR62829/MTU9992. Mapping studies identified RM6100 linked to *Rf4* gene on chromosome at distance of 8.7 cM and 7 cM in the above hybrids. The marker RM6100 identified restorer and maintainer lines with 97.5% efficiency.

DNA marker based assessment of genetic purity of seeds of rice hybrids and parental lines

In practice, seed of different strains is often mixed due to the difficulties of handling large numbers of seed samples used within and between crop breeding programmes. Markers can be used to confirm the true identity of individual plants. The maintenance of high levels of genetic purity is essential in rice hybrid production in order to exploit heterosis. Genetic purity of parental lines and hybrids in case of rice is of crucial importance, as one percent reduction in purity of hybrid seed, results in a reduction of about 100 kg/ha in yield of commercial crop. In hybrid rice, SSR and STS markers were used to confirm purity, which was considerably simpler than the standard 'grow-out tests' that involve growing the plant to maturity and assessing morphological and floral characteristics (Yashitola *et al.*, 2004).

Marker Assisted screening of genotypes for the presence of wide compatibility (WCG) genes

The utilization of wide compatibility genes (WCG) in overcoming the partial hybrid sterility (HS) while exploiting the inter-sub-specific hybridization (indica/japonica) was further speeded up with the development of new set of multiplex marker system at S5 locus. This system called S5-MMS targets the functional nucleotide polymorphisms (FNP) at S5, is suitable for deployment in marker assisted selection (MAS) of wide compatible genotypes and for identification of allelic status at S5 locus in rice varieties. This way many rice genotypes were identified possessing S5-neutral allele (S5n), which could be immediately deployed in hybrid rice breeding for development of superior inter-sub specific rice hybrids.

Thus molecular markers are becoming increasingly useful in enhancing the efficiency in crop improvement. With the availability of molecular markers linked to fertility restoration and wide compatibility, breeding lines can be easily screened and identified as restorers & wide compatible genotypes without any test crosses and complications associated with phenotype based screening. In Directorate of Rice Research 100 breeding lines were screened for the presence of fertility restorer genes and wide compatible *S5* neutral allele. The lines were developed having presence of all *Rf4, Rf3* and *Wc* genes and can be immediately utilized in hybrid rice breeding program to exploit higher level of heterosis. In recent studies, two hybrid sterility genes, S5 and Sa, involved in the *indica–japonica* subspecies hybrid sterility, have been cloned (Chen *et al.*, 2008).

Engineered male sterility

The exploitation of heterosis hinges on the availability of a good male sterility system. In the tropics, over the past decades, many cytoplasmic male sterility (CMS) sources have been successfully developed and used for hybrid seed production in rice. Considerable progress has been made in understanding the organization and expression of plant genes. Efficient methods for transforming many plant species have also been developed. It is now possible to introduce virtually any genetic sequence into a plant genome and modulate its expression with a certain degree of precision. With these tools, several methods have been developed for disrupting normal pollen development (for male sterility) and for restoring normal pollen development (fertility restoration) in the hybrid (Narayanan, 1998).

One of the earliest and successful attempts to induce male sterility by genetic engineering involved the transfer and tissue-specific expression of a "toxin" gene that disrupted normal pollen development. The toxin gene from a fungal source, barnase (an RNase), driven by a tapetum-specific promoter (TA 29), was made to express itself specifically in the tapetal tissue of developing tobacco anthers (Mariani *et al.*, 1990). Fertility restoration in these male sterile transgenic plants could be achieved by crossing with another transgenic tobacco line that expressed the specific RNase inhibitor, barstar (Mariani *et al.*, 1992). Using the same system, male sterility has been successfully engineered in rapeseed, *Brassica napus* (Denis *et al.*, 1993).

Further male gametogenesis is a complex development process that is controlled by the expression of many genes. Genes whose function seems to be specifically associated with male gametogenesis (*MG* genes) have been reported in a few crops. Aarts *et al.*, (1993) isolated a gene from *Arabidopsis*, using transposon mutagenesis, the disrup-tion of which leads to male sterility; this locus was designated *MS2*. With the availability of efficient molecular gene-hunting tools such as gene traps (Sundaresan *et al.*, 1995), it is now relatively easy to identify loci involved specifically in male gemetogenesis. Several molecular methods inhibit the expression of a locus. One of the powerful and widely tested methods is the antisense RNA technique, which involves the expression of the sequence of the gene to be inhibited in antisense (complementary to the mRNA sequence) orientation through transgenes. A plant can thus be made male sterile by inhibiting any of the *MG* genes through this technique (Narayanan, 1998).

In order to develop an alternative novel strategy for hybrid rice production based on genetically engineered nuclear male sterility (NMS), Luo *et al.*, (2006) reported the isolation and characterization of an anther-specific gene, RTS, which is a single-copy gene exclusively expressed in tapetal cells. They showed that specific down regulation of RTS expression leads to pollen abortion, resulting in male sterility in transgenic rice plants. They also demonstrate that the promoter sequence of the RTS gene confers cell-specific expression of a cytotoxic gene, *barnase*, in anthers of transgenic rice, creeping bentgrass and *Arabidopsis*, resulting in disruption of viable pollen grain formation, and thus induction of male sterility in transgenics of different plant species.

Conclusions

Parental lines used for hybrid rice development determines the yield advantage, resistance to biotic and abiotic stresses, adaptability, grain quality, and other traits in rice hybrids. With the advent of transgenic technology, it has now become possible to insert a variety of genes into the plant genome from different sources. Further, the genetic diversity of parental lines can now be better determined by using molecular marker technology, which should aid in selecting parents suitable for making hybrids. The future hybrids are expected to have wider adaptability to different environments, better yield capability, increased insect and disease resistance, and value-added grain with high nutritional content such as iron, vitamin A, and lysine. For the future success of hybrid technology, the integration of transgenic and molecular breeding approaches would play a vital role in parental line improvement.

REFERENCES

Aarts MGM, Dirkse WG, Stiekema WJ and Periera A (1993) Transposon-tagging of male sterility gene in *Arabidopsis. Nature* 363: 715-717

Ahmadikhah A, Alavi M, Kamkar B and Kalateh M (2009) Mapping *Rf3* locus in rice by SSR and CAPS markers. *International Journal of Genetics and Molecular Biology* 1(7): 121-126

Alam MF, Datta K, Abrigo E, Oliva N, Tu J, Virmani SS and Datta SK (1999) Transgenic insect resistant maintainer line (IR68899B) for improvement of hybrid rice. *Plant Cell Rep* 18: 571-575

Bazarkar L, Ali AJ, Babaeian NA, Ebadi AA, Allahghollipour M, Kazemitabar K and Nematzadeh G (2008) Tagging four fertility restorer loci for wild abortive cytoplasmic male sterility system in rice (*Oryza sativa* L.) using microsatellite markers. *Euphytica* 164 (3): 669-677

Chen IS, Zhang IH and Ge MF (2008) A triallelic system of *S5* is a major regulator of the reproductive barrier and compatibility of indica-japonica hybrids in rice. *PNAS* 105 (32): 436-441

Chen IS, Zhang IH, Ge MF, Wan BH and Bai HS (1994) Studies on fertility stability in the anther culture-derived progeny of Xieqingzao *A J Crops* 2: 1-4

Chen S, Lin XH, Xu CG, and Zhang Q (2000) Improvement of bacterial blight resistance of Minghui 63, an elite restorer line of hybrid rice, by molecular marker-assisted selection. *Crop Sci* 40: 239-244

Datta SK (2000) Potential benefit of genetic engineering in plant breeding: rice, a case study. *Agric Chem Biotechnology* 43(4): 197-206

Luo H, Lee JY, Hu Q, Vasilchik KN, Timothy K, Lickwar C, Albert P, Joel MK, Thomas K

and Hodges TK (2006) RTS, a rice anther-specific gene is required for male fertility and its promoter sequence directs tissue-specific gene expression in different plant species. *Plant Mol Biol* 62: 397–408

Mariani C, De Beukeleer M, Truettner J, Leemans J, Goldberg RB (1990) Induction of male sterility in plants by chimeric ribonuclease gene. *Nature* 347: 737-741

Mariani C, Gossela V, De Beukeleer M, De Block M, Goldberg RB, De Greef W, Leemans J (1992) A Denis M, Delourme R, Gourret J-P, Mariani C, Renerd M. (1993) Expression of engineered nuclear male sterility in *Brassica napus*. *Plant Physiol* 101: 1295-1304

Narayanan KK (1998) Developing and using novel sources of male sterility. In: Virmani SS, Siddiq EA, Muralidharan K, editors. Advances in hybrid rice technology. Proceedings of the 3rd International Symposium on Hybrid Rice, 14-16 Novemeber 1996, Hyderabad, India. Manila (Philippines): *International Rice Research Institute*. pp 235- 244

Sattari M, Kathiresan A, Gregorio GB and Virmani SS (2008) Comparative genetic analysis and molecular mapping of fertility restiration genes for WA, Dissi, and Gambiaca cytoplasmic male sterility system in rice. *Molecular Breeding* 168: 305- 315

Sundaresan V, Springer P, Volpe T, Howard S, Dean C and Jones JDG (1995) Patterns of gene action in plant development revealed by the enhancer trap and gene trap transposable elements. *Genes Dev.* 9: 1797-1810. *Plant Mol Biol* 62: 397–408

Tu J, Zhang G, Datta K, Xu C, He Y, Zhang Q, Khush GS and Datta SK (2000) Field performance of transgenic elite commercial hybrid rice expressing *Bacillus thuringiensis* ä-endotoxin. *Nature Biotechnology* 18: 1101-1104

Virmani SS (2003) Advances in hybrid rice research and development in the tropics. In: Virmani SS, Mao CX, Hardy B, editors. Hybrid rice for food security, poverty alleviation, and environmental protection. Proceedings of the 4th International Symposium on Hybrid Rice, Hanoi, Vietnam, 14-17 May 2002. Los Baños (Philippines): *International Rice Research Institute*. pp 7-20

Yashitola J, Sundaram RM, Biradar SK, Thirumurugan T, Vishnupriya MR, Rajeshwari R, Viraktamath BC, Sarma NP and Sonti RV (2004) A sequence specific PCR marker for distinguishing rice lines on the basis of wild abortive cytoplasm from their cognate maintainer lines. *Crop Sci* 44: 920-924

4

Nuclear and Organelle Specific Markers with Specific Emphasis on Chloroplast and its Uses in Genetic Analyses

Amandeep Hora, Jyoti Ushahra and *C.P. Malik*

Immense loss of valuable plant species in the precedent centuries and its adverse impact on environmental and socioeconomic values has triggered the conservation of plant resources. Suitable identification and characterization of plant materials is indispensable for successful conservation of plant resources and to make sure their sustainable use. During the last few decades, the use of molecular markers, revealing polymorphism at the DNA level, has been playing an escalating part in plant biotechnology and their genetic studies. There are different types of markers viz. morphological, biochemical and DNA based molecular markers. Among these DNA based markers are differentiates in two categories: non PCR based (RFLP) and PCR based markers (RAPD, AFLP, SSR, SNP etc.), along with these markers, the micromatellite DNA marker has been used most widely due to its simple use by PCR, followed by a denaturing gel electrophoresis for allele size determination, and highly informative nature provided by its large number of alleles per locus. Despite this, CCMP and ITS marker systems are now on the scene and have gained high popularity. Development of such new and specific types of markers make their importance in understanding the genimic variability and the diversity between the same as well as the different species of the plants. In this review, we have discussed about the nuclear and organellar markers, their advantages, disadvantages and the applications of these markers in comparison with other marker systems.

Introduction

A molecular marker is a specific segment of DNA with a known location on a chromosome that is representative of the differences at the genome level. Genetic polymorphism is classically defined as the simultaneous occurrence of a trait in the same population of two or more discontinuous variants or genotypes. Molecular markers are gradually being recognized as valuable tools for assessing genetic diversity amongst germplasm and developing linkage maps. The major advantage of molecular markers are their dispersal across the genome is complete, expression is unaffected by the environment, assessment is independent of the stage of plant development etc. An ideal molecular marker technique should have the following criteria:

1. Highly polymorphic nature: must be polymorphic and consistently distributed throughout the genome.
2. Codominant inheritance: determination of homozygous and heterozygous states of diploid organisms.
3. Frequent occurrence in genome.
4. Should be easy, fast and cheap to detect.
5. Provide adequate resolution of genetic differences.
6. Need small amounts of DNA samples.
7. Have linkage to distinct phenotypes.
8. Require no prior information about the genome of an organism.
9. High reproducibility.
10. Easy exchange of data between laboratories.

Unfortunately, there is no single molecular marker which fulfills above requirements. Several molecular markers have been developed which detect polymorphism at the DNA level and are being used in plants include: Restriction Fragment Length Polymorphism (RFLP), Amplified Fragment Length Polymorphic (AFLP), Randomly Amplified Polymorphic DNA (RAPD), Simple Sequence Repeats (SSRs) or microsatellites, Inter Simple Sequence Repeats (ISSRs), Diversity Arrays Technology (DArT) Marker, Single-nucleotide polymorphisms (SNPs), Consensus Chloroplast Microsatellite Primers (CCMPs). These markers differ from each other with respect to significant features such as genomic abundance, level of polymorphism detected, locus specificity, reproducibility, technical requirements and cost (Agarwal *et al.*, 2008; Kumar *et al.*, 2009). Basic marker techniques can be classified into two categories:

(1) Non-PCR-based techniques or hybridization based techniques and,

(2) PCR-based techniques

Non-PCR or hybridization-based markers

These markers are based on DNA-DNA hybridization between a DNA or RNA probe and genomic DNA.

Restriction Fragment Length Polymorphism (RFLP)

It is the most extensively used hybridization-based molecular marker. This marker was first used in 1975 to identify DNA sequence polymorphisms for genetic mapping of a temperature-sensitive mutation of adeno-virus serotypes (Grodzicker *et al.*, 1975). RFLP is a technique that exploits variations in homologous DNA sequences (Tharachand *et al.*, 2012). In RFLP, DNA is exposed to restriction digestion followed by hybridization of generated

DNA fragments with radioactively labeled DNA probe (Grover *et al.*, 2012). Polymorphism is generated due to nucleotide substitutions or DNA rearrangements like insertion or deletion or point mutation. The RFLP markers are moderately highly polymorphic, co dominantly inherited and having high reproducibility. The technique is not very commonly used because it is time consuming, involves expensive and radioactive/toxic reagents, labor intensive and also requires large amount of DNA.

PCR based markers

PCR is a technique principally used for amplification of DNA without using a living organism. PCR based markers include *in vitro* amplification of DNA segments in the presence of short oligonucleotides of arbitary sequence and a thermo-stable DNA polymerase like Taq polymerase. PCR-based techniques are of two types on the basis of primers used for amplification:

1) Arbitrary or semi-arbitrary primed PCR techniques that developed without former sequence information (e.g., RAPD, AFLP, ISSR).
2) Site-targeted PCR techniques that developed from known DNA sequences (e.g., EST, CAPS, SSR).

Random Amplified Polymorphic DNA (RAPD)

RAPD marker was developed by Welsh and Mc Clleland (1990). It is based on differential PCR amplification of genomic DNA from short oligonucleotide sequences (10bp). It is a non-radioactive assay that requires small quantity of DNA and applicable to wide range of species. This approach requires no prior knowledge of the genome that is being analyzed. RAPD is quick and easy to assay. The main drawback of RAPDs is their low reproducibility and it requires purified, high molecular weight DNA (Kumar *et al.*, 2009). It is broadly used to analyze differences between the genomes of two organisms. RFLP and RAPD markers exhibited similar distributions throughout the genome, both recognized similar levels of polymorphism. RAPD, however, is more rapidly.

Amplified Fragment Length Polymorphism (AFLP)

To overcome the drawback of reproducibility associated with RAPD, Vos *et al.*, (1995) developed AFLP. This technique is based on the detection of genomic restriction fragments by PCR amplification and can be used for DNAs of any origin or complexity. It combines the power of RFLP with the flexibility of PCR-based technology by ligating primer-recognition sequences (adaptors) to the restricted DNA (Lynch and Walsh, 1998). It is based on a selectively amplifying a separation of restriction fragments from a complex mixture of DNA fragments attained after digestion of genomic DNA with restriction endonucleases. AFLP represents dominant marker system among the different marker systems and also provides multi-locus and genome wide marker profiles. AFLP is highly accurate, reliable, quite cheap and reproducible from lab to lab. It is extremely useful in detection of polymorphism between closely related genotypes. At the same time, it has limitation of highly expensive and requires large amount DNA than RAPD.

Inter Simple Sequence Repeats (ISSRs)

ISSR is a PCR based multi locus marker system. This technique was reported by Zietkiewicz (*et al.*, 1994). It involves amplification of DNA segments (about 100–3000bp) present at an amplifiable distance between oppositely oriented microsatellite repeat regions (Semagn *et al.*, 2008). This method uses microsatellites, regularly 16–25bp long, as primers in PCR reaction. ISSR markers habitually show high polymorphism although the level of polymorphism varies with the detection method used. Polymorphism arises due to rearrangements in microsatellite regions as it alters the distance between repeats. ISSR technique is simple, quick and the use of radioactivity is not essential.

Simple Sequence Repeats (SSRs)

Microsatellite or short tandem repeats or simple sequences repeats are monotonous repetitions of very short (one to five) nucleotide motifs, which occur as interspersed repetitive elements in all eukaryotic genomes and demonstrating high degree of variability. The term microsatellite was coined by Litt and Lutty (1989). These are identical markers for genetic mapping and population studies because of their abundance. Microsatellites have been used successfully to determine the degree of relatedness among individuals or a group of accessions, therefore, favored in population studies and for the identification of closely related cultivars. Microsatellite polymorphism can be detected by Southern hybridization or PCR (Kumar *et al.*, 2009).

Single-nucleotide Polymorphisms (SNPs)

SNPs were first described by Jordan and Humphries (1994). These are DNA sequence variations that occur when a single nucleotide (A, T, G and C) in the genome sequence is altered (Tharachand *et al.*, 2012). They represent a particular location at a chromosomal site at which the DNA sequence of two individuals differs by a single base. They are usually more prevalent in the non-coding regions of the genome (Agarwal *et al.*, 2008). SNPs produce the most abundant of molecular markers and their detection relies on finding differences between two sequences.

Diversity Arrays Technology (DArT) Markers

DArT was established to provide a practical and cost-effective whole-genome fingerprinting tool. This technology offers a hasty and sequence-independent shortcut to medium-density whole genome scans of any plant species. It is a microarray hybridization based method that enables the simultaneous typing of several hundred polymorphic loci spread over the genome. DArT marker is a high throughput, quick, cost effective and highly reproducible method. The main drawback of DArT is it comprises numerous steps, including preparation of genomic representation for the target species, cloning, and data managing and analysis.

Consensus Chloroplast Microsatellite Primers (CCMPs)

Information on the genetics of species would be valuable for scheming suitable plant breeding program, conservation of genetic resources, gene sequencing, gene mining or tagging etc. Genetic preservation strategies for their improved development requires progress of inherited information, inhabitants organization knowledge of plants, acquainted with eminence variability of chloroplast DNA, mitochondria and nuclear genome of plants due to its uniparental inheritance, the absence of recombination and premeditated mutation rates chloroplast genome is used (Provan *et al.*, 2001). Since, the chloroplast genomes in higher plants are more conservative than mitochondrial and nucleic genomes (Wolfe *et al.*, 1987), chloroplast microsatellites have the potentiality in the phylogenetic studies among plants with great taxonomic distances than nuclear or mitochondrial microsatellites (Ishii and McCouch, 2000). Therefore, the chloroplast genetic investigation was considered as a trustworthy technique to mark out the origin, evolution and phylogeny of many plant species. Usually, RFLPs and other genetic markers are the most conventional method for the study of cytoplasmic inheritance in somatic hybrids of higher plants. During last decade, cleaves amplified polymorphic sequence (CAPS) using mitochondrial or chloroplast specific primers are universal primers and are simple, inexpensive, valuable and extra competent in application in comparison to RFLPs (Bastia *et al.*, 2001: Guo *et al.*, 2002: Cheng *et al.*, 2003). Chloroplast microsatellites or chloroplast simple sequence repeats (cpSSRs) are characteristically mononucleotide tandem repeats which has various realistic implications for genetic investigation (Weising and Gardner, 1999). Its amplification protocol has been described by Andrianoelina *et al.*, (2006) who defined a combination of the different alleles as a chlorotype which is established at each locus for the reason of the non recombining quality of the chloroplast genome (chlorotypes were then treated as alleles at a distinct locus). For chloroplast microsatellites, the genetic diversity within *D. monticola* is one of the uppermost among tropical tree species assessed with chloroplast microsatellites (Muller *et al.*, 2009) prevent the species extinction.

Different studies revealed that chloroplast SSR have been employed for information of execution of genetic conservation approach and restoration of populations in Madagascar (Crandall *et al.*, 2000; Lhuillier *et al.*, 2006; Muller *et al.*, 2009).

Chloroplast SSR (cpSSR) markers have demonstrated their utility in studying genetic relationships, to assess the maternal and paternal plastid inheritance (Cato and Richardson, 1996), evaluation of interspecific polymorphism, detection of hybridization, introgression, phylogeny of plant population and are also applicable tool in the research of plant population genetics, understanding crop evolution, domestication, and phylogenetics (Provan *et al.*, 2001). Comparing with RFLPs and CAPS, cpSSR are proved to be more convenient, proficient, simpler and less expensive for organelle analysis of *Citrus* somatic hybrids at a very early regeneration stage. It was useful to verify the chloloroplast genomic origin of citrus somatic hybrids which is the first information on cytoplasmic inheritance analysis hybrids in higher plants by cpSSR. DNA sequence information of the chloroplast genome is necessary for the development of cpSSR primer pairs. These primers were designed according to the conserved nature of intron regions in chloroplast of higher plants. The relatively low numbers of repeats are typical of cpSSRs in other plant species, where long stretches of mono-nucleotide repeats are very exceptional (Powell *et al.*, 1996; Provan *et al.*, 1999a). In spite of these comparatively short repeat lengths, it has been revealed that cpSSR primers designed in one species will generate polymorphic products in other species and even in assorted genera (*e.g.* primers derived from *Nicotiana tabacum* revealing cpSSR polymorphism in *Solanum* spp.; Provan *et al.*, 1999b).

They unanimously disclose intra-specific dissimilarity in repeat number when situated in the non-coding regions of the chloroplast genome (cpDNA). The results revealed that the chloroplast genomes in the somatic hybrids were indiscriminately inherited from either parent, which were documented in the previous reports on *Citrus* somatic hybrids being based on RFLPs analysis (Grosser *et al.*, 1996; 2000; Moreria *et al.*, 2000; Guo and Deng, 2001). In the past several years, cpSSR has turn out to be useful method for evaluating genetic diversity of the cpDNA genome in plant species such as pine (*Pinus contorta)* (Powell *et al.*, 1995), potato (*Solanum tuberosum*) (Bryan *et al.*,1999), barley (*Hordeum vulgare*) (Provan *et al.*, 1999), rice (*Oryza sativa*) (Provan *et al.*, 1997; Ishii and McCouch, 2000), soybean (*Glycine max*) (Powell *et al.*, 1996: Xu *et al.*, 2002) and Kiwifruit (*Actinidia deliciosa)* (Weising and Gardner, 1999) cpSSRs are used for identification (for example *Dalbergia monticola)* that may previously not have been recognized using any other molecular markers such as RFLPs, RAPDs, AFLPs, etc. whether there is any genetic discrepancy present in the chloroplast genome of numerous plants) and to study genetic variation within and among populations and geographical structure in established populations. Chloroplast SSRs have been used in population and systematic studies in a variety of species (Powell *et al.*, 1996; Provan *et al.*, 1999a). This facilitates the perceptive of both chronological and current actions. It provides a balancing observation of gene flow blueprint because in angiosperms chloroplasts are transmitted by seeds (Petit *et al.*, 2005). In *Pinaceae,* to measure discontinuity within an efficiently haploid genome uniparental inheritance of mitochondrial and chloroplast genomes provides opportunities. However, in plants due to a high rate of sequence reformation mitochondrial genomes have not typically been helpful for phylogenetic analyses (Sederoff *et al.*, 1981; Wu *et al.*, 1998). Chloroplast microsatellites are predominantly valuable markers for paternal inheritance of chloroplast genomes in most conifers for studying mating systems, uniparental lineages, and gene flow via both pollen and seeds in most conifers (Neale *et al.*, 1986; Neale and Sederoff, 1981). Predominant paternal inheritance of chloroplast DNA has been established in European Abies with previously categorized (Vendramin *et al.*, 1996) *Pinus thunbergii* primers at two extremely inconsistent microsatellite loci (Vendramin and Ziegenhagen, 1997; Ziegenhagen *et al.*, 1998; Vendramin *et al.*, 1999). Intra-specific diversity and population structure in heterologous amplification with these primers have also been demonstrated (Cato and Richardson, 1996; Morgante *et al.*, 1998). In wild plant species and their varieties the potential of cpSSRs into biological and evolutionary processes has yet to be entirely accepted because of the intensifying outline of studies employing cpSSRs, while studies of economically imperative plants and their varieties remains obscured.

Chloroplast markers are further sensitive to drift, one of the major effects of disintegration. The only limitation is because of the same number of repeats may evolve in two different microsatellite lineages through independent mutational events i.e. its homoplasy effect, (Navascue´s and Emerson, 2005) instead of that recent simulations have verified that chloroplast microsatellites are capable of studying genetic organization and gene flow (Hansen *et al.*, 2005). Consequently to determine levels of inconsistency amongst populations, have great influence in estimating population structure information of mutation rates at SSR loci is significant. The low chloroplast DNA haplotypes is closely related to slow mutation rates (Provan *et al.*, 2001). Till date there have been no statistics published for mutation rates at simple repeat loci in the chloroplast genome.

Variation that can be detected among plant species using cpDNA-specific universal primers can be useful for many applications in plant science. Among these, systematics and evolutionary relationships studies at the different taxonomic levels (e.g. interfamilial, inter-generic, inter-specific, intra-specific and inter-population level) have been by far the most prevalent. Examples of other studies that made use of such primers have been reported in establishing the genetic variation and relationships between two species of *Melilotus indica* and *Melilotus alba*. A set of ten consensus chloroplast microsatellite primers (*ccmp1* to *ccmp10*) (Fig. 1)specific to chloroplast genomes of dicotyledonous angiosperms (Weising and Gardner, 1999) were used for characterization of organelle genome of *Melilotus sp.* All the *ccmp* primers except *ccmp* 8 and 9 yielded a discrete PCR product.

Table 1 : Allele sizes of amplification products generated by consensus chloroplast microsatellite primers (*ccmp1* to *ccmp10*)

Species	Size of amplification product (bp) and GenBank accession numbers							
Primer	*ccmp1*	*ccmp2*	*ccmp3*	*ccmp4*	*ccmp5*	*ccmp6*	*ccmp7*	*Ccmp10*
M. indica	129 140	70	82	130	137	120	160	190
M. alba	140	70	82	130	137	120	160	190
A. mexicana	139	240	110	150	100	95	150	115
Nicotiana tabacum	139	189	112	126	121	103	133	103

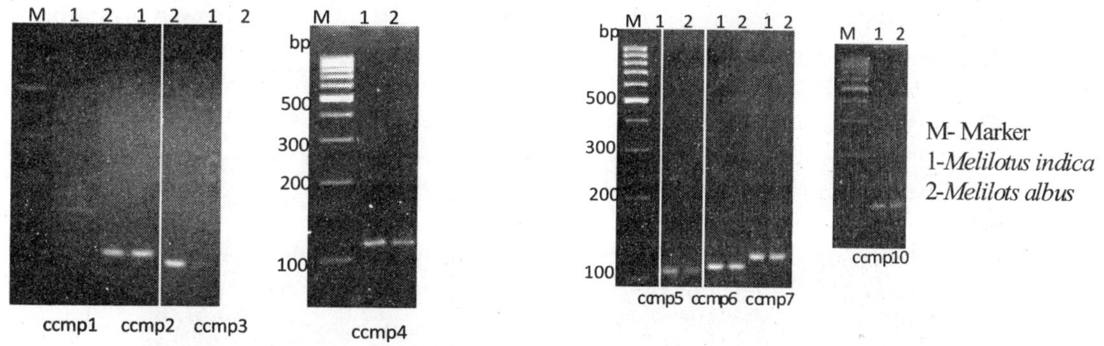

Fig.1 CCMP profiles of the *M. indica* and *M. alba* generated by different primers of *ccmp*.

Very diminutive differences were observed in allele size of two different genera with ccmp1, 3 and 5. The two species appear to be genetically similar. Due to lesser level of genetic variation detected through CCMP it could be a case of gene mutation or minor chromosomal alterations.

Similarly, with a view to assessing genetic relationship among the two species of *Argmone*, to evaluate the organelle specific primer polymorphism, to confirm hybridity of the two taxa i.e. AH and BH, and identification of maternal and paternal parents consensus chloroplast microsatellite primers were used.

Ten primers specific to chloroplast genome gave amplification products. Allele sizes of amplification products (bp) from *Argemone* species and the two hybrids generated by consensus chloroplast microsatellite primers (ccmp1 to 10) are set in Table 2 (Fig. 2). PCR amplification of *Argemone* species and hybrids with ccmp3 primer pair made interesting revelation. Very small difference was observed in allele size of two species with ccmp3. The two hybrids appear to be genetically similar. The ccmp3 primer confirmed that *A. mexicana* is the paternal and *A. ochroleuca* as the maternal parent. Allele length of AH and BH with ccmp3 primer was the same as that of the maternal parent thus, confirming *A. ochroleuca* as the maternal parent.

Table 2 : Allele sizes of amplification products from *Argemone* species generated by consensus chloroplast microsatellite primers (*ccmp1* to 10)

Species	Size of amplification product (bp) and GenBank accession numbers							
Primer	*ccmp1*	*ccmp2*	*ccmp3*	*ccmp4*	*ccmp5*	*ccmp6*	*ccmp7*	*Ccmp10*
A. mexicana	139	240	110	150	100	95	150	115
AH	139	240	110	150	100	95	150	115
BH	139	240	110	150	100	95	150	115
A. ochroleuca	139	240	110	150	100	95	150	115
Nicotiana tabacum	139	189	112	126	121	103	133	103

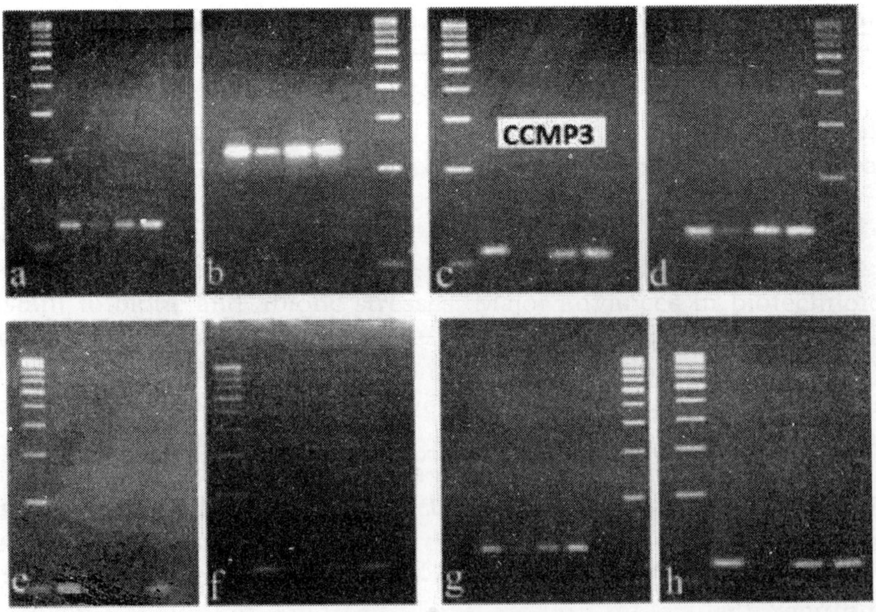

Fig. 2 (a-h) Amplification products from *Argemone* species and the two putative 'hybrids' generaied by consensus chloroplast microsatellite primers (ccmp-1 to 10). M: *A. mexicana* AH, BH: Hybrids O: *A. ochroleuca*. a. ccmp-1, b. ccmp-2, c. ccmp-3, d. ccmp-4, e. ccmp-5, f. ccmp-6, g. ccmp-7, h. ccmp-1C.

For wider development and application of cpSSRs some suggestions may be kept in view:

(i) DNA sequencing of cpSSR alleles is essential, given the complex nature of the genetic variation associated with hyper variable cpDNA regions.

(ii) *De novo* sequencing of non-coding cpDNA is the mainly efficient and resourceful way to develop cpSSR markers in wild species.

(iii) They are potentially constructive, universal cpSSR primers which provide admittance to merely a miniature of variable markers.

(iv) Cross-species intensification can frequently be advantageous when genus-specific cpSSR primers are accessible

(v) The reliability of cpSSR length based genetic assays need to be validated in all studies related to higher plants.

Whole chloroplast genome as a single locus barcode

Use of the whole chloroplast genomes sequence as a single locus barcode suggested by Parks *et al.*, (2009) and Nock *et al.*, (2011) because of its advancement and significant cost reduction in sequencing technology. This approach does not require the availability of universal primers as PCR amplification. Moreover due to the increased matrix length and number of informative sites, resolution would be enormously increased as has been demonstrated and highlighted by an investigation on 32 gymnosperms, where resolving power of the suggested two-locus barcode (rbcL-matK) and whole chloroplast genome where compared (Parks *et al.*, 2009).

Sequence from chloroplast genome

Maturase K (matK)

Among the gene present in chloroplast genome, *matK* has been reported to have a high evolutionary rate. It is about 1550 bp long encodes for an enzyme known as maturase, which splices type II introns from RNA transcripts (Ems *et al.*, 1995). Evolution rate of *matK* in angiosperms is adequate for resolving the inter-generic as well as interspecific relationships (Johnson and Soltis, 1995). It amplification primers can be easily designed from the conserved reason of *trnK* and flank *matK* on both sides (Johnson and Soltis, 1995).

rbcL

It is large subunit of the ribulose-bisphosphate carboxylase gene. Some chloroplast-encoded genes are interrupted by introns; this is not the case of *rbcL* gene. Unambiguous alignment is important advantages of the *rbcL* gene for the study of seed plant evolutionary history. A second major advantage is a conservative rate of nucleotide substitution. (Clegg, 1993).

rpoB- rpoC1

The *rpoB* gene coding for a β like subunit and *rpoC1* gene coding for a β′-subunit of the chloroplast DNA-dependent RNA polymerase. *rpoC* region is divided into *rpoC1* and *rpoC2* and is located within the large single copy region of the plastid genome in most angiosperms. These genes, along with *rpoB*, co-transcribed as a single operon and encode three subunits of the chloroplast RNA polymerase.

accD

The *accD* gene encodes the β carboxyl transferase subunit of acetyl coA carboxylase, which catalyzes the first step in fatty acid biosynthesis in chloroplasts. Its average length of *accD* is 622 bp, and it ranges from 320–784 bp.

ycf5

The chloroplast gene, *ycf5*, encodes a polypeptide containing 313 amino acid residues. (Tsuruya *et al.*, 2006). This polypeptide is involved in c type cytochrome biosynthesis. Although, the gene possesses high proportion of variable sites, it is not much popular for plant identification because of its low amplification success in most groups and problems in alignment. It is a potential barcode for some plants.

NDHA dehydrogenase J Subunit (ndhJ)

It is one of the 11 genes among the plastid *ndh* gene complex, formerly reported from the plastid genome of Tabacco (Shinozaki *et al.*, 1986). It was earlier known as ORF159. These complexes consist of *ndhA-ndhK* and *ndhJ* was found to be linked with *ndhC* and *ndhK* as a single operon (Burrows *et al.*, 1998). It encodes for NADH dehydrogenase, a 30 KDa protein, which is one of the subunits of functional respiratory protein complex of around 550 KDa present inside the mature chloroplast (Burrows *et al.*, 1998).

atpF -atpH inter-genic spacer region

The two genes *atpF* and *atpH* individually encode CFO I and CFO III subunit of ATP synthase (Drager *et al.*, 1993). Its amplification success was high, there was a significant variation (218 - 847bp) in the retrieved length of the sequence, due to which alignment of the sequence was difficult. It has been suggested only as a supplementary locus that can be used in combination with the standard two–locus barcode (*rbcL+matK*), for a better species resolution in certain taxonomic groups.

trnH-psbA inter-genic spacer

It is one of the most variable inter-genic spacer present in the chloroplast genome of angiosperms. Its length varies from 296-1120 bp with an average of approximately 450bp. It has been successfully amplified from a wide range of angiosperm gymnosperm, ferns, mosses and liverworts (lee *et al.*, 2007). In manly angiosperm, the *trnH -psbA* spacer is exceedingly short (~300bp) in length. Thus, its sequence may not have enough number of informative or variable sites to distinguish species in those groups. Besides, the length of this locus is around 1000bp in some monocots (Chase *et al.*, 2007) and conifers (Hollingsworth *et al.*, 2009).

psbK-psbI inter-genic spacer

The *psbK* and *psbI* genes encode two low molecular weight polypeptides, k and I, respectively. These are involved in photosystem II. They obtained high success in amplification and sequencing of this locus. It has been recommended only as a supplementary barcode due to the contradiction observed in obtaining unambiguous bidirectional sequences.

trnL (UAA) -trnF(GAA): intron and inter-genic spacer

The *trnL* and *trnF* code for tRNAs for leucine and phenylalanine, individually. The intron present in *trnL* was first employed in the plant systematic studies by Taberlet *et al.* (1991). Although, it does not exhibit adequate variation as compared to other non-coding regions of the chloroplast DNA.

Sequence from nuclear genome

Nuclear ribosomal cistron internal transcribed spacer (nrITS)

The nuclear genome is bi-parentally inherited. In nuclear genome, only one region i.e., *nrITS* has been tested for barcoding of plants. The reasons for selecting a limited number of genes from the nuclear genome for DNA barcoding could be problems in obtaining high PCR amplification success for single or low-copy nuclear genes (especially from the degraded samples) and the conserved nature of functional genes across a large number of taxa which would have resulted in a low species discrimination success (Vijayan and Tsou, 2010). Due to its bi-parental inheritance, ubiquitous occurrence and moderately higher number of variable sites due to less functional constraints, *ITS* has been considered as a phylogenetic marker of prime importance for both plants and animal (Alvarez and Wendel, 2003).

REFERENCES

Agarwal M, Shrivastava N and Padh H (2008) Advances in molecular marker techniques and their applications in plant sciences. *Plant Cell Rep* 27: 617-631

Alvarez I and Wendel JF (2003) Ribosomal ITS sequences and plant phylogenetic and evolution. *Genetics* 29: 417-434

Andrianoelina O, Rakotondraoelina H, Ramamonjisoa L, Maley J, Danthu P and Bouvet J-M (2006) Genetic diversity of Dalbergia monticola (Fabaceae) an endangered species in the fragmented oriental forest of Madagascar. *Biodiversity Conservation* 15: 1109-1128

Basha SD and Sujatha M (2009) Genetic analysis of *Jatropha* species and interspecific hybrids of *Jatropha curcas* using nuclear and organelle specific markers. *Euphytica* 168: 197-214

Bastia T, Scotti N and Cardi T (2001) Organelle DNA analysis of *Solanum* and *Brassica* somatic hybrids by PCR with universal primers. *Theor Appl Genet* 102: 1265-1272

Bryan GJ, McNicoll J, Ramsay G, Meyer RC and Jong WSD (1999) Polymorphic simple sequence repeat markers in chloroplast genomes of *Solanaceous* plants.*Theor Appl Genet* 99: 859-867

Cato SA and Richardson TE (1996) Inter- and intraspecific polymorphism at chloroplast SSR loci and the inheritance of plastids in *Pinus radiata* D. Don. *Theoretical and Applied Genetics* 93: 587- 592

Chase MW, Cowan RS, Hollingsworth PM, van den Berg S, Petersen G, Seberg O, Jorgsensen T, Cameron KM, Carine M, Pedersen N, Hedderson TAJ, Conrad F, Salazar GA, Richardson JE, Hollingsworth ML, Barraclough TG, Kelly L and Wilkinson M (2007) A proposal for a standardized protocol to barcode all land plants. *Taxon* 56: 295-299

Cheng YJ, Guo WW and Deng XX (2003) Molecular characterization of cytoplasmic and nuclear genomes in phenotypically abnormal Valencia orange (*Citrus sinensis*) + Meiwa kumquat (*Fortunella crassifolia*) intergeneric somatic hybrids. *Plant Cell Rep* 21: 445-451

Clegg MT (1993) Chloroplast gene sequences and the study of plant evolution. *Proc Nati Acad Sci* USA 90: 363-367

Crandall KA, Bininda-Emonds ORP, Mace GM, Wayne RK (2000) Considering evolutionary processes in conservation biology. *Trends in Ecology and Evolution* 15: 290–295

Drager RG and Hallick RB (1993) A novel *Euglena Gracilis* chloroplast operon encoding 4 ATP synthase subunit and 2 ribosomal-protein contain 17 introns. *Current Genetic* 23: 271-280

Ems SC, Morden CW, Dixon CK, Wolfe KH, Depamphilis CW and Palmer J (1995) Transcription, splicing and editing of plastid RNAs in the nonphotosynthetic plant *Epiphagus virginiana. Plant Molecular Biology* 29: 721-733

Grodzicker T, Williams J, Sharp P, Sambrook J (1975) Physical mapping of temperature sensitive mutants of adenovirus. *Cold Spring Harbor Symp Quant Biol* 39: 439-446

Grosser JW, Gmitter FG Jr, Tusa N, Recupero GR, Cucinotta P (1996) Further evidence of a cybridization requirements for plant regeneration from citrus leaf protoplasts following somatic fusion. *Plant Cell Rep*15: 672-676

Grosser JW, Ollitrault P, Olivares-Fuster O (2000) Somatic hybridization in *Citrus*: an effective tool to facilitate variety improvement. *In Vitro Cell Dev Biol-Plant* 36: 434-449

Grover S, Singh Y, Pareek N, Malik CP (2012) DNA fingerprinting: approaches and applications with emphasis on random amplified polymorphic DNA (RAPD) *Jour Pl Sci Res* 28(1): 1-14

Guo WW, Cheng YJ, Deng XX (2002) Regeneration and molecular characteization of intergeneric sonatic hybrids between *Citrus reticulata* and *Poncirus trifoliata. Plant Cell Rep* 20: 829-834

Guo WW and Deng XX (2002) Wide somatic hybrids of *Citrus* with its related genera and their potential in genetic improvement. *Euphytica* 118: 175-183

Hansen OK, Kjaer ED and Vendramin GG (2005) Chloroplast microsatellite variation in Abies nordmanniana and simulation of causes for low differentiation among populations. *Tree Genetics & Genomes* 1: 116–123

Ishii T and McCouch SR (2000) Microsatellites and microsynteny in the chloroplast genomes of *Oryza* and eight other Gramineae species. *Theor Appl Genet* 100: 1257-1266

Johnson LA and Solitis DE (1995) Phylogenetic inference in saxifragaceae sensu strict and gilia (Polemoniaceae) using *matK* sequences. *Annals of the Missouri Botanical Garden* 82: 149-175

Jordan SA and Humphries P (1994) Single nucleotide polymorphisms in exon 2 of the BCP gene on 7. *Hum Mol Genet* 3: 931-935

Kumar P, Gupta VK, Mishra AK, Modi DR and Pandey BK (2009) Potential of Molecular Markers in Plant Biotechnology. *Plant Omics Jour* 2(4):141-162

Lee HL, Yi DK, Kim KJ (2007) Development of plant DNA barcoding markers from the variable noncoding regions of chloroplast genome. In: Abstract, Second International Barcode of life conference, Academia sinica, Taipei, Taiwan, pp. 18-20. http://www.bolin.org/conferences/assets/files/conference_abstract_book.pdf

Lhuiller E, Butaud JF, Bouvet J-M (2006) Extensive clonality and strong differentiation in the insular Pacific tree Santalum insulare: implications for its conservation. *Annals of Botany* 98: 1061–1072

Litt M and Luty JM (1989) A hypervariable microsatellite revealed by in vitro amplification of a dinucleotide repeat within the cardiac muscle actin gene. *Am J Hum Genet* 44:397-401

Lynch M and Walsh B (1998) Genetics and analysis of quantitative traits. Sinauer Associates, Sunderland, MASSACHUSETTS, pp 980

Moreira CD, Chasee CD, Gmitter FG Jr and Grosser JW (2000) Inheritance of organelle genomes in citrus somatic cybrids. *Mol Breeding* 6: 401-405

Morgante M, Felice N and Vendramin GG (1998) Analysis of hypervariable chloroplast microsatellites in Pinus halepensis reveals a dramatic genetic bottleneck. In A. Karp, P. G. Isaac, and D. S. Ingram [eds.], Molecular tools for screening biodiversity: plants and animals, 407–412. Chapman and Hall, London, UK.

Muller F, Voccia M, Ba A and Bouvet J-M (2009) Genetic diversity and gene flow in a Caribbean tree Pterocarpus officinalis Jacq, a study based on chloroplast and nuclear microsatellites. *Genetica* 135: 185–198

Navascue´s M and Emerson BC (2005) Chloroplast microsatellites: measures of genetic diversity and the effect of homoplasy. *Molecular Ecology* 14: 1333–1341

Neale DB, Wheeler NC and Allard RW (1986) Paternal inheritance of chloroplast DNA in Douglas fir. *Canadian Journal of Forest Research* 16: 1152–1154

Nock CJ, Waters DLE, Edwards MA, Bowen SG, Rice N, Cordeiro GM and Henry RJ (2011) Chloroplast genome sequences from total DNA for plant identification. *Plant Biotechnology Journal* 9: 328-333

Parks M, Cronn R, Liston A (2009) Increasing phylogenetic resolution at low taxonomic levels using massively parallel sequencing of chloroplast genomes. *BMC Biology* 7: 84

Petit RJ, Duminil J, Fineschi S, Hampe A, Salvini D and Vendramin GG (2005) Comparative organization of chloroplast, mitochondrial and nuclear diversity in plant populations. *Molecular Ecology* 14: 689–701

Powell W, Morgante M, Andre C, Mcnicol JW, Machray GC, Doyle JJ, Tingey SV and Rafalski JA (1995a) Hypervariable microsatellites provide a general source of polymorphic DNA markers for the chloroplast genome. *Current Biology* 5: 1023–1029

Powell W, Morgante M, Doyle JJ, McNicol JW, Tingey SV and Rafalski AJ (1996a) Genepool variation in genus *Glycine* subgenus *Soja* revealed by polymorphic nuclear and chloroplast microsatellites. *Genetics* 144: 793-803

Powell W, Morgante M, McDevitt R, Vendramin GG and Rafalski JA (1995b) Polymorphic simple sequence repeats regions in chloroplast genomes: applications to the population genetics of pines. *Proc Natl Acad Sci USA* 92: 7759-7763

Powell W, Machray G and Provan J (1996b) Polymorphism revealed by simple sequence repeats. *Trends Plant Sci* 1: 215–222

Provan J, Powell W, Hollingsworth PM (2001) Chloroplast microsatellites: new tools for studies in plant ecology and evolution. *Trends Ecol Evol* 16: 142-147

Provan J, Soranzo N, Wilson NJ, Goldstein DB and Powell W (1999b) A low mutation rate for chloroplast microsatellites. *Genetics* 153: 943-947

Provan J, Soranzo N Wilson NJ, McNicol JW, Morgante M *et al.* (1999a) The use of uniparentally inherited simple sequence repeat markers in plant population studies and systematics, pp. 35–50 in *Molecular Systematics and Plant Evolution*, edited by P. M. Hollingsworth, R. M. Bateman and R. J. Gornall. Taylor and Francis, London.

Sederoff RR, Levings III S, Timothy DH and WWL HU (1981) Evolution of DNA sequence organization in mitochondrial genomes of Zea. *Proceedings of the National Academy of Sciences*, USA 78: 5953–5957

Semagn K, Bjørnstad A and Ndjiondjop MN (2006) An overview of molecular marker methods for plants. *African Jour of Biotech* 5(25): 2540-2568

Shinozaki K, Ohme M, Tanaka M, Wakasugi T, Hayashida N, Matsubayashi T, Zinta N, Chunwongse J, Obokata J, yamaguchi-Shinozaki K, Ohto C, Torazawa K, Meng BY, Sugital M, Deno H, Kamogashira T, Yamada K, Kusuda J, Takaiwa F, Kato A, Tohdoh N, Shimada H and Sugiura M (1986) The complete nucleotide sequence of the tobacco chloroplast genome: its gene organization and expression. *The EMBO journal* 5: 2043-2049

Tarberlet P, Gielly L, Pautou G and Bouvet J (1991) Universal primers for amplification of three non-coding regions of chloroplast DNA. *Plant Molecular biology* 17: 1105-1109

Tharachand C, Immanuel Selvaraj C and Mythili MN (2012) Molecular markers in characterization of medicinal plants: An overview. *Res. Plant Biol* 2(2): 01-12

Tsuruya K, Suzuki M, Plader W, Sugita C and Sugita M (2006) Chloroplast transformation reveals that tobacco *ycf5* is involved in photosynthesis. *Acta physiologiae Plantarum* 28: 365-371

Vendramin GG and Ziegenhagen B (1997) Characterization and inheritance of polymorphic plastid microsatellites in Abies. *Genome* 40: 857–864

Vendramin GG, Degen B,Petit RJ, Anzidei M, Madaghiele A and Ziegenhagen B (1999) High level of variation at Abies alba chloroplast microsatellites loci in Europe. *Molecular Ecology* 8: 1117–1126

Vendramin GG, Lelli L, Rossi P and Morgante (1996) A set of primers for the amplification of 20 chloroplast microsatellites in Pinaceae. *Molecular Ecology* 5: 595–598

Vijayan K and Tsou CH (2010) DNA barcoding in plants: taxonomy in a new perspective. *Current science* 99: 1530-1541

Vos P, Hogers R, Bleeker M, Reijans M, van de Lee T, Hornes M, Frijters A, Pot J, Peleman J, Kuiper M and Zabeau M (1995) AFLP: a new technique for DNA Fingerprinting. *Nucleic Acids Res* 23: 4407-4414

Weising K and Gardner RC (1999) A set of conserved PCR primers for the analysis of simple sequence repeat polymorphisms in chloroplast genomes of dicotyledonous angiosperms. *Genome* 42: 9-19

Wolfe KH, Li WH and Sharp PM (1987) Rates of nucleotide substitution vary greatly among plant mitochondrial, chloroplast and nuclear DNAs. *Proc Natl Acad Sci USA* 84: 9054-9058

Xu DH, Abe J, Gai J and Shimamoto Y (2002) Diversity of chloroplst DNA SSRs in wild and cultivated soyabeans evidence for multiple origins of cultivated soyabean. *Theor Appl Genet* 105: 645-653

Ziegenhagen B, Scholz F, Madaghiele A and Vendramin GG (1998) Chloroplast microsatellites as markers for paternity analysis in Abies alba. *Canadian Journal of Forest Research* 28: 317–321

Zietkiewicz E, Rafalski A and Labuda D (1994) Genome fingerprinting by simple sequence repeats (SSR)-anchored PCR amplification. *Genomics* 20: 176-183

5

Morphological, Physiological and Biochemical response in Plants Subjected to Salt Stress

Ekta Kathuria, Dheera Sanadhya and *C.P. Malik*

Soil salinity causes drastic effects in plants including osmotic stress, ionic and nutritional imbalances. These ultimately limit plant growth and development. A common feature of these effects is the production of reactive oxygen species (ROS) such as singlet oxygen (O_2), superoxide anion (O_2^-), hydrogen peroxide (H_2O_2) and hydroxyl radical (·OH). These cause cellular damage to the plant. To alleviate the damage, plants employ antioxidants both non-enzymatic and enzymatic. Non-enzymatic antioxidants include ascorbate and glutathione which scavenge ROS species. In enzymatic antioxidants, superoxide dismutase (SOD) reacts with superoxide anion (O_2^-) to produce H_2O_2. H_2O_2 is then scavenged by ascorbate peroxidase (APx) and catalase (CAT). Hence, to sustain food production, development of salt tolerant crops is essential. The approach used to improve salt tolerance includes the generation of transgenic plants by introduction of novel genes or by alteration in the expression level of existing genes. This review focuses on the morphological, physiological and biochemical responses that a plant exhibit under salinity and the various strategies employed to combat stress.

Introduction

Abiotic stresses enforce negative impact of environmental factors on the living organisms (Mane *et al.*, 2010). They pose harmful effect on crops concerning the growth and productivity. The problem becomes much more severe when various abiotic stresses occur together. These stresses include heavy metals, high winds, extreme temperatures, salinity, drought, flood and other natural disasters, such as tornado and wild fire. Adaptations to the stressful conditions vary differentially from plant to plant, even if, they are living in the same area

(Munns, 2002). As soil is required to produce crops, vegetables, fruit, timber and other economically important items, so it is a valuable resource in agriculture and forestry (Mane *et al.*, 2010a). A healthy soil imparts a higher chance of survival under stressful conditions. Maintenance of soil fertility, and crop productivity, is therefore one of the fundamental issue.

In one third world population, various aspects of human lives including human health and agricultural productivity are affected by salinity. Salinity has caused degradation of arable land over many hundred-year periods and degradation of cultivated land during less than 100 years. For example, in California 4.5 out of 8.6 mha irrigated agricultural land has become salt affected during the last century. At present, its extent throughout the world is increasing regularly in extent (Schwabe *et al.*, 2006) and it has now become a very serious problem for crop production (Munns and Tester, 2008), particularly in arid and semi-arid regions.

About 1.93 lakh km area has been effected by saline water of electrical conductivity > 4000 mS cm^{-1}. Ground water salinity is so high in Rajasthan and Gujarat that salt is formed by solar evaporation of well water. According to an estimate of FAO (2008; http://www.fao.org/ag/agl/agll/spush accessed on 6th August, 2008) over 6% of the world's land is salt affected. In addition, out of 230 mha of irrigated land, 45 mha (~20%) are salt affected. However, the intensity of salinity stress varies from place to place. 6.73 mha of land is affected by salinity and alkalinity problem in India. Generally, dry land salinity is categorized into three different types: low salinity (ECe 2-4dS/m), moderate salinity (ECe 4-8 dS/m) and high salinity (ECe > 8 dS/m). Further, it can be categorized into two groups on the basis of source:

A. Primary salinization- It results from weathering of minerals and soil derived from saline parent rocks.

B. Secondary salinization- It is caused by human interference such as irrigation, deforestation, overgrazing or intensive cropping (Sposito, 1989; Ashraf, 1994).

According to an estimate of FAO (2008), 32 mha (~2%) out of 1500 mha are affected by secondary salinity to varying degrees depending upon the type of salinity causing factor. Rengasamy (2006) categorized salinity in three groups based on soil and ground water processes causing salinity as: Ground water associated salinity (GAS), Non-ground water associated salinity and Irrigation associated salinity.

Salt response as a multigenic trait

Processes such as seed germination, seedling growth and vigour, vegetative and reproductive growth of the plant are adversely affected under saline conditions (Lianes *et al.*, 2005). Plants are classified as glycophytes or halophytes based on the capacity to grow in salt medium. Glycophytes cannot tolerate salt stress. It causes reduction in the osmotic potential of the soil, severe ion toxicity and nutrient imbalance and deficiencies.

These all cause the molecular damage and death of the plant (McCue *et al.*, 1990). Salt and osmotic stresses present a high degree of similarity with respect to physiological, biochemical, molecular and genetic effects. The halophyte *Mesembryanthemun crystallinum* is a model system for studying effects of salt response at molecular level. The metabolism of plant shifts from

C3 to CAM (Crassulacean acid metabolism) under salt or drought stress as organic acids, malate and oxalate are some important osmolytes in CAM plants.

Major symptoms that appear due to prolonged salt stress exposure are growth inhibition, accelerated development, senescence and then the death of the plant. The synthesis of abscisic acid is induced which closes stomata and cause a decrease in photosynthesis and as a result oxidative stress occur. The assimilation of other ions like potassium, nitrate, phosphate, calcium etc. is effected under salinity. These ions play important role in maintenance of cell turgor, membrane potential and enzyme activities (Arshi *et al.*, 2002).

Cellular mechanism of salt stress survival

Extreme saline conditions cause hyperosmotic stress and change in ion equilibrium (Hasegawa *et al.*, 2000b; Zhu, 2001). There are two types of survival strategies:

a. Either the plant become dormant in the situation

b. Or the plant adjust to tolerate the saline environment

Salinity initially establishes an imbalance of water potential between apoplast and symplast which then cause a decrease in turgor pressure that results in growth reduction (Bohnert *et al.*, 1995). To overcome turgor pressure, osmotic adjustment is done by accumulation of compatible osmolytes and osmoprotectants (Bohnert and Jensen, 1996). The ions (Na^+, Cl^-) move to the vacuole for minimization of cytotoxicity (Blumwald *et al.*, 2000) directly from the apoplast through membrane vesiculation (Hasegawa *et al.*, 2000b) and by ion transport systems located in the plasma membrane and tonoplast.

Need to develop salt tolerant crops

Salt stress affected more than 10 percent of arable land, which results in more than 50% decline in average yields of major crops (Bray *et al.*, 2000). For this reason, there is an increasing demand for new plant cultivars that have potential for higher yield under such abiotic constraints. With great advancements in the field of plant physiology and molecular biology in the present era, there are great expectations that plant breeders will certainly provide salt tolerant lines/cultivars with higher yield. Generally, it is believed that stress tolerant plants have ability to maintain higher rates of growth in saline conditions. However, during the past decade, progress made in this area is very slow because there is a great controversy among plant physiologists, plant breeders, plant molecular biologists about physiological basis of stress tolerance in plants (Yeo, 1988; Hasegawa *et al.*, 2000; Munns, 2002; Serraj and Sinclair, 2002; Wang *et al.*, 2003; Ashraf, 2004; Ashraf and Harris, 2004; Flowers, 2004; Reynolds *et al.*, 2005; Cuartero *et al.*, 2006; Munns and Tester, 2008). Although there is a reasonable consensus on methods and strategies such as screening for stress tolerant individuals, identification of promising traits, conferring stress tolerance in plants, and development of stress tolerant plants through breeding or genetic engineering. There is still no consensus on physiological traits that are primarily responsible for salt tolerance in plants. In a few plant species, salt tolerant plants have evolved specialized complex mechanisms as adaptive measures for the survival. Genes for cellular based mechanisms of stress tolerance are common in genotypes, development of an adaptive mechanism in plants to tolerate abiotic stresses requires the combination of several morphological, physiological and metabolic processes which depends on a multitude of genes and varies within each target environment. However, among various mechanisms of stress tolerance, mechanisms that regulate ion or water homeostasis are of prime importance (Bartels and Sunker, 2005; Munns and Tester, 2008). Thus, nature of various biochemical and physiological characters responsible for determining crop productivity under stress conditions is very complex (Ashraf *et al.*, 2008).

Morphological and Physiological responses

Seed germination

Usually the term seed germination refers to a complex array of successive processes that lead to radical emergence. The soluble salts in high concentration interfere with the balanced absorption of essential nutritional ions by the plants. The main effect is slow and insufficient germination of seeds, physiological drought, wilting, desiccation, stunted growth, reduction in leaf area, reduction in root and shoot lengths, retarded flowering, fewer flowers, sterility, small seeds etc (Mane *et al.*, 2010). High salt stress delays the emergence of nodal roots, leaf and tiller; decrease in relative growth rate, leaf area ratio, specific leaf area; and decrease in stomatal conductance, leaf-level transpiration and internal CO_2 concentration (Hoffman *et al.*, 2006) along with senescence (lutts *et al.*, 1999). It is well evident that a crop species or cultivar with better germination and seedling growth under salt stress will be more stress tolerant at later stages and will produce better crop growth and productivity (Francois, 1994; Ashraf and Mc Neilly, 2004; Ahmadi and Arkedani, 2006). Lower levels of salinity delay germination, whereas higher levels reduce the final percentage of seed germination (Ghoulam and Fares, 2001). Salts of different nature and concentration, when present in soil, increases its water potential, restricting the movement of water towards the seed surface (Tester and Devenport, 2003; Polesskaya *et al.*, 2006; Houimli *et al.*, 2008). Under increasing salinity stress, water is not imbibed which delays the germination percentage. The adverse effect of salinity on germination has been known since 1896 and the work has been reviewed from time to time. High soil salt content reduces the rate of germination and total seed germination percentage (Sinha *et al.*, 2004). Germination in *Zea mays* (*Z.mays* L.) decreases linearly with rising salinity. Different genotypes of maize were shown to differ in their tolerance to soil salinity at seed germination and seedling growth and whole plant level (Tyagi and Rangaswamy, 1993; Sharma and Gill, 1994). A rapid screening methodology has already been developed to evaluate large number of genotypes in solution culture under laboratory conditions (Sinha *et al.*, 2004).

Root length, shoot length and leaf area

The shoot and root length are the most important parameters for salt stress because roots are in direct contact with soil and absorb water from soil and supply it to the rest of the plant. For this reason, root and shoot length provides an important clue to the response of plants to salt stress (Jamil, 2004). The ionic status in the roots remains relatively good, the ion concentration do not increase with time as in leaves. It has a low ionic strength than the external solution (Tattini *et al.*, 1995; Munns, 2002). The inhibition of shoot growth is due to ion accumulation and then to osmotic adjustment (Osorio *et al.*, 1998). A decrease in shoot to root ratio with increase in NaCl concentration was noticed by (Tattini *et al.*, 1995). According to Munns (1993), the first phase of the growth response results from the effect of salt outside the plant that causes a reduction in leaf and root growth. In maize, when salinity is applied to the root medium, elongation of leaf is immediately stopped (Van Volkenburgh and Boyer, 1985). Toxic effects of the NaCl and unbalanced nutrient uptake by the seedlings reduce shoot and root development. In case of high salinity, water uptake is reduced which then inhibit the elongation of root and shoot. Similar observations have been reported in barley (*Hordeum vulgare* L.), Huang *et al.*, (1995); pigeon pea (*Cajanus cajan*), Subbaro *et al.*, (1991); tepary bean (*Phaseolus acutifolius*), Goertz *et al.*, (1991); tomato (*Lycopersicon*), Foolad (1996); and sugar beet and cabbage, (Jamil, 2004). In all these crops, salinity significantly reduced the root and shoots length at all salinity treatments. Salinity in soils restricts water availability to plants as in water stress which causes reduction in growth rate and in production (Munns, 2002). Increase in salinity levels cause a progressive and significant decrease in plant height, dry matter of shoot and root and seed yield in three cultivars of *Vigna aconitifolia*. Mathur *et al.*, (2006) studied that after exposure of the plants to 300 $mmolL^{-1}$ NaCl, root and shoot fresh weight was decreased by half and also considerable morphological changes were observed. Along with growth reduction, the roots became yellow and the leaves appeared dark-green. The biomass accumulation (fresh and dry weight) represents overall growth of the plant. Application of NaCl (50 to 150 mM) to the seeds reduced fresh and dry weight in two cultivars of *Brassica Juncea* and wheat (Rehman, 2005).

Biomass and moisture content

The shoot biomass production is reduced in the plant due to chlorosis and necrosis of the leaves that decrease photosynthetic active area (De Harralde *et al.*, 1998). Jaleel *et al.*, (2008) found a decrease in fresh weight in *Catharanthus roseus* while Cicek and Cakirlar (2002) found similar results in maize. On the contrary, Ceyhan and Ali (2002) reported an increase in dry weight in lettuce plants. The availability of water is the main environmental factor that limits photosynthesis and growth (Mujeeb-ur-Rahman *et al.*, 2008). The plants growing at higher or lower levels of salinity have low concentrations of water in leaves and shoots. Relative leaf water content (RLWC) measures the water content of a leaf relative to the maximum amount that the leaf can take under full turgidity and hence is considered as an appropriate measure of plant water status under stress. While leaf water potential has been used as an estimate of plant water status when dealing with water transport in the soil-plant-atmosphere continuum, it does not account for osmotic adjustment that commonly occurs in plant roots and leaves in response to stress. Osmotic adjustment is a powerful mechanism for conserving cellular hydration under drought stress and can be accounted for measurement of leaf RLWC.

Hence, RLWC is an appropriate estimate of plant water status because it accounts for both leaf water potential and osmotic adjustment. Plant responses to salt stress, as experienced at early seedling stage, and responses to water stress have much in common. This is because under saline condition, there are two phases of plants responses to salt stress;

a. The first phase is a typical water stress response, due to the high osmotic potential of the soil solution caused by accumulation of soluble salts.

b. The second phase is the injury caused by ionic stress, mainly seen as ion toxicity and nutritional imbalances. The salt tolerant cultivars, could therefore, be selected based on their ability to maintain high relative water content during the initial phase of salt stress.

This study was conducted to determine the association between RLWC and salinity tolerance in rice and to investigate the possibility of using RLWC as screening criteria for breeders to incorporate tolerance to salt stress.

Biochemical responses to salt stress

Growth, development and yield

High salt concentrations in the growth medium reduces growth rate and causes poor and spotty growth of crops, uneven or stunted growth and hence poor yields. However, the extent of adverse effects of salt stress on crop plants greatly differs and it depends on the type and duration of salt stress and climate conditions (Shannon *et al.*, 1994). The nature of salt responsible to salinization plays an important role in the growth and development of plants. A decrease in plant height, root length, number of leaves, leaf area, and dry weight of root, stem and leaves in two cultivars of wheat was observed by Malibari, (1993) and Sharma *et al.* (1994). Similarly, Gorham and Bridges (1995) claimed a significant reduction in fresh and dry weight of cotton. While, Grattan and Maas (1988) reported reduction in shoot dry weight in soybean plant. Salinity causes an imbalance in the uptake of mineral nutrients and their accumulation within the plants. Osmotic stress, ion imbalances of Ca^+ and K^+, and toxic effects of Na^+ and Cl^- cause physiological impairments in the plants.

However, these adverse effects of salinity on plant growth occur in two phases (Munns, 2005).

A. Salinity reduces plant's ability to take up water due to low water potential of external growth medium (in the presence of high Na^+). In this phase, cellular and metabolic changes are very similar to drought affected plants (Munns, 1993). Na^+ does not enter the plant and remains outside the plant but causes salt-induced osmotic effect which reduces plant growth. Thus, growth inhibition in this phase is not due to salt specific effect. Furthermore, salt induced growth inhibition is regulated by hormonal signals coming from roots (Munns, 2002).

B. Na^+ is taken up by roots from where it is transported to shoot through transpiration stream in xylem (Tester and Davenport, 2003). A large number of reports are available which indicate that re-circulation of Na^+ from shoot to root through phloem rarely exists. Thus, major contribution of Na^+ transport is unidirectional thereby resulting in a progressive accumulation of Na^+ in the leaves and eventually the leaves die (Tester and Davenport, 2003). However, the extent of leaf injury depends on the ability of leaf cells to compartmentalize salts in the vacuole (Munns, 2002).

High accumulation of Na^+ in cytoplasm inhibits enzyme activity, and other metabolic processes such as protein synthesis and photosynthesis (Ashraf, 2004; Munns, 2005) thereby reducing leaf growth or causing leaf death.

Photosynthesis

Photosynthesis is an important physiological process by which plants can synthesize their own food. Since it is a primary process in plant productivity, growth of plants is mainly dependent on photosynthesis and also primary process to be affected by salinity. However, the rate of photosynthesis varies with the change in environmental factors, thereby affecting plant growth. For instance, the decline in productivity observed for many plant species subjected to excess salinity is often associated with the reduction in photosynthesis capacity (Dubey, 2005; Allakhverdiev *et al.*, 2005; Kao *et al.*, 2006; Tejera *et al.*, 2006). For example, in 34 canola cultivars, an increase in salinity caused reduction in stomatal conductance and photosynthetic capacity, which resulted in reduced growth (Ulfat *et al.*, 2007). The total chlorophyll content decreases under salinity (Jaleel *et al.*, 2008) in salt stressed sorghum and maize. Salinity causes changes in chloroplast ultrastructure and decrease the rate of photosynthesis. However, salt-induced reduction in photosynthetic capacity of plants may have been due to stomatal and non-stomatal factors. Salinity reduces stomatal conductance immediately due to salt-induced osmotic shock and shortly afterward due to local synthesis of ABA (Fricke *et al.*, 2004). In tomato plants, salinity causes reduction of plant leaf area and stomatal density that result in reduction in photosynthesis (Romero-Aranda *et al.*, 2001). While analyzing various factors affecting net CO_2 assimilation rate in durum wheat, James *et al.* (2002) found that rates of photosynthesis per unit leaf area in salt-treated plants are often unchanged, even though stomatal conductance is reduced. This contradiction could be explained by the changes in cell anatomy such as smaller but thicker leaves that result in a higher chloroplast density per unit leaf area. For example, both epidermal and mesophyll cell thickness and intercellular spaces were decreased in NaCl treated leaves of the mangrove *Bruguiera parviflora* (Parida *et al.*, 2004). Photosystem II is a relatively sensitive component with respect to salt stress (Allakhverdiev *et al.*, 2000). A significant decrease in the efficiency of PSII, ETC and assimilation of CO_2 occurs under salinity. No change in the photosynthetic potential of the plants has been observed by Mathangi *et al.*, (2006). Usually the values of chlorophyll a are dominant over chlorophyll b, but under salinity, the values become closer (Mane *et al.*, 2010). Salinity also reduced intercellular spaces in the leaves of spinach (Delfine *et al.*, 1998). It is also reported that decline in leaf expansion due to salt stress results in accumulation of unused photosynthates in growing tissues and consequently down-regulates photosynthesis. However, the reduction in leaf area due to salinity means that photosynthesis per plant is reduced (Munns and Tester, 2008). Membrane deterioration is also observed along with the stress (Mane *et al.*, 2010). The plant has to effectively compartmentalize the ions in the vacuole, cytoplasm and chloroplast to make the photosynthetic systems resistant to salinity.

Photosynthetic Pigments

Net photosynthesis, transpiration rate and stomatal conductance are significantly affected by salt stress due to changes in chlorophyll content and chlorophyll fluorescence, damage of photosynthetic apparatus and chloroplast structure (Baki *et al.*, 2000; Fidalgo *et al.*, 2004; Kao *et al.*, 2003; Pinheiro *et al.*, 2008). However, Sepehr and Ghorbali (2006) and Siddiqi *et al.* (2009) reported that both chlorophyll a (Chl-a) and chlorophyll b (Chl-b) amounts decreased with NaCl application in *Zea mays* and *Carthamus tinctorius* plants. However, in *Hibiscus esculentus* plants exposed to salt stress, significant increases were only found in Chl-a content but Chl-b was not significantly affected by salt stress (Ashraf *et al.*, 2008). Pinheiro *et al.* (2008) reported that Chl-a, Chl-b, Chl-a/b and carotenoid (Car) contents showed increase and decrease depending on exposure of NaCl in *Ricinus communis.*

Carbohydrate metabolism

Carbohydrates play numerous roles in living things, as structural components (cellulose, chitin) and storage components (starch, glycogen). These are important component of dry matter production and energy relations of the cells. Any change in them is of utmost importance as it is directly involved in various physiological processes such as photosynthesis, translocation and respiration. Carbohydrate metabolism is affected by increase in salinity level as well as by the type of ions present in the ambient solution (Polonenko *et al.*, 1983). For the various organic osmotica, sugars contribute up to 50% of the total osmotic potential in glycophytes, when exposed to salt conditions. Binzel *et al.* (1987) reported an increase in total soluble sugar in tobacco cells adapted to NaCl and reported that organic soluble also may take significant contribution in osmotic adjustment. Weimberg (1987) reported that sucrose is the main form of translocatable carbohydrate from leaves to root and a marked increase in sucrose content cause a reduction in content of reducing sugar under saline conditions. He has found 30 fold increase in sucrose level at -1.2 MPa of NaCl stress in *Triticum* species. The increase in sugar may or may not be associated with a change in starch content. Basra *et al.*, (1991) reported that pre-soaking of maize seed in potassium salts (K_2HPO_4 or KNO_3) each at (0.25%) at 45°C showed increased activity of acid invertase and accumulation of soluble sugars, especially the non reducing fraction. An increase in glucose, fructose and maltose content (except sucrose) has been reported in maize. Similarly, imposition of salt in the growth medium significantly reduced the total sugars in the leaves of all 8 lines of canola (*B. napus*) except cv. Oscar, in which soluble sugars were slightly increased with increasing supply of NaCl (Qasim, 2000). Al-Sobhi (2006) noticed an increase in insoluble and total carbohydrates in shoot and root seedlings of *Calotropis procera* with increase in age and salt.

The accumulation of soluble carbohydrates in plants is a response to salinity or drought, along with a considerable decrease in net CO_2 assimilation rate (Murakeozy *et al.*, 2003). Sugars and starch accumulate under salt stress (Parida *et al.*, 2002) and play a leading role in osmoprotection, osmotic adjustment, carbon storage, and radical scavenging. Dkhil *et al.*, (2010) observed that okra seeds treated with salinity treatment showed lower level of starch and high level of total sugars with increasing salinity. Dubey and Singh (1999) reported that the sugar contents increased more in sensitive than in the tolerant rice cultivars. An increase in soluble sugar has been correlated with the salt tolerance in many plants (Rathert, 1983). A decrease in starch content and an increase in both reducing and non-reducing sugars and polyphenol levels have been reported in leaves of *Bruguiera parviflora* (Parida *et al.*, 2002). Katsuhara *et al.* (1997) observed that within one hour of initiation of salt stress, level of glucose-6-phosphate and UDP glucose decrease markedly and suggested that such decrease might lead directly or indirectly to death of barley root cell under salinity. Cayuela *et al.* (1996) reported an increase in sugar content in tomato leaves under salinity. The content of soluble sugars and total saccharides increase significantly in the leaves of tomato, but starch content is not affected (Khavarinejad and Mostofi, 1998). Kuznetsov and Shevyakova (1997) observed level of soluble sugars and decrease level of starch in tobacco cells under saline conditions.

Imposition of salt in the growth medium significantly reduced the total sugars in the leaves of all 8 lines of canola (*B.napus*) except cv. Oscar, in which soluble sugars were slightly increased with increasing supply of NaCl (Qasim, 2000). Neera and Jasleen (2004) observed a gradual and progressive accumulation of total soluble sugars in chickpea cultivars. However, the quantum of accumulation was highest in the tolerant cultivars than sensitive ones. Similar results have been noticed in cumin (Garg *et al.*, 2002). Trehalose, a disaccharide gets accumulated under abiotic stresses and protects membrane proteins in the cells. It helps to reduce the aggregation of denatured proteins (Singer and Lindquist, 1998) and has a suppressive effect on apoptotic cell death.

Nitrogen metabolism

Nitrogen metabolism of plants changes under saline conditions but the data obtained by different workers are rather controversial. Nitrogen metabolism is also affected by various types of salinity and the plant species tested.

Protein

Stress induces change in the confirmation of protein, thus causing their denaturation. The increase in soluble protein at low salinity and decreases at high salinity has already been observed in mulberry cultivars (Agastian *et al.*, 2000). It could be predicted that plants under stress would have a powerful protein turnover machinery to degrade stress-damaged and environmentally regulated proteins (Abdel *et al.*, 2003). Soluble protein content decreased significantly with increasing salinity in the medium except tolerant genotype of cucumber. A higher content of soluble proteins has been observed in salt tolerant variety of barley, sunflower, finger millet, and rice (Ashraf and Harris, 2004).

It has been known to decrease the protein content in plant tissue either by decreasing amino acid availability for the synthesis of proteins or by increasing their rates of degradation (Polonenko *et al.*, 1983). An increase in the protein contents was due to the slower depletion of protein rather than the synthesis. Several reports had indicated that salinity decreased the synthesis of proteins and increased their degradation by increasing the activity of hydrolytic protein (Sheoram and Nainawatee, 1990). Maintenance of higher protein and free amino acids concentration under salt stress in tolerant as compared to sensitive genotype is in concurrence with earlier reports in Mustard (Garg *et al.*, 1999; Burman *et al.*, 2001) Pearl millet (Sharma and Gill, 1991) and Cluster bean (Lahiri *et al.*, 1996). A decrease in protein and nucleic acid content was reported in roots, leaves and embryo axis of mung bean, chickpea and cowpea under NaCl and Na_2SO_4 salinity. However, Basra *et al.*, (1991) reported an accumulation of soluble protein in germinating embryo under potassium salt solutions. The protein synthesis is very sensitive to fall in osmotic potential. In higher plants, osmotic stress induces several proteins in vegetative tissues, related to late-embryogenesis-abundant (LEA) proteins. This correlation provides a protective role under dehydration stress (Ingram and

Bartels, 1996). Engineered plants with an over expressing barley LEA gene, *HVA1* proved to be stress tolerant than other wild type (Xu *et al.*, 1996). Increase in salinity level progressively and significantly decrease the total soluble protein in three cultivars of *Vigna aconitifolia* (Mathur *et al.*, 2006).

Free amino acids

Amino acids represent one of the important classes of metabolites in the cell and so, are the building blocks of proteins. These form the chemical basis of life and play important role in metabolism. Amino acids (alanine, arginine, glycine, serine, leucine, and valine), imino acid (proline), non-protein amino acids (citrulline, ornithine) and amides (glutamine, asparagine) accumulate in plants under salt stress (Mansour, 2000). Salinity has been reported to increase in total free amino acids (Jones and Gorham, 1983; Le Rudulier *et al.*, 1984; Weimberg, 1987; Sheoram and Nainawattee, 1990). Weimberg *et al.* (1984) and Dubey and Rani (1987) have demonstrated that as salinity increases, free amino acids generally increased to a maximum of about twice that of non stressed plants. Binzel *et al.* (1987) found an increase of total free amino acid content under salt stress conditions which are positively correlated with cell osmotic potential. Amino acids such as proline, asparagine and aminobutyric acid provide osmotic adjustment to the plant under saline conditions. Levels of all 20 amino acids increase under salinity in both root and shoot, with shoot maintaining a high level of all these amino acids in both absence and presence of salinity in two cultivars of green gram (Mishra *et al.*, 2006). However, Mathur *et al.* (2006) studied that with increase in salinity level, there is decrease in free amino acids in three cultivars of *Vigna aconitifolia.*

Proline

Proline is an imino acid and its accumulation as a compatible osmolytes occurs in most plant species belonging to different plant families in response to various abiotic stresses including salinity (Rhodes *et al.*, 1999; Hsu *et al.*, 2003; Ashraf, 2004; Kishore *et al.*, 2005; Ashraf and Foolad, 2007). The extent of proline accumulation in plants under abiotic stress conditions depends on the type of species and the extent of stress (Kishore *et al.*, 2005). Proline is an important compatible solute that plays a role in osmotic adjustment, stabilizes sub-cellular structures (e.g. membranes and proteins), scavenges free radicals, and buffers cellular redox potential under stress conditions (Ashraf and Foolad, 2007). It occurs widely in higher plants and its accumulation, is in larger amounts than other amino acids (Abraham *et al.*, 2003) regulates the accumulation of useable N. The proline content showed an increasing trend in all genotypes of chickpea under salt stress (Singh, 2004). Its accumulation occurs in cytosol, contributing to cytoplasmic osmotic adjustment (Ketchum *et al.*, 1991), membrane stability and mitigates the effect of NaCl on cell membrane disruption (Mansour, 1998). Maggio *et al.* (2002) viewed that proline act as a signalling molecule which is able to activate multiple responses required for adaptation.

There are two ways for proline biosynthesis:

a. L-ornithine pathway

b. L-glutamate pathway

Delauney *et al.* (1993) showed that two enzymes: pyrroline-5-carboxylate synthetase (P5CS) and pyrroline-5-carboxylate reductase (P5CR) are important in proline biosynthetic pathway. It has also been reported that upon relief of stress, proline is broken down to provide sufficient reducing agents that support mitochondrial oxidative phosphorylation and generation of ATP for recovery from stress and repairing of stress-induced damages (Hare and Cress, 1997; Hare *et al.*, 1998). Increased level of free proline is an indicator of protein degradation. In addition, proline is also known to induce expression of salt stress responsive genes (Satoh *et al.*, 2002; Oono *et al.*, 2003; Chinnusamy *et al.*, 2005). It is also one of the important indicators of salt tolerance in some *Brassica spp eg*. exposure of seedlings of salt sensitive and salt tolerant cultivars of Swede rape (*B. napus*) to NaCl stress increased the free proline concentration and the results let the authors to conclude that proline could be used as one of the important selection criteria of salt tolerance in swede rape (Borodina, 1991). It was observed that with a NaCl concentration of 300 mM L^{-1} treatment in nutrient solution, the proline concentration in roots and shoots increased two folds value compared to control in *Canola spp.*

Proline is known to serve as compatible osmolytes, protector of macromolecules and also as scavengers of reactive oxygen species under stressful conditions (Hellman *et al.*, 2000; Ashraf and Foolad, 2007). Proline accumulation in the plant tissue due to salinity stress is reported in many studies, e.g. in rape (Kundu and Paul, 1997), durum wheat (Bajji *et al.*, 2001), and alfalfa (Irrigoyen *et al.*, 1992) in pistachio (Hokmabadi *et al.*, 2005).

ROS

Molecular oxygen in its ground state (triplet oxygen) is essential to life on earth. When it receives extra energy or electrons, it generates a variety of ROS that cause oxidative damage to living cells including lipids, proteins and nucleic acids. Triplet oxygen has two unpaired electrons with parallel spin present in different orbitals. On receiving extra energy, these show anti-parallel spin, a change that increases the oxidizing power of oxygen (Krieger-Liszkay, 2004). When it receives an electron, it forms O_2^- which is converted into H_2O_2 and OH^- (Apel *et al.*, 2004). In photosynthesis, light energy captured by PS II and PS I excite electrons, which is finally accepted by oxygen as a final acceptor, leading to formation of O_2^- (Foyer *et al.*, 2000). ROS is also formed in photorespiration, respiration, and from NADPH oxidases, amino oxidases, and cell wall bound oxidases (Mittler, 2002). The most important ROS formed under normal condition are O_2^- and H_2O_2.

In certain situations, the plant uses AO S in a beneficial way. AO S play an important in the induction of protection mechanism during biotic and abiotic stresses. It plays an important role in the activation of resistance responses during incompatible plant-pathogen interactions. On infection, an NADPH oxidase of plasma membrane is activated, that produce superoxide radicals. These on spontaneous dismutation or via superoxide dismutase converted into H_2O_2.

H_2O_2 has various defensive properties:

a. High levels of H_2O_2 are toxic for both pathogen and plant cells.

b. It acts as a substrate in peroxidative cross-linking reactions of lignin precursors in cell wall formation.

c. It acts as a signaling molecule.

d. It induces several defense genes and a programmed cell death pathway.

These different effects of H_2O_2 are also regulated by other potential signals, such as salicylic acid, superoxide and nitric oxide. Studies on maize hypocotyls and on potato and mustard seedlings have shown that H_2O_2 mediates heat and cold tolerance.

Plants possess both enzymatic and non-enzymatic mechanism for scavenging ROS (Azevedo Neto *et al.*, 2008).

a. Non-enzymatic defense include antioxidants such as MDA, Ascorbate, ROS, H_2O_2.

b. Enzymatic defense include Superoxide dismutase (SOD), Catalase (CAT), Guaiacol peroxidase (GPX), Ascorbate peroxidase (APX) and other enzymes (Chinnusamy *et al.*, 2005; Azevedo Neto *et al.*, 2008).

Generation and scavanging of ROS under salt stress in relation to the type of photosynthesis

Three types of photosynthesis are found in plants (C3, C4 and CAM). In C3 plants, CO_2 is first fixed to 5-C sugar to form a 3-C compound by Rubisco. In C4 and CAM plants, CO_2 is first fixed into a 3-C sugar by PEPc to form 4-C compound, which then releases CO_2 with Rubisco. In maize as a C4 plant, fixation of CO_2 and release into the action of Rubisco are spatially separated due to dimorphic chloroplast (leaf mesophyll and bundle sheath cells).

Salt stress encourages the formation of ROS in plants.

a. Stomatal conductance decreases to avoid excessive water loss. This decreases the internal CO_2 concentration (Ci) and slows down the reduction of CO_2 by Calvin cycle. This cause the depletion of oxidized $NADP^+$, a final acceptor of electrons in PSI and increases the leakage of electrons to O_2 and so, O_2^- is formed (Hsu *et al.*, 2003).

b. The decrease in Ci slows down the reactions of Calvin cycle and induces photorespiration in C3 plants leading to formation of H_2O_2 in peroxisome (Leegood *et al.*, 1995; Ghannoum, 2009).

c. The cell membrane bound NADPH oxidase activate salt stress and generate ROS (Hernandez *et al.*, 2001; Mazel *et al.*, 2004; Tsai *et al.*, 2005).

Stress increases the rates of respiration (Moser *et al.*, 1991; Jeanjean *et al.*, 1993). Photorespiration is strongly induced by salt stress in C3 but not in C4 and CAM plants (Cushman *et al.*, 1997). Stepien and Klobus (2005) reported that maize suffered less oxidative stress under salt stress. Moreover, antioxidative activities of SOD, APX and GR were higher in maize. So, it is still unclear that whether the higher tolerance of maize to oxidative stress is due to lower production of ROS or due to higher activities of antioxidative enzymes.

Polyamime

These are low molecular weight polycations of high biological activity. The most commom are Put, Spd, Spm, and cadaverine. These play important role in diverse fundamental processes of the cell eg. growth and development, cell division and elongation, embryogenesis, morphogenesis, fruit development, enzyme regulation, replication, transcription and translation as these are present in all compartments of the cell (Bouchereau *et al.*, 1999; Kaur-Sawhney *et al.*, 2003; Galston *et al.*, 1997; Walden *et al.*, 1997). These occur in a much high concentration (10^{-9}-10^{-5} M) as compared to that of other endogenous hormones (10^{-13}-10^{-7} M). Under environmental stressed conditions, polyamines get accumulated in the cell. These polycations interact with the negatively charged macromolecules and stabilize DNA, RNA, membranes and some proteins (Bachrach, 2005).

Role as non-enzymatic antioxidant

The exogenous treatment of free PAs on injury of tobacco leaf caused by O_3 cause an increase in conjugated PAs (Langabartels *et al.*, 1991). Free PAs also function as ROS scavenger (Ha *et al.*, 1998; Kuznetsov *et al.*, 2007). An interesting analysis has been made out on the content of free PAs- perchloric PCA soluble and PCA insoluble in mature leaves and roots of C3-CAM species of *M.crystallinum* under saline conditions (Shevyakova *et al.*, 2006a). The grown leaves contain PCA bound forms- Put, Spm and Spd which increase with increasing duration and concentration of salt treatment. The roots contain both the forms. This supports the involvement of high-molecular polyamine as a mechanism for salt-tolerance. PAs scavange ROS by forming soluble conjugates with various phenol derivatives. PAs play important function in oxidative stress which is mediated by H_2O_2 formed during oxidative degeneration.

Role as enzymatic antioxidant

PAs regulate the expression of genes encoding antioxidant enzymes. PAs induce the expression of antioxidant genes for Spd in peroxidase of tobacco plants (Hiraga *et al.*, 2000) and for Cad in S OD in roots of *M.Crystallinum* (Aronova *et al.*, 2005). The data (Hiraga *et al.*, 2000; Aronova *et al.*, 2005) suggest that PAs not only act as ROS scavenger but also as activators of expression of genes encoding antioxidant enzymes.

Antioxidants (Non- Enzymatic)

Hydrogen peroxide

Most environmental stresses affect the production of active oxygen species. The electron transport chain of mitochondria and chloroplast are the main site of active oxygen species (AO S) production. Electron passing through the transport chain in the photo systems and mitochondria react with the oxygen to form superoxide radicals (O_2^-) and hydrogen peroxide (H_2O_2). Micro bodies are also the source of AOS. H_2O_2 is also formed in glyoxisomes during fatty acid degradation and in peroxisomes during photorespiration. H_2O_2 is a vital cellular component and play role in development, metabolism, homeostasis of aerobic organism (Bienert *et al.*, 2006). It is one of the stable and non-radical ROS in plants and regulates acclimation, defense and development (Slesak *et al.*, 2007). It acts as a signaling molecule and a translocating secondary messenger that trigger Ca^{2+}, protein modification and gene expression (Bienert *et al.*, 2006). This in turn induces various transcription factors which then induce antioxidative enzymes (Agrawal *et al.*, 2005).

Ascorbic acid

It exists in high concentrations in many cellular environments such as stroma of chloroplast- 2.3×10^{-3} M. It possesses high antioxidative activity (Smirnoff, 1996). It reduces two equivalents of O_2^- and produce H_2O_2 and dehydroascorbic acid. It reacts with singlet oxygen at a faster rate (Noctor *et al.*, 1998). In physiological conditions, ascorbic acid exists mostly in the reduced form in leaves and chloroplast (Smirnoff, 2000). In cytosol, it can be present upto 20 mM and in chloroplast stroma, upto 20-300 mM. Ascorbic acid is a main RO S-detoxifying compound in the aqueous phase as it donates electrons. It has the ability to scavange superoxide, hydroxyl radicals and singlet oxygen. It reduces H_2O_2 to water via ascorbate peroxidase reaction. It is involved in the regulation of cell division, cell cycle progression from G1 to S phase (Smirnoff, 1996) and cell elongation (De Tullio *et al.*, 1999). It is found at concentration of 1-2 mM in legume nodules (Dalton *et al.*, 1986; Matamoros *et al.*, 1999). It is essential for ASC-GSH pathway.

Salt stress increases the accumulation in roots, stems and leaves by interactions with damaging AO S (Shalate and Neumann, 2001). Its isolation from Arabidopsis mutant provided the first genetic evidence for its importance in stress tolerance.

Glutathione (GSH)

GSH, a reduced form of Glutathione, is a tripeptide (gammaglu-cys-gly) present in plants, animals and microorganisms. Other homologs are also present with some replacements in gly at C-terminus such as homoglutathione- gamma-glu-cys-ß-ala, hydroxymethylglutathione- gamma-glu-cys ser (Klapheck *et al.*, 1992). The reduced (GSH) and oxidized form (GSSG) of glutathione are capable to interchange. The oxidized compound is sulfur-sulfur linked compound called as glutathione disulfide. The ratio of the oxidized and reduced glutathione is a sensitive indicator of oxidative stress. It plays important role in:

a. Sulfur metabolism (Rennenberg *et al.*, 1982)

b. Protection from stress

It is an important non-protein thiol that acts as an antioxidant and is a precursor of phytochelatins. The decrease in glutathione content cause oxidative stress to the plant (Reed, 1985). GSH has been considered as a signalling molecule against different stresses.

The synthesis of glutathione occurs in two constitutive steps (Hell *et al.*, 1988, 1990):

a. Cysteine and glutamate are combined by gamma glutamyl cysteinyl synthase, with cysteine as a limiting factor.

b. Next is the reaction of glycine with gamma glutamyl cysteine to form GSH. Reaction is catalyzed by GSH synthetase.

If it is not converted to GSH, it is converted to 5-Oxyproline catalyzed by gamma glutamyl cyclotransferase.

GSH pool acts as a:

a. Cofactor G-S-S-G transferase

b. Substrate for gamma glutamyl transpeptidase which catalyzes the transfer of glutamine from GSH to other amino acids.

c. Cofactor for antioxidative enzymes.

GSH plays an important role in cellular homeostasis, protein synthesis, enzyme catalysis, transmembrane transport, receptor action, intermediary metabolism, and cellular maturation. Under stress conditions, oxidation of GSH is accompanied by net glutathione degradation (Foyer *et al.*, 1976). Glutathione accumulates in response to increased AOS generation (May *et al.*, 1996; Schmidt *et al.*, 1986; Smith *et al.*, 1985). Regulation of GSH synthesis is significant due to enhanced AOS generation.

It can be synthesized in the cytosol and the chloroplast. In maize, glutathione is more abundant in bundle sheath cells than in mesophyll cells (Doulis *et al.*, 1997). The sulfate assimilation in C4 cells is confined to the bundle sheath cells (Burnell *et al.*, 1984; Gerwick *et al.*, 1980) although cys synthesis may occur in both types of the cells (Burnell *et al.*, 1984).

Antioxidants (Enzymatic)

Since higher plants are immobile, they can't escape from environmental stresses. The plants scavenge the toxic effects of AOS by antioxidative enzymes, peptides and metabolites (Tanaka, 1994). High antioxidant activities are inferred as a symptom of oxidative damage. These enzymes (Catalase, Peroxidase, APX and SOD) perform a specific role in ROS detoxification. Increase in activities of SOD, APX, CAT, PX under drought, high temperature and salinity, and comparatively higher activity in tolerant wheat genotypes has been reported by Sairam *et al.*, 1997; 1998; 2000. Sairam and Srivastava (2002) reported comparatively higher Cu/Zn-SOD, Fe-SOD, APX, and GR activity in chloroplastic fraction and Mn-SOD in mitochondrial fraction in tolerant wheat genotypes in response to salinity. Hernandez *et al.*, 2001 reported NaCl-induced enhanced mRNA expression and activity of Mn-SOD, APX, GR and MDHAR in tolerant pea cv. Granda, while in salinity-sensitive cv. Chills. Roxas *et al.* (2000) reported over expression of a tobacco glutathione-S-transferase (GST) and glutathione peroxidase (GPX) in transgenic tobacco seedlings under a variety of stresses. Doulis *et al.* (1997) reported on the differential localization of antioxidants in maize. Total SOD and APX activity in maize is only detected in bundle sheath cells, whereas GR and DHAR activity are exclusively present in the mesophyll tissues. Each cell type plays an important role in destroying ROS. The mesophyll cells contain chloroplast with both PSI and PSII, whereas the bundle sheath cells are deficient in PSII. These exhibit little oxygen evolution. As a result, the majority of ROS species are formed in the mesophyll cells.

Superoxide Dismutases (SOD)

SOD (Superoxide oxidoreductase; EC 1.15.1.1) is considered key enzymes of the antioxidative stress defence mechanism (Van Camp *et al.*, 1994a). They determine the cellular concentrations of O_2^- and H_2O_2 since they dismutate superoxide into O_2 and H_2O_2.

$$2H^+ \quad + \quad 2O_2^- \quad ---\rightarrow \quad H_2O_2 \quad + \quad O_2.$$

They are found in all aerobic organisms and sub cellular compartments where active oxygen species are formed. Various iso-enzymes are found on the basis of metal cofactor such as (Cu/Zn), Fe, Mn. FeSOD and MnSOD are structurally similar (Stallings *et al.*, 1984), Cu/Zn show no structural relationship to each other (Bowler *et al.*, 1991). The three enzymes are located in different subcellular compartments, having different molecular properties and differential sensitivity to inhibitors. Recently, a new type of SOD with Ni in the active centre has been described in *Streptomyces* (Kim *et al.*, 1996). All SOD enzymes are nuclear encoded, and SOD genes are sensitive to environmental stresses as a result of ROS formation. This is proved in an experiment with *Zea mays,* where a 7-day flooding treatment cause an increase in TBARS content, membrane permeability and formation of superoxide anion radical and hydrogen peroxide in the leaves (Yan *et al.*, 1996). Four different cytosolic and one chloroplastic Cu/ZnSODs are identified. The cytoplasmic SOD-2, 4, 4a and 5 are made up of two subunits of mol.wt 17KDa each (Baum and Scandalios, 1981; Cannon and Scandalios, 1989).

Ascorbate peroxidase

Ascorbate peroxidases (APX; E C 1.11.1.11) destroy harmful H_2O_2 via the ascorbate-glutathione pathway in chloroplasts and cytosol of plants, algae, and some cyanobacteria. The pathway provides protection against oxidative stress by a series of coupled redox reactions in photosynthetic tissues, mitochondria and peroxisomes (Foyer and Halliwell, 1976).

APX uses ascorbate as an electron donor and eliminates H_2O_2.

$$2\ \text{Ascorbate} + H_2O_2 \dashrightarrow 2\ \text{Monodehydroascorbate} + 2\ H_2O$$

Monodehydroascorbate reductase (MDHAR) catalyzes the regeneration of ascorbate from monodehydroascorbate (MDA) in stroma and in chloroplast membrane by ferredoxin. MDA can also dissociate into ascorbate and dehydroascorbate (DHA). Dehyroascorbate reductase (DHAR) catalyzes the reduction of DHA. DHAR involves reduced glutathione as an electron donor. Oxidized glutathione is recycled by NADPH-consuming glutathione reductase. In pea and radish, APX activities are induced by drought and salt stress. APX levels increase during germination along with an increase in ascorbate levels (de Gara *et al.*, 1997). Seven different APXs are found in plants: two soluble cytosolic forms, three membranes bound cytosolic forms, one in stroma of chloroplast and one in membrane bound thylakoid. These differ in molecular weight; pH optima; affinity to ascorbate and H_2O_2; electron donor specificity (Creissen *et al.*, 1994). The chloroplast isoforms use ascorbate as an electon donor and cytosolic isoforms use pyrogallol.

Catalases

Plants have multiple forms of catalase (H_2O_2: H_2O_2 oxidoreductase; EC 1.11.1.6), mainly found in peroxisomes, glyoxisomes and in mitochondria in maize. It oxidizes substrate such as methanol, ethanol, formaldehyde and formic acid with utilization of H_2O_2. It catalyzes the decomposition of H_2O_2 to water and oxygen.

$$H_2O_2 \xrightarrow{\text{Catalase}} 2H_2O + O_2$$

$$H_2O_2 + RH_2 \xrightarrow{\text{Catalase}} 2H_2O + R$$

Catalase is a tetramer of four polypeptide chains, each over 500 amino acids. Catalase is of three types based on their gene expression. Class I catalases are most prominent in photosynthetic tissues involved in removal of photo respiratory H_2O_2. Class II catalases are expressed in highly vascular tissues that play a role in lignification. Class III catalases are only abundant in seeds and young seedlings and play an important role in removal of excessive H_2O_2 formed during fatty acid degradation in glyoxisomes. Catalase is the most effective antioxidant enzyme that prevents oxidative damage. Catalase activity in maize (Azevedo *et al.*, 2006), wheat (Sairam *et al.*, 2002) and *Sesamum indicum* (Koca *et al.*, 2007) is affected under salt stress.

Peroxidase

Peroxidase (POD) is widely distributed in higher plants and involved in various processes such as lignifications, auxin metabolism, salt and heavy metal tolerance. It catalyzes a reaction of the form:

$$ROOR' + \text{electron donor } (2e^-) + 2H^+ \dashrightarrow ROH + R'OH$$

Higher POD activity in kikuyu plants under salt stress is a result of minimization of oxidative stress. Mechanical properties of cell wall are changed under salt stress. This could be related to salt adaptation (Sancho *et al.*, 1996).

Under anoxia, a differential response of the peroxidase system is observed in coleoptiles and rice seedlings.

The different protective systems operate in the plant to protect the plant from stressed condition (Fig.1). These include:

- Proline and free amino acids were significantly accumulated in seedlings in response to salinity. Their accumulation possibly contributed to osmotic adjustment at the cellular level and enzyme protection stabilizing the structure of macro molecules and organelles. Besides osmotic adjustment, they aided in maintaining the integrity of plasma membrane, acted as a source of carbon and nitrogen and scavenged hydroxyl radicals.
- Increased salt levels caused an enhanced production of ROS (H_2O_2 and O_2^-) which in turn affected the membrane integrity causing an increased lipid peroxidation, and percentage membrane injury. Higher lipid peroxidation reflected on the extent of oxidative injury to cell membranes. Lipid peroxidation was measured as the amount of MDA produced when polyunsaturated fatty acids in the membrane underwent oxidation by the accumulation of free oxygen radicals. This feature could be taken as an indicator of membrane damage.
- In response to increased ROS and membrane injury, L-Ascorbic acid, a precursor for oxalate and tartarate synthesis in plants, act as a non-enzymatic antioxidant. It participated in a variety of processes including photosynthesis, cell cycle, cell wall growth, and cell expansion, synthesis of ethylene, gibberellins, anthocyanins, and hydroxyprolines and was resistant to environmental stresses.
- Activity of H_2O_2 scavenging enzymes changed in response to NaCl stress. APX,

CAT and SOD played significant role in detoxification. Catalase catalyzes the decomposition of H_2O_2 into less reactive gaseous oxygen and water molecules. Activity of PX is not as significant against salt stress as compared with other oxidases e.g. ascorbate peroxidase, a H_2O_2 scavenger that belongs to the ascorbate glutathione cycle.

- The major substrate for the reductive detoxification of H_2O_2 is ascorbate, which was continuously regenerated from its oxidized form.
- GSH provided protection against oxidative stress by reduction of ascorbate via ascorbate-glutathione cycle. Its content enhanced with increased levels of stress.

Changes in the minerals

Essential minerals are required by the plants to grow and survive. Salt stress largely affects macronutrients as compared to micronutrients (El-Fouly and Salma, 1999). Excessive ions are accumulated in the plants due to vacuolation (Huang and Stevninck, 1990) and so, the concentration of micronutrients depends on the type of crop species (Sharpley, 1992) and the levels of salinity and organs of the plant reflect the concentration of macronutrients (Hu and Schmidhalter, 2001). High ion concentration in the solution is taken by the roots and so, the ions are accumulated in the tissue. These ions then inhibit the uptake of other ions (K^+ and Ca^{2+}) through the roots. So, other nutrients become deficient due to high concentration of Na^+. Different species differ in their ability to either accumulate or exclude Na ions. Leaves are more susceptible than roots to Na^+ as these ions accumulate more in shoots than in roots (Tester and Devenport, 2003). Cl^- is used by pepper plants for osmotic adjustment in saline conditions (Martinez- Ballesta *et al.*, 2004). In saline conditions, Na^+ cause membrane depolarization which then activates Ca^{2+} channels and so, Ca^{2+} oscillations are generated which then signal salt stress (Sanders *et al.*, 1999). Glycophytes show improved salt tolerance when Ca is added to root growth medium (Lauchli, 1990).

The source of Mg^{2+} is the Mg-centered porphyrin ring in chlorophyll. Mg^{2+} decreased with increasing salinity in halophyte, *Suaeda fruticosa* (Khan *et al.*, 2000a), *Sporobolus virginicus* (Marcum and Murdoch, 1992), alfaalfa (Eschie and Rodriguez, 1999), cowpea genotypes (Murillo-amador *et al.*, 2002), bittlet almond (Shilbi et al., 2003) and in *Pinica granatum* (Naeini *et al.*, 2004).

Salinity decrease iron content in pea plants (El-Hamdaoui *et al.*, 2003), alfalfa (Eschie and Rodriguez, 1999) and bitter almond (Shilbi *et al.*, 2003).

Ion homeostasis

Salt stress disturbs homeostasis and distribution of ion (Serrano *et al.*, 1999; Zhu, 2001). Ion concentration should be maintained in the cell by accurate ion flux. The plant growth and development is restricted by high Na^+ and Cl^- in soil and water (Wang *et al.*, 2003). Researchers have already proved that Na^+ homeostasis is modulated by Ca^{2+} and high $[Na^+]$ ext negatively affects K^+ uptake (Rains and Epstein, 1967). High Na^+ interrupt K^+ uptake by roots through common transport system, so it must be compartmentalized into vacuoles (Hassegawa *et al.*, 2000). Cellular homeostasis in the plants under saline conditions is

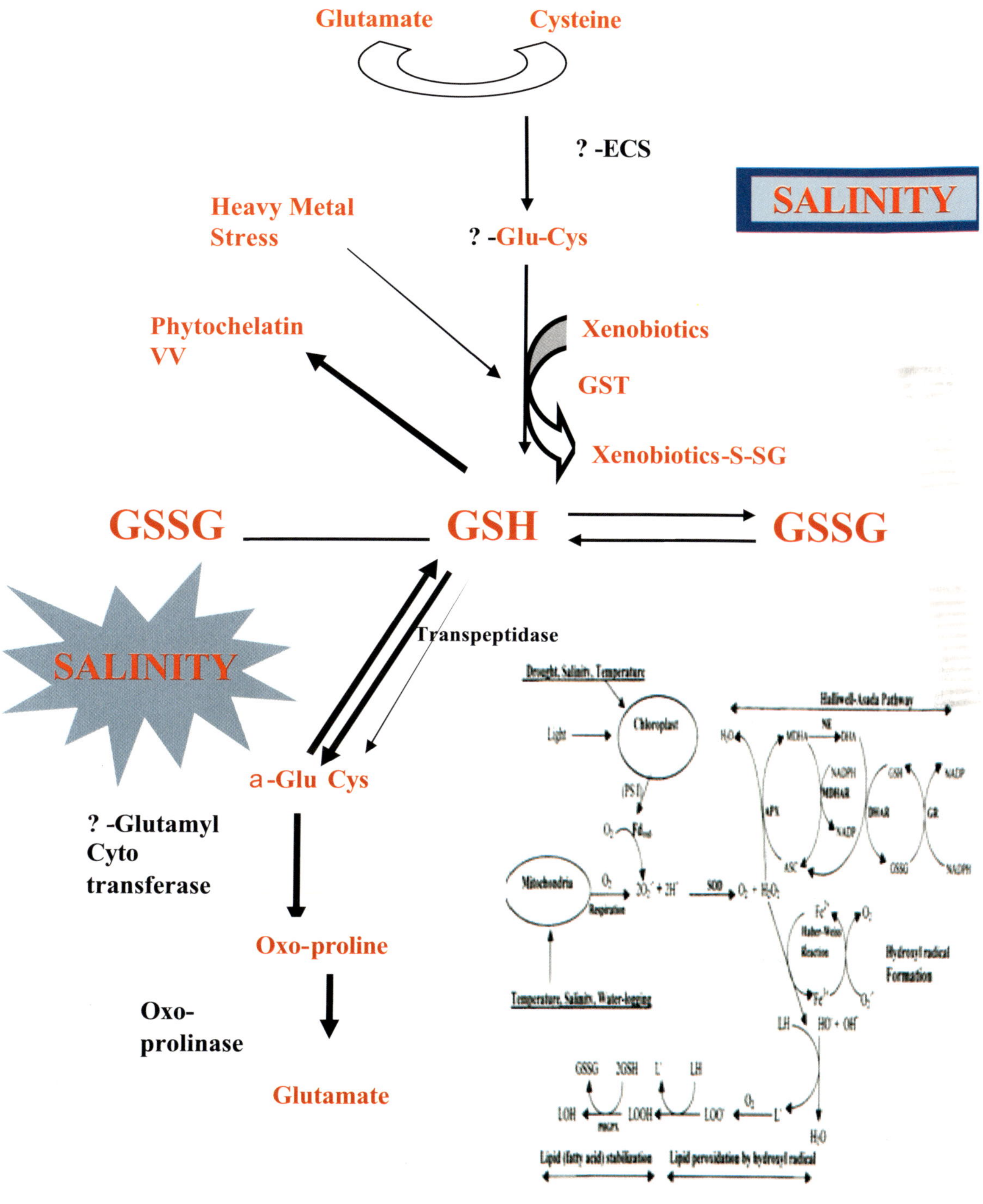

Fig. 5.1. Different protective systems operating in the plant to protect from stressed condition

maintained by ATPases, channels and co-transporters. The exchange of Na^+ for H^+ across the membrane is done with Na^+/H^+ antiporter. This then maintains cytoplasmic pH, sodium levels and cell turgor pressure (Serrano *et al.*, 1999).

Maintenance of Na^+ homeostasis

Pumps

Cytotoxic ions are compartmentalized from the cytoplasm by H^+ pump (Maeshima, 2000; Maeshima, 2001). These pumps provide the driving force for secondary active transport and so, electrochemical ion flux is facilitated. A P-type ATPase H^+ pump maintain large pH and membrane potential gradient across the membrane (Morosomme and Boutry, 2000). A vacuolar-type H^+ ATPase and a vacuolar pyrophosphatase maiantains pH and membrane potential across the tonoplast (Drozdowicz and Rea, 2001). A new plasma membrane H^+-ATPase is identified as a salt tolerant determinant by *aha*4-1 mutation (Vitart *et al.*, 2001).

Channels

Na^+ enters the plant cells by affinity of some non-selective cation channels (Amtmann and Sanders, 1999). These can be voltage-dependant or voltage-independant. VIC is the major route for entry of Na^+ into plant cells whereas voltage-dependant cation channels are the minor routes. Ca^{2+} inhibits Na^+ entry into the roots by blocking non-selective cation channels, as these are sensitive to Ca^{2+}. Other divalent cations e.g., Mg^{2+}, Ba^{2+} and Zn^{2+} maintain Na^+/K^+ ratio in the cytosol by regulating K^+ transporters across the membrane (Shabala *et al.*, 2005).

Na^+ efflux

SOS I gene is an exchanger of Na^+/H^+ that facilitates transport of Na^+ encode Na^+/H^+ antiporter in the plasma membrane catalyze Na^+ efflux (Shi *et al.*, 2000,2002; Qiu *et al.*, 2002; Quintero *et al.*, 2002). Its activity is detected in all tissues of salt- stressed plants, but it was highest in root epidermal cells and vascular tissue (Shi *et al.*, 2002). *SOS I* gene control Na^+ transport between roots and leaves by loading and unloading of Na^+ between xylem and phloem. This is mediated by H^+-ATPases- AHA4 (Vitart *et al.*, 2001). In salt-stressed plants, transcript level of H^+-ATPases increases (Niu *et al.*, 1993). An overexpression of H^+ pyrophosphatase found in *Arabidopsis, AVP*1 improved both salt and drought tolerance (Gaxiola *et al.*, 2001). *AtHKT1* (High affinity K transporter) is the only member of the *Arabidopsis* gene family, which works as an effector of Na^+ influx (Rus *et al.*, 2001). Rice *OsHKT1* and *OsHKT2* were identified based on sequence similarity with wheat *HKT* (Horie *et al.*, 2001). *OsHKT1* functions as a Na^+ influx system like *AtHKT1* but *OsHKT2* is a Na^+/K^+ symporter.

Na^+ vacuolar compartmentalization

An important mechanism that reduces accumulation of Na^+ in the cytosol is the compartmentalization of Na^+ in the vacuole. It seperates Na^+ from cytosolic enzymes, help in the osmotic adjustment and maintain water uptake from saline conditions. The change in pH across the tonoplast facilitates vacuolar compartmentalization of the cation cause the activation of Na^+/H^+ antiporter. The *Arabidopsis AtNHX1* has sequence similarity to mammalian

NHE transporters (Apse *et al.*, 1999; Gaxiola *et al.*, 1999; Quintero *et al.*, 2000). Salt tolerance is enhanced in the transgenic plants that over express *AtNHX1*. This is due to the abundant accumulation of transporter in the tonoplast (Apse *et al.*, 1999; Quintero *et al.*, 2000; Zhang and Blumwald, 2001). So, this proves the importance of *AtNHX* family in vacuolar compartmentalization of Na^+.

K^+ Homeostasis

To maintain cellular metabolism, high K^+/Na^+ is essential. Na^+ competes with the K^+ for uptake in the roots. The level of *AtKC1*, a K^+ transporter gene is increased under salt stress in *Arabidopsis* roots (Pilot *et al.*, 2003). K^+ channels are regulated by protein kinase (Li *et al.*, 1998) and phosphatases (Cherel *et al.*, 2002). K^+ flux across the membrane is maintained by K^+ channels. In *Arabidopsis, AtCNGC10*, a member of a cyclic nucleotide gated family and calmodulin is isolated.

Regulation of ion homeostasis

SOS (salt overly sensitive) pathway

Under stressed condition, plants produce some signals which are then transduced into different proteins and help in tolerance mechanism through signalling cascade and regulates ion homeostasis (Hasegawa *et al.*, 2000a; Sanders, 2000; Zhu, 2000; 2001; Chinnusamy *et al.*, 2005). This pathway is induced by high intracellular or extracellular Na^+ which then triggers Ca^{2+} signal in the cytosol (Zhu, 2001). This results into changes in both activity and expression of transporters for Na^+, K^+ and H^+. This change in concentration of ions is sensed by trans-membrane protein present on the surface of plasma membrane or inside the cell (Zhu, 2003). SOS1 also plays an important role in sensing excess Na^+ through channels or kinases e.g., histidine and wall associated kinases (Urao *et al.*, 1999; Kreps *et al.*, 2002; Seki *et al.*, 2002). SOS3 encodes a myristoylated Ca binding protein which senses cytosolic Ca changes and translates it to downstream responses (Ishitani *et al.*, 2000). This protein showed similarity to β-subunit of calcineurin and neuronal Ca^{2+} sensors (Liu and Zhu, 1998; Ishitani *et al.*, 2000). SOS3 then interacts and activates ser/thr protein kinase, SOS2 in a calcium dependant way (Liu *et al.*, 2000). SOS3 interacts at 21st amino acid of SOS2 and form SOS3-SOS2 kinase complex. This then phosphorylates and activates Na^+/H^+ exchanger encoded by SOS1 gene (Shi *et al.*, 2000). Along with SOS2, the activity of SOS1 is also regulated by SOS4. SOS4 catalyzes the formation of pyridoxal-5-phosphate, a cofactor for SOS1 (Shi *et al.*, 2002). Apse *et al.*, 1999 have discovered a single endogenous gene (*AtNHX1*) that encodes Na^+/H^+ antiporter protein. Cheng *et al.*, (2004) and Qiu *et al.*, (2003) suggested that SOS2 kinase regulates the activity of *AtNHX1* and *CAX1* positively and *AtHKT1* negatively. In yeast, high calcium concentration activates PP2B phosphatase calcineurin which then transcript ENA1 leading to encoding of P-type ATPase. This ATPase is responsible for efflux of Na^+ across the plasma membrane (Nakamura *et al.*, 1993; Mendoza *et al.*, 1994). So, it was suggested that this is basically a Calmodulin-dependant pathway which gets activated by Ca^{2+} (Ehlers and Augustine, 1999; Sanders *et al.*, 1999). Calmodulin then follows calcineurin pathway and activates signalling (Yokoi *et al.*, 2002).

Changes in gene expression-Transcription factors

On salt exposure, changes occur in the gene expression of plants. Salt-tolerant species are used to isolate genes that are involved in salt tolerance which provide a further clue that how these differ from salt-sensitive species. Experimental approach involved is screening cDNA libraries constructed from mRNA isolated from species under salt stress.

Protein profiling of the plant under stress is done to elucidate the changes in gene expression at the transcriptional and post-transcriptional levels. The results supported both qualitative and quantative changes in the pattern of polypeptides formed under salt stress (Ericson *et al.*, 1984; Moons *et al.*, 1997). These studies were further extended to study the changes in mRNA (Gulick *et al.*, 1987; Robinson *et al.*, 1990).

Signalling pathways are controlled by a large number of transcriptional factors. Moons *et al.* (1997) firstly discovered the relation between salt-tolerance and level of endogenous ABA in salt-treated roots of both salt-tolerant and salt-sensitive varieties in rice. Some transcription factors are independent of ABA regulation and are called CBFs-C-Repeat binding proteins or DREBs- Dehydration Responsive Element Binding Proteins. Two transcription factors have been isolated from Arabidopsis- *DREB1* (*DREB1A*, *DREB1B* and *DREB1C*) and *DREB2* (*DREB2A* and *DREB2B*). Salt stress and osmotic stress induce *DREB2A* and *DREB2B* (Liu *et al.*, 1998; Thomshow *et al.*, 1999; Shinzaki and Yamaguchi-Shinozaki, 2000). *ABREB1* and *ABREB2* are the two transcription factors from *Arabidopsis* bZIP and, are upregulated by salinity, drought and ABA. Transgenic plant produced from transformation of *Arabidopsis* with *DREB1* gene showed improved tolerance to salinity and drought. Sigma factors such as RopS and SigB in heterotrophic bacteria (Hecker *et al.*, 1996) and SigF in phototrophs regulate the expression of salt-inducible genes.

Transgenic strategies to combat stress

Recent studies in gene mapping have offered new opportunities in understanding the genetics of stress-resistance genes. The information thus gained will be useful for the breeders to develop stress-tolerant crops. Breeders use molecular genetic markers to keep track of a genetic locus that is responsible for salt resistance. One strategy involved is marker-assisted selection and the other involved introduction of selected genes into selected crops. Selected genes are first analysed through either targeted or non-targeted strategy. Targeted strategy includes complete information about the components and reaction. Non-targeted approach includes the processes like differential hybridization and shortgun cloning. The selected genes whose products ameliorate the stressed conditions are upregulated. Then, these salt-resistant varieties are developed to combat the stressed condition. A single gene manipulation can increase tolerance to stressed condition by SOS1 (Shi *et al.*, 2003).

Various salt tolerant transgenic plants express genes involved in synthesis/over expression of ion transporters and compatible solutes - organic compounds such as amino acids (e.g., Proline) Sugars and their alcohol (e.g., trehalose, mannitol) and ammonium compounds (e.g., glycinebetaine and polyamines). Pommerrenig *et al.*, (2007) introduced *PmSDH1* in *Plantago major* that overexpress Sorbitol dehydrogenase and make the transgenic plant tolerant to salt stress. For more trehalose production, introduction of *TPS* + *TPP* fusion (Jang *et al.*, 2003); *otsA* +*otsB*(*TPSP*) (Garg *et al.*, 2002) in rice and rice (PB1) respectively is being done, which cause overproduction of trehalose-6- phosphate synthase and T-6-P phosphatase. Trehalose levels in seeds and leaves were more and transgenic plants showed enhanced tolerance both during and after stress and had longer and thicker roots. Transgenic plants had high levels of glycine betaine and grew faster compared to wild types on removal of stress e.g., transgenic rice (Nipponbare) is produced by incorporation of *ch-codA*, which activates Choline oxidase A (Sakamoto *et al.*, 1998). Transgenics engineered for over expression of polyamines have also been developed (Roy and Wu, 2001; 2002; Kumria and Rajam, 2002; Waie and Rajam, 2003; Anderson *et al.*, 1998; Capell *et al.*, 2004). Overproduction of osmolyte helps to minimize the undesirable effects and osmoregulation would be the best strategy for abiotic stress tolerance.

Transgenic plant formed had enhanced P-ATPase hydrolytic activity, increased photosynthesis and root proton exportation capacity, and reduced ROS generation. Plants have developed complex antioxidant defense system (enzymatic and non-enzymatic) to scavange ROS. Enzymes involved in oxidative protection e.g., glutathione peroxidase, SOD, APx, and glutathione reductase are over expressed to produce transgenic plants (Zhu *et al.*, 1999; Roxas *et al.*, 1997). Bowler *et al.* (1991) and Van Camp *et al.* (1996) had developed transgenic tomato by over expressing *SOD* in chloroplast, mitochondria and cytosol. Photosynthetic performance of tobacco (Sen Gupta *et al.*, 1993) and potato (Perl *et al.*, 1993) has been improved by over expression of Cu/Zn SOD in chloroplast under chilling stress. Transgenic plants exhibited greater salt tolerance by introduction of *Oscdpk7* Rice (Notohikari) which encodes for Calcium-dependent protein kinase (Saijo *et al.*, 2000) and by incorporation of *Mn-SOD in* Rice (Sasanishiki) which activates Superoxide dismutase (Tanaka *et al.*, 1999). Same is observed when *ProDH is introduced in Arabidopsis which activates* Proline hydrogenase (Nanjo *et al.*, 1999).

Introduction of *AtNHX1* in *A. thaliana* cause Na^+ sequestration and make the plant tolerant to salinity (Luming *et al.*, 2006). Zhao *et al.* (2006) introduced *SOD 2 gene in* Rice (Zhonghua no. 11). He introduced *SsNHX1in* Rice (Zhonghua no. 11) which encoded for Na^+/H^+ antiporter, resulted in Na^+ sequestration; plants had an increase in H^+-ATPase and H^+-PPase activity, reduced ROS generation and increased photosynthesis. Transgenic plant overexpressing Na^+/H^+ antiporter gene was produced by overexpression of nhaA in Rice (Zhongzou 321) (Wu *et al.*, 2005). Introduction of *HbNHX1* gene responsible for vacoular Na^+/H^+ antiporter in Tobacco Transgenic plants showed increased tolerance to salt stress (Lu *et al.*, 2005). Same as in, introduction of *OsNHX1* responsible for Na^+/H^+ antiporter in rice (Nipponbare) make the transgenic plants tolerant to salt stress (Fukuda *et al.*, 2004). Shi *et al.*, (2003) had introduced *AtSOS1* and Gao *et al.*, (2003) introduced *SOD2* in *Arabidopsis* responsible for Plasma membrane Na^+/H^+ antiporter and the transgenic plants showed increased tolerance to salt stress due to Na^+ extrusion. Transgenics survived under conditions of 300 mM NaCl for 3 days by overexpression of Na^+/H^+ antiporter gene in Rice (Kinuhikari) on introduction of *AgNHX1*(Ohta *et al.*, 2002). Over expression of vacoular Na^+/H^+ antiporter is done by introduction of *AtNHX1* in *Brassica napus* (Zhang *et al.*, 2001) and *AtNHX1in Arabidopsis* (Apse *et al.*, 1999). Over expression of vacuolar H^+-pyro-phosphatase make the transgenic plant tolerant to salt stress due to vacuolar acidification by incorporation of *AVP1* in *Arabidopsis* (Gaxiola *et al.*, 2001). Transgenic tomato formed showed increased tolerance to salt stress by introduction of *BADH* which encodes Betaine dehydrogenase (Jia *et al.*, 2002) and *AtNHX1* which encodes vacoular Na^+/H^+ antiporter (Zhang and Blumwald, 2001).

Mohanty *et al.* (2002) produced transgenic rice (PB 1) by introduction of codin which encodes for Choline oxidase A and the 50% R1 plants formed survived after exposure to salt stress for 1 week. Holmstrom *et al.* (2000) reported incorporation of *beta in* Tobacco exhibited greater salt tolerance due to overexpression of Choline dehyrogenase and Nakayam *et al.* (2000) introduced *EctA* in Tobacco and obtained the same results. Hoshida *et al.* (2000) reported that over-expression of *c-GS2* in rice (Kinuhikari) encode for Chloroplastic glutamine synthetase reduced the increase of Na^+ content at high salinity.

LEA (late embryogenesis abundance) proteins are high molecular wt, ABA inducible proteins which accumulate during seed desiccation under salinity and osmotic stress (Galau *et al.*, 1987). On the basis of amino acid sequence homology, these are placed in different groups. These proteins have 11-mer amino acid motifs with the consensus sequence TAQAAKEKAGE which is being repeated 13-times (Dure, 1993). The Group 1 LEA proteins contain high proportion of Gly and charged amino acid and, so are highly hydrophilic. These enhance water binding capacity and provide aqueous environment for cellular components. *Em* gene encodes wheat EM protein which is more hydrated than other globular proteins (McCubbin *et al.*, 1985; Bostock *et al.*, 1992). The members of Group 2 LEA proteins are called as dehydrin. These have a conserved, lys-rich amino acid domain located in the C-terminus and, one at upstream position. In vegetative tissues, these genes are responsive to ABA under salinity and cold stresses (Mundy *et al.*, 1988; Gilmour *et al.*, 1992). Tandemly repeated 11-mer amino acid motifs are repeated many times in Group 3 and 5 LEA proteins (Dure, 1993). These sequester ions during water loss. A Group 3 LEA protein, *HVA*1 from barley, on constitutive expression provide tolerance to salt and water in transgenic rice (Xu *et al.*, 1996). Group 4 LEA proteins contain Gly and -OH group rich amino domain at C-terminus that forms an unstructured random coil and a conserved N-terminal domain that forms an α-helix (Dure, 1993). In cotton leaves, LEA D95 class of proteins is found which is structurally similar to cDNA, pcC27-45 from *Craterostigma plantigineum* in response to salt and water in leaves and callus (Galau *et al.*, 1999).

A typical halophyte with high resistance to cold, drought, oxidative stresses and salinity has been identified as salt cress. Salt cress could serve as an important genetic model system for botanist, geneticists, and breeders for better understanding of the genetic mechanisms of abiotic stress tolerance as the plant is of small size, having short life cycle, copious seed production, small genome size, and an efficient transformation. When the genome of salt cress (Shandong ecotype) was sequenced using the paired-end Solexa sequencing technology, a draft sequence of about 134-fold coverage was identified. The final length of the assembled sequences was found to be 233.7 Mb, covering about 90% of the estimated size (~260 Mb). A total of 28,457 protein-coding regions were predicted in the sequenced salt cress genome. The average exon length of salt cress and *A. thaliana* genes was similar, whereas the average intron length of salt cress was about 30% larger than that of *A. thaliana* as determined by the researchers. A dramatically different lifestyle, a unique gene complement, significant differences in the expression of orthologs, a larger genome size with higher content of transposable elements (TEs) especially retrotransposons.

Concluding remarks

Salt stress causes huge losses of agricultural productivity. Therefore, the effects of severe environmental stresses need to be completely removed. Strategy to regulate the biochemical indicators at cellular level could be of great help for improving the tolerance of plants against salt stress. The length of exposure and concentration of salt treatment affect the morphological, physiological and biochemical aspects of the plant. A low salt concentration could only affect the physiological mechanisms where as a high salt concentration affect the morphology of the plant. Plant adaptation includes the tissue tolerance to accumulated Na^+ and Cl^- and osmotic tolerance through exclusion of Na^+.

A decrease in biomass and moisture content of the plant is associated with the chlorosis and necrosis of the leaves. The ions are not effectively compartmentalized in the vacuole, cytoplasm and chloroplast and so, the activity of photosystems get declined. Reducing sugars, amides and imino acid such as proline provide osmoticum to the plant. Increased levels of polyphenols under salinity serve as an important defense mechanism by induction of secondary metabolism. Membrane disorganization, ROS, metabolic toxicity and inhibition of photosynthesis are effected under salinity. To scavenge these ROS, antioxidants both enzymatic and non-enzymatic constitute a defensive team and protect the cell from oxidative damage. The major substrate detoxification of ROS is Ascorbate, which is continuously formed from its oxidized form. An important function of GSH to protect the cell from oxidative stress is through reduction of Ascorbate via Ascorbate-glutathione cycle.

REFERENCES

Abd EL, Baki GK, Siefritz F, Man HM, Welner H, Kaldenhoff R, Kaiser WM (2000) Nitrate reductase in *Zea mays* L. under salinity. *Plant Cell Environ* 23: 15-521

Abdel HA, Khedr MAA, Amal A, Abdel W, Quick WP, Abogadallah GM (2003) Proline induces the expression of salt stress responsive proteins and may improve the adaptation of *Pancratium maritimum* L to salt stress. *J Exp Bot* 54(392): 2553-2562

Abraham E, Rigo G, Szekely G, Nagy R, Koncz C, Szabados L (2003) Light-dependent induction of proline biosynthesis by abscisic acid and salt stress is inhibited by brassinosteroid in *Arabidopsis. Plant Mol Biol* 51: 363–372

Agarwal S, Sairam RK, Srivastava GC, Tyagi A, Meena RC (2005) Role of ABA, salicylic acid, calcium and hydrogen peroxide on antioxidative enzymes induction in wheat seedlings. *Plant sci* 169: 559-570

Agastian P, Kingsley SJ, Vivekanandan M (2000) Effect of salinity on photosynthesis and biochemical characteristics in mulberry genotypes. *Photosynthetica* 38: 287-290

Ahmadi SH and Ardekani J N (2006) The effect of water salinity on growth and physiological stages of eight Canola (*Brassica napus*) cultivars. *Irri Sci* 25(1): 11-20

Allakhverdiev SI, Sakamoto A, Nishiyama Y, Murata N (2000) *Plant Physiology* 122:1201-1208

Allakhverdiev SI, Klimov VV, Hagemann M (2005) Cellular energization protects the photosynthetic machinery against salt-induced inactivation in *Synechococcus. Biochemica Biophysica Acta* 1708: 201–208

Al-Sobhi OA, Al-Zahrani HS, Al-Ahmadi (2006) Effect of salinity on chlorophyll & Carbohydrate contents of *Calotropis procera* seedlings. *Scientific Journal of King Faisal University* 7: 114-127

Amtmann A and Sanders D (1999) Mechanisms of Na+ uptake by plant cells. In: eds) Advances in Botanical Research. *Academic Press.* 76-114

Anderson SE, Bastola DR, Minocha S C (1998) Metabolism of polyamine in transgenic cells of carrot expressing a mouse ornithine decarboxylase cDNA. *Plant Physiology* 116: 299-307.

Apel K and Hirt H (2004) Reactive oxygen species: metabolism, oxidative stress and signal transduction. *Annu Rev Plant Biol* 55: 373-399

Apse MP, Aharon GS Snedden WA, Blumwald E (1999) Salt tolerance conferred by overexpression of a vacuolar Na+/H+ antiport in *Arabidopsis. Science* 285: 1256-1258

Aronova EE, Shevyakova NI, Sretsenko LA, Kuznetsov VIV (2005) Cadaverine- induced induction of superoxide dismutase gene expression in *Mesembryanthemum crystallinum* L. *Doklady of biological Sciences* 403: 1-3

Arshi A, Abdin MZ, Iqbal M (2002) Growth and metabolism of senna as affected by salt stress *Biol Plant*, 45: 295-298

Ashraf M (1994) Organic substances responsible for salt tolerance in *Eruca sativa. Biol Plant 36:* 255–259

Ashraf M (2004) Some important physiological selection criteria for salt tolerance in plants. *Flora* 199: 361–376

Ashraf M and Harris PJS (2004) Potential biochemical indicators of salinity tolerance in plants. *Plant Sci* 166: 3-16

Ashraf M and McNeilly T (2004) Salinity tolerance in *Brassica* oilseeds. *Crit. Rev.Plant Sci* 23(2): 157-174

Ashraf M and Foolad MR (2007) Roles of glycinebetaine and proline in improving plant abiotic stress resistance. *Environ Exp Bot* 59: 206-216

Ashraf MHR, Harris PJS, Kwon TR (2008) Some prospective strategies for improving crop salt tolerance. *Adv Agron* 97: 45-110

Azevedo NAD, Prico JT, Eneas-Filho J, Braga De Abreu CE, Gomes-Filho E (2006) *Environ Exp Bot* 56: 235-241

Azevedo NAD, Gomes-Filho E, Prico JT (2008) Salinity and oxidative stress. *In: Khan N A Sarvajeet S, Abiotic stress and Plant Responses* 58-82

Bajji MS, Lutts, Kient J M (2001) Water deficts effects on solute contribution to osmotic adjustment as a function aging in three durum wheat (*Triticum durum Defs.*) cultivars performing differently in arid conditions. *Plant Science* 86: 115-123

Bartels D and Sunker R (2005) Drought and salt tolerance in plants. *Crit Rev Plant Sci* 24: 23-58

Basra AS, Grewal RD, Malik CP (1991) Germination performance of maize seed at unfavourable temperature by potassium salts. *Indian J Plant Physiology* 34(4): 378-381

Baum JA and Scandalios JG (1981) Isolation of cytosolic and mitochondrial superoxide dismutases of maize. *Arch Biochem Biophysics* 206: 249-264

Bienert GP, Schjoerring JK, Jahn TP (2006) Membrane transport of hydrogen peroxide. *Biochem Biophysics Acta* 1758: 994-1003

Binzel ML, Hasegawa PM, Rhodes D, Handa S, Handa AK and Bressan RA (1987) Solute accumulation in tobacco cells adapted to NaCl. *Plant Physiol* 84: 1408-1415

Blumwald E, Aharon GS and Apse MP (2000) Sodium transport in plant cells. *Biochemica Biophysica Acta* 1465: 140-151

Bohnert HJ and Jensen RG (1996) Strategies for engineering water stress tolerance in plants. *Trends in Biotechnology* 14: 89-97

Bohnert HJ, Nelson DE, Jensen RG (1995) Adaptations to environmental stresses. *Plant Cell* 7: 1099-1111

Borodina RA (1991) Accumulation of free proline in seedlings of swede rape under salt stress. *Sel'skokhozyaistvennaya-Biologiya* 1: 119–124

Bostock RM and Quairano RS (1992) Regulation of *Em* gene expression in rice. *Plant Physiol.* 98:1356-1363

Bouchereau A, Aziz A, Larher F, Martin-Tanguy J (1999) Polyamines and environmental challenges: recent development. *Plant Science* 140:103-125

Bowler C, Montagu MV, Inze D (1992) Superoxide dismutases and stress tolerance. *Annu Rev Plant Physiol and Plant Mol Biol* 43: 83-116

Bowler C, Slooten L, Vandenbraden S, Rycke RD, Botterman J, Sysbesma C, Van Montagu M, Inze D (1991) Manganese superoxide dismutase can reduce cellular damage mediated by oxygen radicals in transgenic plants. *EMBO J* 10:1723-1732

Bray EAJ, Bailey-Serres and Weretilnyk E (2000) Responses to abiotic stress In: Buchanan B Gruissem W Jones R. (eds.). Biochemistry and Molecular Biology of Plants American Society of Plant Physiology Rockville, MD 1158–1200

Burman UBK, Garg, Kathju R (2001) Genotype variation in growth, mineral composition, Photosynthesis and leaf metabolism of Indian mustard under salinity stress. *Ind J Plant Physiology* 6: 374-380

Burnell JN (1984) Sulfate assimilation in C4 plants- Intercellular and intracellular location of ATP sulfurylase, cysteine synthase and cysthionine b-lyase in maize leaves. *Plant Physiol* 75: 873-875

Cannon RE and Scandalios JG (1989) Two cDNAs encode nearly identical Cu/Zn superoxide dismutase proteins in maize. *Mol Gen Genet* 219: 1-8.

Capell T, Bassie L, Christou P (2004) Modulation of the polyamine biosynthetic pathway in transgenic rice confers tolerance to drought stress. *Proc Natl Acad Sci USA* 101: 9909-9914

Cayuela E, Pervez -Alfocea Caro M, Bolrin MC (1996) Priming of seeds with NaCl induces in tomato plants grown under salt stress. *Physiol. Plant* 96: 231-236

Ceyhan T and Ali I (2002) *Journal of Plant Nutrition* 25:27-41

Cheng NH, Pittman JK, Zhu JK and Hirschi KD (2004) The protein kinase SOS2 activates the *Arabidopsis* H+/Ca2+ antiporter *CAX1* to integrate calcium transport and salt tolerance. *J Biol Chem* 279: 2922-2926

Cherel I, Michard E, Platet N, Mouline K, Alcon C, Sentanac H, Thibaud JB (2002) Physical and functional interaction of the Arabidopsis KR channel *AKT2* and phosphatase *AtPP2CA*. *Plant Cell* 14:1133-1146

Chinnusamy VA, Jagendorf, Zhu J K (2005) Understanding and improving salt tolerance in plants *Crop Sci* 45: 437–448

Cicek N, Cakirlar H (2002) The effect of salinity on some physiological parameters in two maize cultivars. *Bulgaria J Plant Physiol* 28(1-2): 66-74

Creissan GP, Edwards EA, Mullineaux PM (1994) Glutathione reductase and ascorbate peroxidase In C H Foyer and P M Mullineaux (eds), Causes of photooxidative stress and Amelioration of defense Systems in Plants, CRC Press, Boca Raton 343-364

Cuartero JMC, Bolarin MJ, Asins, Moreno V (2006) Increasing salt tolerance in the tomato. *J Exp Bot* 57(5): 1045–1058

Cushman JC, Bohnert HJ (1997) Molecular genetics of Crassulacean acid metabolism. *Plant Physiol* 113: 667-676

Dalton DA, Russell SA, Hanus FJ, Pascose GA, Evans HJ (1986) Enzymatic reactions of ascorbate and glutathione that prevent peroxide damage in soyabean root nodules. *Proc Nail Acad Sci* 83: 3811-3815

De Gara L, de Pinto MC, Arrigoni O (1997) Ascorbate synthesis and ascorbate peroxidase activity during the early stage of wheat germination. *Physiol Plant* 100: 894-900

De Herralde F, Biel C, Save R, Morales MA, Torecillas A, Alacron JJ, Sanchez-Blanco MJ (1998) *Plant Science* 139: 9-17.

De Tullio MC, Paciolla C, Dalla Vecchia F, Rascio N, D'Emerico S, De Gara L, Liso R, Arrigoni O (1999) Changes in onion root development induced by the inhibition of peptidyl-prolyl hydroxylase and influence of the ascorbate system on cell division and elongation. *Planta* 209: 424-434

Delauney AJ and Verma DPS (1993) Proline biosynthesis and osmoregulation in plants. *Plant J* 4: 215-223

Delfine SA, Alvino M, Zacchini, Loreto F (1998) Consequences of salt stress on conductance to CO2 diffusion, Rubisco characteristics and anatomy of spinach leaves. *Aust J Plant Physiol* 25: 395–402

Dkhil Ben Besma and Denden Mounir (2010) Salt stress induced changes in germination, sugar, starch and enzyme of carbohydrate metabolism in *Abelmoschus esculentus* (L.) Moench seeds. *African J Agriculture Research* 5(6): 406-415

Doulis AG, Debian N, Kingston-Smith AH, Foyer CH (1997) Differential localization of antioxidants in maize leaves. *Plant Physiol* 114: 1031-1037

Drozdowicz YM and Rea PA (2001) Vacuolar (H+) pyrophosphatases: from the evolutionary backwaters into the mainstream. *Trends Plant Sci.* 6: 206-211

Dubey RS and Singh AK (1999) Salinity induces accumulation of soluble sugar and alters the activity of sugar metabolizing enzyme in rice plants. *Biologia Plant* 42: 233

Dubey RS (2005) Photosynthesis in plants under stress full conditions In: *Handbook photosynthesis*, M. Pessarakli (ed.). 2nd (Ed), C. R. C. Press, New York, USA. 717-718

Dubey RS and Rani M (1987) Protease and protein in germination rice seed in relation to salt tolerance *Plant Physiol Biochem India* 14: 174-182

Dure L III (1993) A repeating 11-mer aminoacid motif and plant dessication. *Plant J* 3: 363-369

Dure LS (1993) the Lea proteins of higher plants. In *Control of Plant gene Expression* (ed. Verma D P S) CRC Press, Boca Raton, USA 325-335

Ehlers MD and Augstine GJ (1999) Cell signaling *Calmodulin* at the channel gate. *Nature* 399: 107-108

El-Fouly MM and Salama ZH (1999) International Symposium: Nutrient Management under Salinity and Water Stress. 1-4 March, Technion-ITT, Haifa

El-Hamdaoui A, Redondo-Nieto M, Torralba B, Rivilla R, Bonilla I and Bolanos L (2003) *Plant soil* 251:93-103

Ericson MC and Alfinito SH (1984) Proteins produced during salt stress in tobacco cell culture. *Plant Physiol* 74: 506-509

Eschie HA and Rodriguez V (1999) *Journal of Agronomy and Crop Science* 182:273-278

Fidalgo F, Santos A, Santos I, Salema R (2004) Effects of long-term salt stress on antioxidant defence systems, leaf water relations and chloroplast ultrastructure of potato plants. *Annals of Appl Bio* 145: 185-192

Flowers TJ (2004) Improving crop salt tolerance *J Exp Bot* 55(96): 307-319

Foolad MR (1996) Response to selection for salt tolerance during germination in tomato seed derived from PI174263. *J Am Soc Hort Sc.* 121: 1001–1006

Foyer CH and Halliwell B (1976) The presence of glutathione and glutathione reductase in chloroplast : a proposed role in ascorbic acid metabolism. *Planta,* 133: 21-25

Foyer CH and Noctor G (2000) Oxygen processing in photosynthesis: Regulation and signalling. *New Phytol* 146: 359-388

Francois LE (1994) Growth, seed yield and oil content of canola growth under saline conditions. *Agron J* 86: 233–237

Fricke WG, Akhiyarova D, Veselov and Kudoyarova G (2004) Rapid and tissuespecific changes in ABA and in growth rate in response to salinity in barley leaves. *J Exp Bot* 55: 1115–1123

Fukuda A (2004) Function, intracellular localization and the importance in salt tolerance of a vacuolar Na+/H+ antiporter from rice. *Plant Cell Physiol.* 45: 146-159

Galau GA, Bijaisoradat N, Hughes DW (1987) Accumulation kinetics of cotton late embryogenesis-abundant (Lea) mRNAs and storage protein mRNAs:coordinate regulation during embryogenesis and role of abscisic acid. *Dev Biol* 123: 198-212

Galau GA, Wang HY-C, Hughes DW (1999) Cotton Lea5 and Lea4 encode atypical late embryogenesis-abundant proteins. *Plant Physiol* 101: 695-696

Galston AW, Kaur-Sawhney R, Atabella T, Tiburcio AF (1997) Plant polyamines in reproductive activity and response to abiotic stress. *Botanica Acta* 110: 197-207

Gao X (2003) Overexpression of *SOD2* increases salt tolerance of Arabidopsis. *Plant Physiol* 133: 1873-1881

Garg BKS, Kathuju Vyas, Lahiri AN (1999) Genotypic difference to salt stress in Indian mustard In: Management of arid ecosystem (Eds. Faroda A S N L Joshi S Kathju and Amal Kar)Arid Zone Research Assosiation and Scientific Publishers (India),Jodhpur 295-302

Garg BK, Burman U, Kathju S (2002) Response of cumin (*Cuminum cyminum L.*) to salt stress. *Indian Journal of Plant Physiology* 16: 70-74

Garg Neera and Jasleen (2004) Variability in response of chickpea (*Cicer arietinum* L.) cultivars to salt stress in germination and early growth of seedlings *Indian Journal of Plant Physiology* 9(1): 21-28

Gaxiola RA, Li J, Undurraga S, Dang LM, Allen GJ, Alper SL, Fink GR (2001) Drought and salt tolerant plant result from over expression of the *AVP1* H+-pump. *Proc Natl Acad Sci USA* 98: 11444-11449

Gaxiola RA, Rao R, Sherman A, Grisafi P, Alper SL, Fink GR (1999) The *Arabidopsis thaliana* proton transporter, *AtNHX1*and *Avp1*, can function in cation detoxification in yeast. *Proc Natl Acad Sci* USA 96: 1480-1485.

Gelburd Diane E (1985) Managing salinity: Lessons from the past. *Journal of soil and water conservation* 40(4): 329-331

Gerwick BC, Ku SB, Black CC (1980) Initiation of sulfate activation: a variation in C4 photosynthesis plants. *Science* 209: 513-515

Ghannoum O (2009) C4 photosynthesis and water stress. *Ann Bot* 103: 635-644.

Ghassemi F, Jakeman AJ, Nix HA (1995) Salinisation of land and Water Resources: Human Causes, Extent, Management and Case Studies. *Australian national University, Canberra, Australia and CAB International, Wallingford, Oxon, UK*

Ghoulam C and Fares K (2001) Effect of salinity on seed germination and early seedling growth of sugar beet. *Seed SciTechnol* 29: 357-364

Gilmour SJ, Artus NN, Thomshow MF (1992) cDNA sequence analysis and expression of two cold-regulated genes of *Arabidopsis thaliana. Plant Mol Biol* 18: 13-21

Goertz SH, Coons JM (1991) Tolerance of tepary and navy beans to NaCl during germination and emergence *Hort Sci* 26: 246–249

Gorham and Bridges J (1995) Effects of calcium on growth and leaf ion concentration of Gossypium Hirsutumgrown in saline hydroponics culture. *Plant Soil* 176(2): 219-227

Grattan SR and Maas EV (1988) Effect of salinity on phosphate accumulation and injury in soybean II role of sulphate, Cl and Na. *Plant Soil* 109(1): 65-71

Gulick PJ and Dvorak J (1987) Gene Induction and repression by salt treatment in roots of the salinity sensitive Chinese Spring wheat and the salinity-tolerant Chinese Spring x

Elytriga elongate amphiploid. *Proc Natl Acad Sci USA* 84: 99-103

Ha HL, Sirisoma NS, Kuppusamy P, Zweller JL, Woster PM, Casero RA (1998) The natural polyamine spermine functions as a free radical scavenger. *Proceedings of the National Academy of Sciences, USA* 95: 11140-11145

Hare PD and Cress WA (1997) Metabolic implications of stress-induced proline accumulation in plant. *Plant Growth Regul* 21: 79-102

Hare PD, Cress WA, Staden JA (1998) Dissecting the roles of osmolyte accumulation during stress. *Plant Cell Environ* 21: 535-553

Hasegawa PM, Bressan RA, Pardo JM (2000a) The dawn of plant to tolerance genetics. *Trends Plant Sci* 5: 317-319

Hasegawa PM, Bressan RA, Zhu JK, Bohnert HJ (2000) Plant cellular and molecular responses to high salinity. *Annu Rev Plant Physiol Plant Mol Biol* 51: 463-499

Hasewaga PM, Bressan RA, Zhu JK, Bohnert HJ (2000b) Plant cellular and molecular responses to high salinity. *Annu Rev Plant Physiol Plant Mol Biol* 51: 463-499

Hecker M, Schumann W, Volker U (1996) Heat shock and general stress response in *Bacillus subtilis*. *Mol Microbiol* 19: 417-428

Hell R and Bergmann L (1988) Glutathione synthase in tobacco suspension cultures: catalytic properties and localization. *Physiol Plant* 72: 70-76

Hell R and Bergmann L (1990) Gamma- glutamylcysteine synthase in higher plants: catalytic properties and subcellular localization. *Planta* 180: 603-612

Hellman HD, Funk D, Rentsch, Frommer WB (2000) Hypersensitivity of an Arabidopsis sugar signaling mutant towards exogenous proline application. *Plant Physiol* 123: 779-790

Hernandez JA, Ferrar MA, Jimenez A, Barcelo AR, Sevilla F (2001) Antioxidative systems and O2/ H2O2 production in the apoplast of pea leaves. Its relation with salt-induced necrotic lesions in minor veins. *Plant Physiol* 127: 817-831

Hiraga S, Ito H, Yamakawa H, Ohtsubo N, Seo S, Mitsuhara I, Matsui H, Honma M, Ohashi Y (2000) An HR-induced tobacco peroxidase gene is responsive to spermine, but not to salicylate, methyljasmonate and ethaphon. *Molecular Plant-Microbe Interactions* 13: 210-216

Hoffman LL, Jeanne LDN, Monroe IE, Shaftel R, Anten NPR, Martinez Ramos M, Ackerly DD (2006) Mangrove Seedling Net Photosynthesis, Growth and Survivorship are Interactively Affected by Salinity and Light. *Biotropica* 38: 606-616

Hokmabadi H, Arzani K, Grierson PF (2005) Growth, chemical composition, and carbon isotopes discrimination of pistachio (*Pistacia vera L.*) rootstock seedlings in response to salinity. *Aust J Agri Res* 56: 135-144

Holmstrom KO, Somersalo S, Mandal A, Palva TE, Welin B (2000) Improved tolerance to salinity and low temperature in transgenic tobacco producing glycine betaine. *J Exp Bot* 51: 177-185

Horie T, Yoshida K, Nakayama H, Yamada K, Oiki S, Shinmyo A (2001) Two types of *HKT* transporters with different properties of Na+ and K+ transport in *Oryza sativa*. *Plant J*

27: 129-138

Hoshida H, Tanaka Y, Hibino T, Hayashi Y, Tanaka A, Takebe T (2000) Enhanced tolerance to salt stress in transgenic rice that overexpress chloroplast glutamine synthetase. *Plant Mol Biol* 43: 103-111

Houimli SIM, Denden M, Hadj SBE (2008) Induction of salt tolerance in pepper (*Capsicum annuum*) by 24-epibrassinolide. *Eur.Asia.J.Bio.Sci* 2: 83-90.

Hsu SYY, Hsu T, Kao CH (2003) The effect of polyethylene glycol on proline accumulation in rice leaves. *Biol Plant* 46: 73–78

Hu Y and Schmidhalter U (2001) Effects of salinity and macronutrients levels on micronutrients in wheat. *Journal of Plant Nutrition* 24: 273-281

Huang J and Redmann RE (1995) Salt tolerance of *Hordeum* and *Brassica* species during germination and early seedling growth. *Can J Plant Sci* 75: 815–819

Huang XC and Steveninck RFM (1990) Salinity induced structural changes in meristematic cells of barley roots. *New Phytologist* 115: 17-22

Ingram J and Bartels D (1996) The molecular basis of dehydration tolerance in plants. *Annu. Rev. Plant Physiol. Plant Mol Biol* 47: 377-403

Irrigoyen JJD and Sanchez-Diaz W (1992) Water stress induced damage in concentration of proline and total soluble sugars in nodulated alfalfa plant. *Physiol Plant* 84: 55-60

Ishitani M, Liu J, Halfter U, Kim CS, Shi W, Zhu JK (2000) *SOS3* function in plant salt tolerance requires N- myristoylation and calcium binding. *Plant Cell* 12: 1667-1677

Jaleel CA, Sankar B, Sriaharan R, Paneerselvam R (2008) Soil salinity alters growth, chlorophyll content and secondary metabolite accumalation in *Catharanthus roseus. Turk J Biol* 32: 79-83

James RAA, Rivelli R, Munns R, Von Caemmerer S (2002) Factors affecting CO2 assimilation, leaf injury and growth in salt stressed durum wheat. Funct. *Plant Biol* 29: 1393–1403

Jamil M and Rha ES (2004) The effect of salinity (NaCl) on the germination and seedling of sugar beet (*Beta vulgaris* L.) and cabbage (*Brassica oleracea* L.). *Korean J plant Res* **7**: 226-232

Jang IC, Oh SJ, Seo JS, Choi WB, Song SI, Kim CH, Kim YS, Seo HS, Choi YD, Nahm BH, Kim JK (2003) Expression of a bifunctional fusion of the *Escherichia coli* genes for the trehalose-6-phosphate synthase and trehalose-6-phosphate phosphatase in transgenic rice plants increases trehalose accumulation and abiotic stress tolerance without stunting growth. *Plant Physiol* 131: 516-524

Jeanjean R, Matthijs HCP, Onana B, Havaux M, Joset F (1993) Exposure of the cyanobacterium synechocystis PCC6803 to salt stress induces concerted changes in respiration and photosynthesis. *Plant Cell Physiol* 34: 1073-1090

Jia W, Wang Y, Zhang S, Zhang J (2002) Salt-stress-induced ABA accumulation is more sensitively trigerred in roots than in shoots. *J Exp Bot* 53: 2201-2206

Joshi SS (1984) Effect of salinity stress on organic and mineral constituents in leaves of pigeon pea (*Cajanus cajan* L. Var.C-II). *Plant and Soil* 872: 676-98

Kao WY, Tsai TT, Tsai HC, Shih CN (2006) Response of three glycine species to salt stress. *Environmental and Experimental Botany* 56: 120–125

Kao WY, Tsai TT, Shih CN (2003) Photosynthetic gas exchange and chlorophyll *a* fluorescence of three wild soybean species in response to NaCl treatments. *Photosynthetica* 41: 415-419

Katsuhara M, Yazaki Y, Sakano K, Kawaski T (1997) Intracellular pH and proton –transport in barley root cells under salt stress:in vivo 35P-nmr Study. *Plant Cell Physiology* 38: 155-160

Kaur-Sawhney R, Tiburcio AF, Altabella T, Galston A (2003) Polyamines in plants: an overview. *Journal of Cell and Molecular Biology* 2:1-12

Kavi Kishore PB, Sangam S, Amrutha RN, Laxmi PS, Naidu KR, Rao KRSS, Rao SKJ, Theriappan P, Sreenivasulu N (2005) Regulation of proline biosynthesis, degradation, uptake and transport in higher plants: its implications in plant growth and abiotic stress tolerance *Curr Sci* 88: 424–438

Ketchum REB, Warren RC, Klima IJ, Lopez- Gutierrez F, Nabros MW (1991) The mechanism and regulation of proline accumulation in suspension cultures of the halophytic grass *Distichlis spicata* L. *J Plant Physiol* 137: 368-374

Khan MA, Ungar IA, Allan M S (2000 a) The effects of salinity on the growth, water status and ion content of a leaf succulent perennial halophyte, Suaeda Fruticosa (L.) Forssk. *Journal of Arid Environments* 45: 73-84

Khavarinejad RA, Mostofi Y (1998) Effects of NaCl on photosynthetic pigments, saccharides, and chloroplast ultrastructure in leaves of tomato cultivars. *Photosynthetica* 35: 151-154

Kim FJ, Kim HP, Hah YC, Roe JH (1996) Differential expression of superoxide dismutases containing Ni and Fe/Zn in *Streptomyces coelicolor. European Journal of Biochemistry* 241: 178-185

Klapheck S, Chrost B, Stark J, Zimmerman H (1992) Gamma-glutamyl-cysteinylserine- a new homologue of glutathione in plants of the family *Poaceae. Botanica Acta* 105: 174-179

Koca H, Bor M, Ozdemir F, Turkan Y (2007) The effects of salt stress on lipid peroxidation, antioxidative enzymes and proline contents of Sesame cultivars. *Environmental Experimental Botany*. 60: 344-351

Kreps JA, Wu Y, Chang HS, Zhu T, Wang X, Harper JF (2002) Transcriptome changes for *Arabidopsis* in response to salt, osmotic and cold stress. *Plant Physiol* 130: 2129-2141

Krieger-Jiszkay A (2004) Singlet oxygen production in photosynthesis. *J Exp Bot*. 56: 337-346

Kumria R and Rajam MV (2002) Ornithine decarboxylase transgene in tobacco affects polyamines, in vitro-morphogenesis and response to salt stress. *J Plant Physiol* 159: 983-990

Kundu PB and Paul NK (1997) Effect of water stress on chlorophyll, proline and sugar accumulation in rape *(Brassica Compestris L.)*. *Bang J Bot* 26: 83-85

Kuznetsov VV and Shevyakova NI (1997) Stress response of tobacco cells to high temperature and salinity, Proline accumulation and phosphorylation of polypeptides. *Physiol Plant* 100: 320-326

Kuznetsov VV, Shorina M, Aronava E, Stesenko L, Rakitin VY, Shevyakova NI (2007) NaCl- and ethylene-dependant cadaverine accumulation and its possible protective role in the adaptation of the common ice plant to salt stress. *Plant Science* 172: 363-370

Lahiri AN, Garg BK, Vyas SP, Kathju S, Mali PC (1996) Genotypic differences to soil salinity in clusterbean Arid Soil. *Res Rehabilit* 10: 333-345

Langebartels C, Kerner KJ, Leonardi S, Schraudner M, Trost M, Heiller W, Sanderman H (1991) Biochemical plant response to ozone. Differential induction of polyamine and ethylene biosynthesis in tobacco. *Plant Physiology* 95: 882-887

Lauchi A (1990) Calcium in Plant Growth and Development. *American Society of plant Physiologists* 4: 26-35

Le Rudulier D, Strom AR, Dandekar AM, Smith LT, Lalentine RC (1984) Molecular biology of osmoregulation. *Science* 224: 1064-1068.

Leegood RC, Lea PJ, Adcock MD, Hausler RR (1995) The regulation and control of photorespiration. *J Exp Bot* 46: 1397-1414

Li J, Lee YRJ, Assmann SM (1998) Guard cells possess a calcium dependant protein kinase that phosphorylates the *KAT1* potassium channel. *Plant Physiol* 116: 785-795

Liu J and Zhu JK (1998) A calcium sensor homolog required for plant salt tolerance. *Science* 280: 1943-1945

Liu J, Ishitani M, Halfter U, Kim CS, Zhu JK (2000) The *Arabidopsis thaliana* SOS2 gene encodes a protein kinase that is required for salt tolerance. *Proc Natl Acad Sci USA* 97: 3730-3734

Lu SY (2005) Antiporter gene from *Hordum brevisubulatum* (Trin.) Link and its overexpression in transgenic tobaccos. *J Integr Plant Biol* 47: 343-349

Lutts S, Majerus V, Kinet JM (1999) NaCl effects on proline metabolism in rice seedlings. *Physiol Plant* 105: 450-458

Maggio A, Miyazaki S, Veronese P, Fujita T, Ibeas JI, Damsz B, Narasimhan ML, Hasegawa PM, Joly RJ, Bressan RA (2002) Does proline accumulation play an active role in stress-induced growth reduction?*Plant Journal* 31: 699-712

Malibari AA (1993) The interactive effects between salinity, Abscissic acid and kinetin on respiration, chlorophyll content and growth of wheat plant. *Indian J Plant Physiol* 36: 232-235

Mane AV, Karadge BA, Samant JS (2010) Salinity induced changes in photosynthetic pigments and polyphenols of Cymbogon Nardus (L.) Rendle. *Journal of Chemical and Pharmaceutical Research* 2: 338-347

Mane AV, Saratale GD, Karadge BA, Samant JS (2010a) Studies on the effects of salinity on growth, polyphenol content and photosynthetic response in *Vetiveria zizanioides*(L.) Nash. *Emiratus Journal of food and Agriculture* 23: 59-70

Mansour MMF (1998) Protection of plasma membrane of onion epidermal cells by glycinebetaine and proline against NaCl stress. *Plant Physiol Biochem* 36: 767-772

Mansour MMF (2000) Nitrogen containing compounds and adaptation of plants to salinity stress. *Biologia Plantarum* 43: 491–500

Marcum KB and Murdoch CL (1992) Salt tolerance of the coastal salt march grass, Sporobolus virginicus (L.) Kunth. *New Phytologist* 120: 281-288

Martinez-Ballesta MC, Martinez V, Carvajal M (2004) *Environmental Experimental Botany* 52: 161-174

Matamoros MA, Baird LM, Escuredo PR, Dalton DA, Minchin FR, Iturbe-Ormaetxe I, Rubio M, Moran J, Gordon A, Becana M (1999) Stress-induced legume root nodule senescence Physiological, biochemical, and structural alterations. *Plant Physiol* 121: 97-111

Mathangi R, Yoav W, Marcelo S (2006) Kikuyu Grass: A valuable Salt-Tolerant Fodder. *Communications in Soil Science and Plant Analysis* 37: 1269-1279

Mathur NSJ and Sachdeva B (2006) Biomass production productivity and physiological changes in moth bean genotypes at different salinity level. *American J Plant Physiology* 1(2): 210-213

May MJ, Hammond-Kosack KE, Jones DG (1996) Involvement of reactive oxygen species, glutathione metabolism and lipid peroxidation in the cf-gene-dependant defense response of tomato cotyledons induced by race-specific elicitors of *Cladosporium fulvum*. *Plant Physiol* 110: 1367-1379

Mazel A, Leshem Y, Tiwari BS, Levine A (2004) Induction of salt and osmotic stress tolerance by overexpression of an intracellular vesicle trafficking protein AtRab7. *Plant Physiol* 134: 118-128

McCubbin WD, Kay CM, Lane BG (1985) Hydrodynamic and optical properties of the wheat germ Em protein. *Can J Biochem CellBiol* 63: 803-811

McCue KF and Hanson AD (1990) Salt-inducible betaine-aldehyde dehydrogenase fron sugar beet: cDNA cloning and expression. *Trends Biotechnol* 8: 358-362

Mendoza I, Rubi F, Rodriguez NA, Pardo JM (1994) The protein phosphatase calcineurin is essential for NaCl tolerance of *Saccharomyces cerevisae*. *J Biol Chem* 269: 8792-8796

Mittler R (2002) Oxidative stress, antioxidants and stress tolerance. *Trends plant Sci* 7: 405-410

Mohanty A, Kathuria H, Ferjani A, Sakamoto A, Mohanty P, Murata N, Tyagi A (2002) Transgenics of an elite indica variety Pusa Basmati 1 harbouring the *codA* gene are highly tolerant to salt stress. *Theor Appl Genet* 106: 51-57

Moons A, Gielen J, Vandekerckhove J, Van Der Straeten D, Gheysen G, Van Montagu M (1997) An abscisic acid and salt stress responsive rice cDNA from a plant gene family. *Planta* 202: 443-454

Morsomme P and Boutry M (2000) The plant plasma membrane (H+)-ATPase: structure, function and regulation. *Biochem Biophys Acta* 1465: 1-16

Moser D, Nicholls P, Wastyn M, Peschek GA (1991) Acidic cytochrome c6 of unicellular cyanobacteria is an indispensible and kinetically competent electron donor to cytochrome oxidase in plasma and thylakoid membranes. *Biochem Int* 24: 757-768

Mujeeb-Ur-Rahman Soomro UA, Zahoor-Ul-Haq M, Gul S (2008) *World journal of Agriculture Sciences* 4: 398-403

Mundy J and Chua NH (1988) Abscisic acid and water stress induce the expression of a novel gene. *EMBO J* 7: 2279-2286

Munns R (1993) Physiological process limiting plant growth in saline soil: some dogma and hypothesis. *Plant, Cell and Environment* 16: 15-24

Munns R (2002) Comparative physiology of salt and water stress *Plant Cell Environ.* 25: 239-250

Munns R (2005) Genes and salt tolerance: bringing them together. *New Phytol* 167(3): 645- 663

Munns R and Tester M (2008) Mechanisms of salinity tolerance *Annu Rev Plant Biol* 59: 651-681

Murakeozy EP, Nagy Z, Duhaze C, Bouchereau A, Tuba Z (2003) Seasonal changes in the levels of compatible osmolytes in three halophytic species of island saline vegetation in *Hungary. J Plant Physiol* 160: 395–401

Murillo-Amador B, Troyo-Dieguez E, Lopez Aguilar R, Lopez-Cortes A, Tinoco-Ojanguren CL, Jones HG, Kaya C (2002) *Australian Journal of Agriculture Research* 53:1243-1255

Naieni MR, Khoshgoftarmanesh AH, Lessani H, Fallashi E (2004) Effects of sodium chloride-induced salinity on mineral nutrients and soluble sugars in three commercial cultivars of pomegranate. *Journal of Plant Nutrition* 27:1319-1326.

Nakamura T, Liu Y, Hirata D, Namba H, Harada S, Hirokawa T, Miyakawa T (1993) Protei phosphatase type B (calcineurin) mediated *Fk506* sensitive regulation of intracellular ions in yeast is an important determinant for adaptation to high salt stress conditions. *EMBO J* 12: 4063-4071

Nakayam H, Yoshida K, Ono H, Murooka Y, Shimyo A (2000) Ectoine, the compatible solute of *Halomonas elongata* conferes hyperosmotic tolerance in cultured tobacco cells. *Plant Physiol* 122: 1239-1248

Nanjo T, Kobayashi TM, Yoshida Y, Kakubari Y, Yamaguchi-Shinozaki K (1999) Antisense suppression of proline degradation improves tolerance to freezing and salinity in *Arabidopsis thaliana.* FEBS *Lett* 461: 205-210

Niu X, Zhu JK, Narasimham ML, Bressan RA, Hasegawa PM (1993) Plasma membrane HR-ATPase gene expression is regulated by NaCl in halophyte (*Atriplex nummularia* L.) cell cultures. *Planta* 190: 433-438

Noctor G and Foyer C (1998) Ascorbate and glutathione: keeping active oxygen under control. *Annual Review of Plant Physiology and Plant Molecular Biology* 49: 249-279

Ohta M (2002) Introduction of a Na+/H+ antiporter gene from *Atriplex gmelini* confers salt tolerance in rice. *FEBS Lett* 532: 279-282

Oldeman LR, Van EVWP, Pulles JHM (1991) World Map of the Satus of Human-Induced Soil Degradation: an Explanatory note. Eds. Oldeman L R Hakkeling R T A Sombroek W G International Soil Reference and Information Centre (ISRIC), Wageningen, 37.

Oono YC, Ooura A, Rahman E, Aspuria T, Hayashi K, Tanaka A, Uchimiya H (2003) *p*-clorophenoxyisobutyric acid impairs auxin response in *Arabidopsis* root. *Plant Physicl* 133: 1135–1147

Osorio J, Osorio ML, Chaves MM, Pereira JS (1998) *Tree physiology* 18: 363-373.

Parida AK and Das AB (2004) Effects of NaCl stress on nitrogen and phosphorus metabolism in a true mangrove *Bruguiera parviflora* grown under hydroponic culture. *J. Plant Pysiol* 161: 921-928

Parida AK, Das AB, Das P (2002) NaCl stress causes changes in photosynthetic pigments, proteins and other metabolic components in the leaves of a true mangrove, *Bruguiera parviflora,* in hydroponic cultures. *J. Plant Biol* 45: 28–36

Perl A, Perl-Treves R, Galili S, Aviv D, Shalgi E, Malkin S, Galun E (1993) Enhanced oxidative stress defence in transgenic potato expressing tomato Cu, Zn superoxide dismutases. *Theor Appl Genet* 85: 568-576

Pilot G, Gaymard F, Mouline K, Cherel I, Sentenae H (2003) Regulated expression of Arabidopsis share K+ channel genes involved in K+ uptake and distribution in plant. *Plant Mol Biol* 5:1-27

Pinheiro HA, Silva JV, Endres L, Ferreira VM, Camara CA, Cabral FF, Oliveira JF, de Carvalho L, dos Santos WT, dos Santos JM, Filho BG (2008) Leaf gas exchange, chloroplastic pigments and dry matter accumulation in castor bean (*Ricinus communis* L) seedlings subjected to salt stress conditions. *Ind Crop Prod* 27: 385-392

Polesskaya OG, Kashirina EI, Alekhina ND (2006) Effect of salt stress on antioxidant system of plants as related to nitrogen nutrition. *Russ J Plant Physiol* 53: 186-192

Pommerrenig B, Papini-Terzi FS, Sauer N (2007) Differential regulation of sorbitol and sucrose loading into the phloem of *Plantago major* in response to salt stress. *Plant Physiol* 144: 1029–1038

Qasim M (2000) Physiological and biochemical studies in a potential oilseed crop canola (*Brassica napus* L.) under salinity (NaCl) stress Ph.D thesis Department of Botany, University of Agriculture, Faisalabad, Pakistan.

Qiu QS, Barkla BJ, Vera-Estrella R, Zhu JK, Schumaker KS (2003) NaR/HR exchange activity in the plasma membrane of *Arabidopsis thaliana* by the *SOS Arabidopsis thaliana. Plant Physiol* 132: 1041-1052

Qiu QS, Guo Y, Dietrich MA, Schumaker KS, Zhu JK (2002) Regulation of SOS1, a plasma membrane Na+/H+ exchanger in Arabidopsis thaliana, by *SOS*2 and *SOS*3. *Proc Natl Acad Sci* USA 99: 8436-8841

Quintero FJ, Blatt MR, Pardo JM (2000) Functional conservation between yeast and plant endosomal Na+/H+ antiporters. *FEBS Lett* 471: 224-228

Quintero FJ, Ohta M, Shi H, Zhu JK, Pardo JM (2002) Reconstitution in yeast of the *Arabidopsis SOS* signaling pathway for Na+ homeostasis. *Proc Natl Acad Sci* USA 99: 9061-9066

Rains D and Epstein E (1967) Sodium absorption by barley roots. Its mediation by mechanisms 2 of alkali cation transport. *Plant Physiol* **42**: 319-323

Rathert G (1983) Effect of high salinity stress on mineral and carbohydrate metabolism of two cotton varieties. *Plant Soil* 73:247-56

Reed DJ (1985) In Oxidative stress (ed) H Sies *London Academic Press* 115-130

Rengasamy P (2006) World salinization with emphasis on Australia *J Exp Bot* 57: 1017-1023.

Rennenberg H (1982) Glutathione metabolism and possible biological roles in higher plants. *Phytochemistry* 21: 2771-2781

Reynolds MP, Mujeeb-Kazi A, Sawkins M (2005) Prospects for utilizing plant adaptive mechanisms to improve wheat and other crops in drought- and salinity prone environments. *Ann Appl Biol* 146: 239–259

Rhodes D, Verslues PE, Sharp RE (1999) Role of amino acids in abiotic stress resistance In: B.K. Singh (ed). Plant Amino Acids: Biochem. Biotech. Marcel Dekker, NY 319-356

Robinson NL, Tanaka CK, Hurkman WJ (1990) Time dependant changes in polypeptide and translatable mRNA levels caused by NaCl in barley roots. *Physiol Plant* 78: 128-134

Romero-Aranda R, Soria T, Cuartero J (2001) Tomato plant–water uptake and plant–water relationships under saline growth conditions. *Plant Sci* 160: 265–272

Roy M and Wu R (2001) Arginine decarboxylase transgene expression and analysis of environmental stress tolerance in transgenic rice. *Plant Sci* 160: 869-875

Roy M and Wu R (2002) Overexpression of S-adenosylmethionine decarboxylase gene in rice increases polyamine level and enhances sodium chloride-stress tolerance. *Plant Sci* 163: 987-992

Roxas VP, Smith RK, Jr Allen ER, Allen RD (1997) Overexpression of glutathione S-transferase/glutathione peroxidase enhances the growth of transgenic tobacco seedlings during stress. *Nat Biotechnol* 15: 988-991

Roxas VP, Sundus AL, Garrett DK, Mahan JR, Allen RD (2000) Stress tolerance in transgenic tobacco seedlings that over express glutathione-S-transferase/glutathione peroxidase. *Plant Cell Physiol* 41: 1229-1234

Rus A, Yokoi S, Sharkhuu A, Reddy M, Lee BH, Matsumoto TK, Koiwa H, Zhu JK, Bressan RA, Hasegawa PM (2001) *AtHKT1* is a salt tolerance determinant that controls Na+ entry into plant roots. *Proc Natl Acad Sci* USA 98: 14150-14155

Saijo Y, Hata S, Kyozuka J, Shimamoto K, Izui K (2000) Overexpression of a single Ca2+ dependant protein kinase confers both cold and salt/drought tolerance on rice plants. *Plants J* 23: 319-327

Sairam RK, Deshmukh PS, Shukla DS (1997) Tolerance to drought and temperature stress in relation to increased antioxidant enzyme activity in wheat. *J Agron Crop Sci* 178: 171-177

Sairam RK, Deshmukh PS, Saxena DC (1998) Role of antioxidant systems in wheat genotypes tolerance to water stress. *Biol Plant* 41: 384-389

Sairam RK, Rao KV, Srivastava GC (2002) Differential response of wheat genotype to long term salinity stress in relation to oxidative stress, antioxidant activity and osmolyte concentration. *Plant Sci* 163: 1037–46

Sairam RK, Srivastava GC, Saxena DC (2000) Increased antioxidative activity under elevated temperature: a mechanism of heat stress tolerance in wheat genotypes. *Biol Plant* 43: 245-251

Sakamoto A, Alia H, Murata N (1998) Metabolic engineering of rice leading to biosynthesis of glycinebetaine and tolerance to salt and cold. *Plant Mol Biol* 38: 1011-1019

Sancho MA, Milrad DFS, Pliego F, Valpuesta V, Quesda MA (1996) *Plant Cell Tissue Organ Culture* 44: 161-167

Sanders D, Brownlee C, Harper JF (1999) *Plant Cell* 11: 691-706

Sanders D (2000) Plant biology: The salty tale of Arabidopsis. *Curr Biol* 10: 486-488

Satoh R, Nakashima K, Seki M, Shinozaki K and Yamaguchi-Shinozaki K (2002) ACTCAT, a novel cis-acting element for proline- and hypo-osmolarity-responsive expression of the *ProDH* gene encoding proline dehydrogenase in *Arabidopsis. Plant Physiol* 130: 709–719

Schmidt A and Kunert KJ (1986) Lipid peroxidation in higher plants: the role of glutathione reductase *Plant Physiol* 82: 700-702

Schwabe KA, Iddo K, Knap KC (2006) Drain water management for salinity mitigation in irrigated agriculture. *Am J Agric Ecol* 88: 133-140

Seki MM, Narusaka J, Ishida Nanjo T, Fujita M, Oono Y, Kamiya A, Nakajima M, Enju A, Sakurai T, Satou M, Akiyama K, Taji T, Yamaguchi-Shinozaki K, Carninci P, Kawai J, Hayashi-zaki Y, Shinozaki K (2002) Monitoring the expression profiles of 7000 Arabidopsis genes under drought, cold and high salinity stresses using a full-length cDNA microarray. *Plant J* 31:279-292.

Sepehr MF and Ghorbanli M (2006) Physiological responses of *Zea mays* seedlings to interactions between cadmium and salinity. *J Integr Plant Biol* 48(7): 807-813

Sen Gupta A, Heinen JL, Holady AS, Burke JJ, Allen RD (1993) Increased resistance to oxidative stress in transgenic plants that over-express chloroplastic Cu/Zn superoxide dismutase. *Proc Nat Acad Sci USA* 90: 1629-1633

Serraj R and Sinclair TR (2002) Osmolyte accumulation: can it really help increase crop yield under drought conditions? *Plant Cell Environ* 25: 333-341

Serrano R, Mulet JM, Rios G, Marguez JA, de Larrinoa IF, Leube MP, Mendizabal I M, Pascual-Ahuir A, Proft M, Ros R, Montesinos C (1999) A glimpse of the mechanisms of ion homeostasis during salt stress. *J Exp Bot* 50:1023-1036

Shabala S, Shabala L, Volkenburgh F, Newman I (2005) Effect of divalent cations on ion fluxes and leaf photochemistry in salinized barley leaves. *J Exp Bot* 56: 1369-1378

Shalata A and Neumann PM (2001) Exogenous ascorbic acid increases resistance to salt stress and reduces lipid peroxidation. *J Exp Botany* 52: 2207-2211

Shannon MC, Grieve CM, Francois LE (1994) Whole plant response to salinity In:Plant-Environ Interaction 199-244 Wilkins, R.E.Ed., Marcel Dekker,New York.

Sharma A and Saran K (1994) Effect os salinity and germination and seedling growth in Black gram. *Neo botanica* 2: 52-57

Sharma PC and Gill KS (1994) Salinity induces effect on biomass yield, yield attributing characters and content in genotype of Indian mustard (*Brassica juncea*). *Indian J Agric Sci* 64: 785-788.

Sharma PC and Gill KS (1991) Effect of salinity on yield and ion distributrion in pearl millet genotypes. *Arid Soil Research Rehabilit* 6: 253-260

Sharpley AN, Meisinger JJ, Power JF, Suarez DL (1992) Advances in Soil Science. Ed.Stewart B *Spinger*: New York 151-217

Sheoram IS and Nainawatee HS (1990) Metabolic changes in relation to environment stress. In: Plant Biochemistry New Delhi 157-178

Shevyakova NI, Rakitin V, Yu Stetsenko LA, Aronova EE, Kuznetsov VIV (2006a) Oxidative stress and fluctuations of free and conjugated polyamines in the halophyte *Mesembryanthemum crystallinum* L. Under Nacl salinity. *Plant growth Regulation* 50: 69-78

Shi H, Ishitani M, Kim CS, Zhu JK (2000) The *Arabidopsis thaliana* salt tolerance gene *SOS* 1 encodes a putative Na+/H+ antiporter. *Proc Natl Acad Sci* USA 97: 6896-6901.

Shi H, Quintero FJ, Pardo JM, Zhu JK (2002) The putative plasma membrane Na+/H+ antiporter *SOS*1 controls long distance Na+ transport in plants. *Plant Cell* 14: 465-477

Shi H (2003) Overexpression of a plasma membrane Na+/H+ antiporter gene improves salt tolerance in *Arabidopsis thaliana*. *Nat Biotechnol* 21: 81-85

Shinozaki K and Yamaguchi-Shinozaki K (2000) Molecular responses to dehydration and low temperature: differences and cross-talk between two stress signaling pathways. *Curr Opin Plant Biol* 3: 217-223

Shilbi RA, Shatnuwi MA, Waldat SIQ (2003) *Communications in Soil Sciences* and *Plant Analysis* 34:1969-1979

Siddiqi EH, Ashraf M, Hussain M, Jamil A (2009) Assessment of intercultivar variation for salt tolerance in safflower (*Carthamus tinctorius* L.) using gas exchange characteristics as selection criteria. *Pak J Bot* 41(5): 2251-2559

Singer MA and Lindquist S (1998) Multiple effects of trehalose on protein folding *in vitro* and *in vivo*. *Mol cell* 1: 639-648

Singh AK (2004) The physiology of salt tolerance in four genotypes of chickpea during germination. *J Agric Sci Technol* **6**: 87-93

Sinha TS, Sharma D, Singh PC, Sharma HB (2004) Rapid screening methodology for tolerance during germination and seedling emergence in Indian mustard (*Brassica juncea*). *Indian J Plant Physiol*. (special issue)

Slesak L, Libik M, Karpinska B, Karpinski S, Miszalski Z (2007) The role of hydrogen peroxide in regulation of plant metabolism and cellular signaling in response to environmental stresses. *Acta biochem Pol* 54: 39-50

Smirnoff N (1996) The function and metabolism of ascorbic acid in plants. *Annu Bot* 78: 661-669

Smirnoff N (2000) Ascorbic acid: metabolism and functions of a multifaceted molecule. *Current opinion in Plant Biology* 3: 229-235

Smith IK (1985) Stimulation of glutathione synthesis in photo respiring plants by catalase inhibitors. *Plant Physiol* 79: 1044-1048

Sposito G (1989) *The Chemistry of Soils* Oxford University Press, Inc. 277

Stallings WC, Pattridge KA, Strong RK, Ludwig ML (1984) Manganese and iron superoxide dismutases are structural homologs. *J Biol Chem* 259: 10695-10699

Stepien P and Klobus G (2005) Antioxidant defense in the leaves of C3 and C4 plants under salinity stress. *Physiol Plant* 125: 31-40

Subbarao GV, Johansen C, Jana MK, Kumar Rao JVDK (1991) Comparative salinity responses among pigeonpea accessions and their relatives. *Crop Sci* 31: 415–418

Tanaka K (1994) Tolerance to herbicides and air pollutants In: Causes of photoxidative stress and amelioration of defense system in plants CRC Press, *Boca Raton* 365-378

Tanaka Y, Hibino T, Hayashi Y, Tanaka A, Kishitani S, Takabe T (1999) Salt tolerance of transgenic rice overexpressing yeast mitochondrial Mn-SOD in chloroplasts. *Plant Sci* 148: 131–138

Tattini M, Gucci R, Coradeschi MA, Ponzio C, Everard JD (1995) *Physiologia Plantarum* 98: 117-124.

Tejera NA, Soussi M, Lluch C (2006) Physiological and nutritional indicators of tolerance to salinity in chickpea plants growing under symbiotic conditions. *Environmental and Experimental Botany* 58: 17–24

Tester M and Davenport R (2003) Na+ tolerance and Na+ transport in higher plants. *Ann Bot* 91: 503–550

Thomashow MF (1999) Plant cold acclimation: freezing tolearance genes ang regulatory mechanism. *Annu Rev Plant Biol* 50: 571-599

Tsai YC, Hong CY, Liu LF, Kao CH (2005) Expression of ascorbate peroxidase and glutathione reductase in roots of reice seedlings in response to NaCl and H_2O_2. *J Plant Physiol* 162: 291-299

Tyagi RK and Rangaswamy NS (1993) Screening of pollen grains vis-à-vis whole plant of oilseed brassicas for tolerance to salt. *Theor Appl Genet* 87: 343-346

Ulfat MHR, Athar M, Ashraf N, Akram A, Jamil A (2007) Appraisal of physiological and biochemical selection criteria for evaluation of salt tolerance in canola (*Brassica napus* L.). *Pak J Bot* 39(5): 1593-1608

Urao T, Yakubov B, Satoh R, Yamaguchi-Shinozaki K, Seki M, Hirayama T, Shinozaki K (1999) A transmembrane hybrid type histidine kinase in Arabidopsis functions as an osmosensor. *Plant cell* 11: 1743-1754

Van Camp W, Van Montagu M, Inze D (1994a) superoxide dismutases. In C H Foyer and P M Mullineaux (eds). Causes of photooxidative stress and Amelioration of defense systems in Plants *CRC Press. Boca Raton* 317-341

Van Camp W, Capiau K, Van Montagu M, Inze D, Slooten L (1996) Enhancement of oxidative stress tolerance in transgenic tobacco plants overproducing Fe- superoxide dismutase in chloroplast. *Plant Physiol* 112: 1703-1714.

Van Volkenburgh E and Boyer JS (1985) Inhibitory effects of water deficit on maize leaf elongation. *Plant Physiol* 77(1): 190-194

Vitart V, Baxter I, Doerner P, Harper JF (2001) Evidence for a role in growth and salt resistance of a plasma membrane H+-ATPase in the root endodermis. *Plant J* 27: 191-201

Waiev B and Rajam MV (2003) Effect of increased polyamine biosynthesis on stress response in transgenic tobacco by introduction of human S-adenosylmethionine gene. *Plant Science* 164: 727-734

Walden R, Cordeiro A, Tiburcio F (1997) Polyamines: small molecules triggering pathways in plant growth and development. *Plant Physiology* 113: 1009-1013

Wang W, Vinocur B, Altman A (2003) Plant responses to drought, salinity and extreme temperatures: towards genetic engineering for stress tolerance. *Planta* 218: 1–14

Weimberg R (1987).Solute adjustment in leaves of two species of wheat at two different stages of growth in response to salinity. *Physiol Plant* 70: 381-388

Wu L (2005) Overexpression of the bacterial *nha*A gene in rice enhances salt and drought tolerance. *Plant Sci* 168: 297-302

Wyn Jones RG and Gorham J (1983) Aspect of Salt and drought tolerance in higher plants In:Genetics engineering of plants. An agriculture prospective eds. T.Kosuge, C.P. Meredith and A. Hollaender Plenum Press, New York. 255-370

Xu DP, Duan XL, Wang BY, Hong BM, Ho THO, Wu R (1996) Expression of late embryogenesis abundant protein gene HVA1 from Barley confers tolerance to water deficit and salt stress in transgenic rice. *Plant Physiol* 110: 249-257

Yan B, Dai Q, Liu X, Huang S, Wang Z (1996). Flooding induced membrane damage, lipid oxidation and activated oxygen generation in corn leaves. *Plant and Soil* 179: 261–268

Yeo AR, Yeo ME, Flowers TJ (1988) Selection of lines with high and low sodium transport from within varieties of an inbreeding species: rice (*Oryza sativa*). *New Phytol* 110: 13-19

Yokoi S, Quintero FJ, Cubero B, Ruiz MT, Bressan RA, Hassegawa PM, Pardo JM (2002) Differential expression and function of *Arabidopsis thaliana NHX* Na+/H+ antiporters in the salt stress response. *Plant J* 30: 529-539

Zhao F, Shanli G, Zhang H, Zhao Y (2006) Expression of yeast SOD2 in transgenic rice result in increased salt tolerance. *Plant Science* 170: 216-224

Zhang HX and Blumwald E (2001) transgenic salt-tolerant tomato plants accumulate salt in foliage but not in fruit. *Nat Biotechnol* 19: 765-768

Zhang HX, Hodson JN, Williams JP, Blumwald E (2001) Engineering salt-tolerant Brassica plants:characterization of yield and seed oil quality in transgenic plants with increased vacuolar sodium accumulation. *Proceedings of the National Academy of Sciences, USA* 98: 12832-12836

Zhu J K (2000) Genetic analysis of plant salt tolerance using *Arabidopsis. Plant Physiol* 124: 941-948

Zhu J K (2001) Plant salt tolerance. *Trends Plant Sci.,* 6: 66–72.

Zhu J K (2003) Regulation of ion homeostasis under salt stress. *Curr Opin Plant Biol* 6: 441-445

Zhu L, Tang GS, Hazen SP, Kim HS, Ward RW (1999) RFLP-based genetic diversity and its development in Shaanxi wheat lines. *Acta Bot Boreali Occident Sin* 19: 13

6

Recent Advances in Photosynthesis for Enhancing Crop Productivity

Pushp Sharma

Improving crop yield to meet the demands of an increasing world population for food and fuel is a central challenge for plant biology. This goal must be achieved in a sustainable manner (i.e. with minimal agricultural inputs and environmental impacts) in the face of elevated levels of CO_2 and more extreme conditions of water availability and temperature. Agricultural yields have generally kept pace with demand in the recent past as a result of the gains made through breeding programs and farming practice, but crops yields are now reaching a plateau. One fundamental component of plant productivity that has not been used to select for increased yield is photosynthesis. There is now the opportunity to exploit our extensive knowledge of this fundamental process for the benefit of humankind. Increase in the yield potential of the major food crops has considerably contributed to the rising food demand/supply over the past decades. Improvement in the photosynthetic efficiency has played only a minor role in the remarkable increase in productivity achieved; further increase in the yield potential will rely in large part on the improved photosynthesis. Critical examination of the inefficiencies in photosynthetic energy transduction in crops from light interception to carbohydrate synthesis, classical breeding, systems biology and synthetic biology are providing new avenues for developing more productive germplasm. Opportunities which could be exploited in near future includes improving display of leaves in crop canopies to avoid light saturation of individual leaves and further investigation of the photorespiratory bypass that has already improved the production of modern cultivars. Long term opportunities include engineering into plants carboxylase that are better adopted to current and future elevated CO_2 concentration and the use of models to guide molecular optimization of resource investment among the components of photosynthetic apparatus for maximizing carbon gain without increasing crop inputs. All these changes collectively have the potential to enhance almost double the yield of the major crops.

Crop plant evolution

Changes that have occurred between wild species and modern cultivar may be grouped into four categories viz. (i) changes conferring adaptedness to cultivation and systematic harvesting; (ii) modification of day length and vernalization requirements; (iii) quality improvement by selection against undesirable and for desirable components and (iv) increase in yield potential. So far, increase in yield potential has been achieved by direct selection for ability to yield the organs of interest under progressively improved systems of agronomic inputs; changes in individual physiological characteristics have resulted only indirectly from such selection. Recently, however, there has been increased interest in the possibility of raising yield by direct selection for important physiological attributes.

Marked increase in the size of seeds, pods, fruits or inflorescences has occurred in many crops during their evolution. Despite these changes, the relative growth rate (RGR) of young plants does not appear to have increased in the course of domestication, even when comparison is made with seedlings of the same size. This has been shown for wheat (Evans and Dunstone, 1970; Khan and Tsunoda, 1970), maize (Duncon and Hesbeth, 1968), tomato (Wilson, 1972) and cowpea (Lush and Wien, 1980).

The relative leaf growth rate is an important determinant of the time taken for the crop canopy to close. Before the canopy achieves full interception of the light, variation in leaf area is a much more powerful determinant of variation in crop growth rate then is variation in photosynthesis rate per unit leaf area. After canopy closure, photosynthetic CO_{-2} exchange per unit leaf may become an important determinant of canopy photosynthesis. Yet there is no evidence till-date of any indirect selection for increase in the maximum light-saturated CO_2 exchange rate per unit leaf area (CER) during the domestication and improvement of wheat, maize, sorghum, pearl millet, sugarcane, cotton or cowpea. Indeed, the highest CERs recorded for wheat, sorghum, pearl millet and cotton have been found in the wild relatives, not in the modern cultivars. In wheat, the higher CER of the more primitive species were associated with higher stomatal and residual conductances and with higher Hill-reaction activity per unit of chlorophyll (Zelanski *et al.*, 1978). The lack of advance in leaf CER during crop evolution may have been due in part to negative relationship between CER and leaf area and the persistence of photosynthetic activity. Also the need for increase in the genetic potential of CER may have been mitigated in some cropping environments by the use of nitrogenous fertilizers. In the absence of genetic increases in photosynthesis and growth, past improvements in yield potential have derived largely from increase in the proportion of accumulated dry weight which is invested in the organs harvested by man i.e. the harvest index. Both the size and the number of these organs have been increased, as have the rate and the duration of their growth in many species. Increase in duration has been associated with greater longevity of leaves or with the storage phase making up a larger proportion of the crop life cycle. Increased rate of storage on the other hand has been shown, at least in wheat, to involve a greater allocation of photosynthetic assimilates to the grain during grain filling. Associated with this greater allocation has been a parallel increase in the cross-sectional area of phloem in the peduncle. In sugarcane also, the phloem cross sectional area serving the storage tissues has been increased in parallel with the latter (Evans and Dunstone, 1970).

Many coordinated changes have occurred in the course of evolution from wild plant to modern cultivar. In general, developmental processes - such as those concerned with germination, flowering, plant form and composition dehiscence, shedding etc. appear to have been more readily modified by empirical selection than the photosynthesis. Two lines of argument may be followed from the observation of improved carbon partitioning during crop evolution. The scope for further improvement in carbon allocation has been discussed by Sharma (2010), so this review aims at increasing photosynthetic and growth rates.

PHOTOSYNTHETIC EFFICIENCY

About 90 per cent of a plant's dry matter is the result of photosynthesis. In other words, there must be a correlation between the net photosynthesis and dry matter content of the whole plant, which vary among cultivars. This relationship is very complex and for many crops a positive correlation is not obtained. The reasons for a poor correlation between leaf photosynthesis and yield are varied. Before a crop canopy achieves full interception of light, leaf area is more important than photosynthetic rate per unit leaf area in determining crop growth rate. After canopy closure, the carbon exchange rate (CER) per unit leaf area becomes the most important factor. High CER has always been considered a desirable characteristic. Still the evidences are needed which can reveal that CER per unit leaf area has increased during the domestication of important crops like wheat, maize, sugarcane and cowpea. Actually, the highest measured CER for wheat, cotton and pearl millet are found in the wild relatives, not in modern cultivars (Gifford and Evans, 1981). However, a correlation seems to exit between yield of forage crops and their respective CER. Several attempts have been made to select for specific photosynthetic component that would result in a higher CER. For example, increasing chlorophyll content does not seem to have much impact on CER. No good correlation has been worked out between Hill reaction activity or phosphorylation per unit leaf area and CER. Nevertheless, a correlation between ribulose bisphosphate (RUBP) carboxylase activity and CER does seem to exist for some crops. Efforts to breed out photo-respiration have failed so far i.e. there is no clear evidence that a C3 plant can have low oxygenase activity with normal carboxylase activity. Considering the low success rate for all these approaches, an alternative might be to breed directly for overall CER, rather than any one specific photosynthetic component.

Crop photosynthesis

Seedling begin as heterotrophic plants depending totally on food mobilized from seed endosperm and pass through a transition phase when photosynthesis commences even while endosperm mobilization reactions continue. Finally, seedlings depend entirely on photosynthesis and become autotrophic. Nelson and Larson (1984) generalized that duration of the pre-autotrophic phase to be 12-days (heterotrophic phase being six days and transition phase being an additional six days). Duration of preautotrophic phases of maize has been reported to range between 14 and 26 days (Cooper and MacDonald, 1970), variations occur because of temperature differences. De-Dutta (1981) reported that rice seedlings become autotrphic at the 3 to 4 leaf stage, somewhere in the range of 14 to 22 days after seedlings emergence. When photosynthesis occurs, dark respiration is assumed to take place simultaneously. Hence, measured photosynthesis consists of difference between true photosynthesis and dark respiration. On the basis of this concept, gross photosynthesis refers to the sum of net photosynthesis and dark respiration (Yoshida, 1981). Crop photosynthesis values collected from the field are primarily determined by photosynthetic capacity per unit leaf area of whole plant canopy, leaf area index and light interception efficiency of leaf canopies (Mae, 1997). It has been concluded that leaf area index and amount of light interception for many crops are closely associated with grain yields.

Photosynthesis from an agricultural perspective

There are theoretical limits to productivity, which are set by the thermodynamic properties of the crop and its environment. In this context (of theoretical maximum yield), the limitations are set by the efficiency of absorption (capture) of light energy and the efficiency of its transduction into biomass. The vital question is whether these limits have been reached already within crop systems or whether there is potential for improvements that have not yet been exploited obviously, improvements in the capture and conversion of light energy have been a central part of crop improvement during the last century. For example, the increase in the erect nature of leaves has permitted a higher leaf area per unit ground area (leaf area index (LAI)), allowing crop canopies to be extremely efficient at absorbing radiation. The rate of conversion has also been improved by input management: the application of fertilizers increases leaf area and also the rate of photosynthesis per unit leaf area. Additionally, enhanced resistance to pests and diseases and a multitude of adaptations to local requirements (photoperiod, growing season duration, and temperature) have resulted in yield progress in a range of agro ecological zones (Evans, 1993). However, many of these features of crop plants, which played such a prominent role in creating the ideotypes of the green revolutions (harvest index, nitrogen responsiveness, stature and canopy architecture) may already be optimized or close to optimization (Long *et al.*, 2006). Therefore, the question remains as to whether the crop biomass production rate is similarly optimized. It might be assumed that yield progress has been associated with an improvement in total biomass production, but some studies show a non significant relationship between yield and biomass (Calderini *et al.*, 1995). Nevertheless, evidence suggests that a critical component of crop production is increasingly dependent upon the capacity to produce more biomass (Hubbard *et al.*, 2007). In the case of rice, the rate of biomass production only seems to have correlated with yield since the early 1980s (Peng *et al.*, 2000). In the UK, wheat yield progress in recent decades has been shown to involve an improvement in biomass production (Shearman *et al.*, 2005). Direct experimental evidence is found in the higher biomass and yield of crops grown in elevated CO_2 infield experiments (Ainsworth *et al.*, 2004; Long *et al.*, 2006). Classical crop physiology tells us that total dry matter content at harvest is closely and linearly correlated with accumulated intercepted solar radiation. The slope of this relationship gives the radiation use efficiency (*E*); in other words, the amount of dry matter produced per unit radiation intercepted (measuring dry matter (DM). Total biomass production can be described by the following equation (Monteith, 1977):

$$\text{Total Biomass} = \sum_{\text{sowing}}^{\text{harvest}} Q \times I \times E$$

(*Q*, the solar radiation over the duration of the crop period; *I*, the interception of the solar radiation by the crop canopy). It follows that the ways to increase total biomass are:

- to increase the duration of crop photosynthesis;
- to increase the interception of the solar radiation by the crop canopy, and
- to increase the efficiency of the conversion of the light energy into plant dry matter.

The duration of crop photosynthesis is of critical importance. It is thought that both selection for higher yields and the increased input of nitrogen fertilizer have acted to increase leaf lifespan in modern crops (Hay and Porter, 2006). Temperature determines the rate of development and therefore the time available for radiation capture. In tropical regions, it is possible to obtain more than one harvest per year and one focus has been on shortening crop duration to allow rapid harvest (Peng *et al.*, 2000). In temperate regions, high temperatures can reduce the grain-filling duration, reducing the biomass production rate during this period. There has been interest in manipulating the timing of senescence to obtain higher crop yields. However, senescence is an essential physiological process that remobilizes nutrients for grain production (Gan and Amasino, 1995) and its manipulation has to be integrated into the regulation of reproductive physiology and appropriate responses to environmental factors (Murchie and Horton, 2007). The efficiency of solar radiation interception (defined as the proportion of incident irradiance absorbed corrected for reflection) by fully formed crop canopies is considered to be generally high. In the case of many cereals, this is a result of a high LAI combined with erect leaves, which increase light penetration into the canopy. The introduction of semi-dwarf growth habit with reduced height genes (*Rht*) had a massive impact on yield in cereals, greatly increasing the harvest index (dry weight of product: dry weight of plant; Austin *et al.*, 1980). It is often assumed that this also improves radiation interception, but the semi-dwarf lines have similar radiation use efficiencies to tall lines (*rht*) (Miralles and Slafer, 1997). The genes responsible for dwarfing in other major crops have been identified (Spiel Meyer *et al.*, 2002) and there may be further potential for improvement. There may be scope for further improvements in canopy architecture and development so that the formation of a full canopy coincides with periods when radiation intensities are highest. For example, more rapid formation of a crop canopy may be important in colder temperate regions where developmental processes such as leaf emergence are temperature-limited (Hay and Porter, 2006). The duration of crop photosynthesis and the interception of solar radiation are the two components of the Monteith equation that have contributed most to the increase in yield of most important crops. Because biomass has on average 40% carbon by dry weight, any improvement in total biomass production means an improvement in photosynthetic carbon fixation. There is general agreement that an improvement in carbon fixation during the second half of last century was an essential ingredient of the increase in crop yield. Surprisingly, this enormous increase in carbon fixation was largely achieved not by increasing the CO_2 assimilation per unit leaf area, but by increasing CO_2 assimilation per unit land area. For this, improved agronomic practices (in terms of plant density, nutrition, water supply, pesticides, herbicides, etc.) created favorable micro environments for plant growth, effectively mitigating the negative impacts of external constraints (biotic and abiotic). Thus, there has been little or no increase in the intrinsic conversion efficiency of light energy into plant dry matter (photosynthetic efficiency) by individual leaves. In fact, studies across a range of crops show that increase in photosynthetic rate per unit leaf area rarely coincides with yield progress (Evans, 1993). In many cases, there may be a negative relationship. In wheat there seems to have been a decline in assimilation rate following domestication (Austin *et al.*, 1982), although one example shows yield progress linked to an improvement in leaf assimilation rate (Fischer *et al.*, 1998). In the case of rice, photosynthesis tended to be lower in the wild *Oryza* species than in *O. sativa* (Cook and Evans, 1983), and among varieties there is evidence for a trend towards higher rates of assimilation in more recent varieties (Zhang and Kokubun, 2004). By contrast, other studies show either no trend or even a lower rate of assimilation in cultivated varieties (Yeo *et al.*, 1994). However, it should be pointed out that yield improvement has targeted traits and practices that did not necessarily depend on increasing or even maintaining the rate of leaf photosynthesis, meaning that the increases in photosynthesis per unit ground area could have happened with no change (or paradoxically even a decline) in photosynthesis per unit leaf area. Attempts to improve yield by directly selecting/breeding for crop plants with high rates of leaf photosynthesis have been rather limited and have had mixed success (Gutierrez-Rodriguez *et al.*, 2000). It has therefore been tacitly assumed that photosynthetic efficiency is a constant in crop systems, already optimized and therefore not a modifiable determinant factor for the increase of crop yield. This assumption was considered to be consistent with research into the mechanisms of the photosynthetic process itself and demonstrations that fundamental photosynthetic parameters such as quantum yield are highly conserved among higher plants (e.g. Björkman and Demmig, 1987). However, although the radiation use efficiency of crops (*E*) is claimed to be fairly consistent for a given crop species, much variation has also been reported (Long *et al.*, 2006; Hay and Porter, 2006), and in the field *E* may fall well below the theoretical maximum (Mitchell *et al.*, 1998; Zhu *et al.*, 2008). To determine the impact of photosynthesis per unit leaf area on yield it is necessary to minimize the effect of other variables. Thus, experiments in which the background genetic variation is reduced have had greater success (Gutierrez-Rodriguez *et al.*, 2000). Similarly, leaf photosynthesis exerts a greater control on biomass production and grain yield when variation in factors such as partitioning, nutrient responsiveness and LAI is minimized (Long *et al.*, 2006; Hubbart *et al.*, 2007). Interestingly, an increase in yield observed in varieties of rice released after 1980 is more closely correlated with an increase in biomass than with an increase in harvest index (Hubbart *et al.*, 2007). The fact that these varieties present higher light-saturated rates of photosynthesis (*P*max) than many older varieties released before 1980 suggests that an increase in leaf-level photosynthesis in rice is more likely to be observed in circumstances where an increase in biomass production dominates. This implies that selection, either directly or indirectly, for improved biomass production has effects on leaf photosynthetic physiology. For many authors the positive response in yield for crops grown under elevated atmospheric CO_2 is strong evidence that an increased rate of leaf photosynthesis can induce more yields. The arguments in this section reveal the potential for increasing crop biomass production through alteration of leaf photosynthesis. There is a need to identify the specific targets that will directly improve leaf photosynthesis, so that further increases in yield will be realized. However, the mechanism slinking whole-plant events to leaf-level events are poorly understood. Photosynthesis in agriculture should be viewed holistically, as an integrated part of a much more complex process, which includes not only the primary events of light harvesting and carbon fixation, but also carbohydrate synthesis, partitioning of biomass (sink size and activity) and harvest of yield, together with the transport efficiencies of water, assimilates and nutrients. Two potential ways forward will be presented: firstly, the direct improvement of the mechanism of photosynthesis itself, resulting in increased photosynthetic capacity and/or efficiency; and secondly, by exploring and analyzing the complexity of photosynthetic regulation in the field ,more effective use of existing photosynthetic capacity by optimizing dynamic responses to the environment can be made. In either case, the resulting improvements in crop yield should not depend upon increased nitrogen (N) fertilization or water supply, but instead should increase the N use and water use efficiencies.

Photosynthesis in the field: potential and actual rates

Many studies clearly show that maximum photosynthetic potential is rarely realized, under field conditions. An analytical approach to describing such losses in potential photosynthesis was first presented by Cheeseman *et al.* (1991). When 'spot measurements' of photosynthetic rate were taken on a large number of leaves in a plant population, and the data points were plotted against irradiance, a high degree of scatter was found, with a theoretical irradiance curve represented as the ceiling below which all data were located. Measurements on a rice crop produced the same picture (Murchie and Horton, 2007). Confirmation of the under performance of the majority of leaves was obtained by measurements of leaf photosynthesis during a diurnal cycle, which showed that peak photosynthetic rates were maintained for only a few hours each day (Black *et al.*, 1995; Murchie *et al.*, 1999). The reasons for this are many and varied (Horton, 2000), but there are two major influences- firstly, the direct limitation imposed by environmental stresses, even under apparently favorable conditions; and secondly, the presence in the plant's genotype of regulatory processes that reduce photosynthetic activity (Fig.1b). There has predictably been a great deal of interest in the former, and increasing the stress tolerance of crops has become a major goal in agricultural improvement (Bohnert *et al.*, 2006, Valliyodan and Nguyen, 2006). By contrast, little attention has been paid to the effect of regulatory mechanisms. These are particularly intriguing, and will be explored at length. From an agricultural perspective, photosynthesis includes all the events from light interception to the export of photosynthate for biomass accumulation and grain production (Fig. 1a). As yield is an integration of this process over time, inevitably it has to include developmental aspects and most importantly the continual changes in environmental factors. The description of photosynthesis in these terms represents a challenge: not only does photosynthesis comprise a large number of integrated reactions and processes, but it is exceedingly complex, and hence difficult to describe in a way that is useful from an agricultural viewpoint. Equally, it is this complexity that provides so many unforeseen opportunities. Photosynthetic complexity arises from several sources: heterogeneity - the leaf consists of a range of cell types in which photosynthetic capacities and even biochemical pathways are different; intracellular co-operation - photosynthesis depends upon the interaction of pathways in the chloroplast, cytosol, vacuole and mitochondria; metabolic regulation - many individual component parts of the process are regulated by metabolic signals generated in other parts, creating a network of feedback and feed forward regulatory networks which provide fine control over flux; acclimation - the capacities of component processes respond to changing environmental conditions, providing course control, and, together with regulatory mechanisms, establish balance in the face of differing environmental and development constraints. An important consideration is that one function of this complexity is to contain the potentially lethal cocktail in the plant cell – light, oxygen and photosensitizing pigments – which serves as a warning that intervention in this system has the possibility of increasing the likelihood of photo-oxidative damage (Horton *et al.*, 2001). Although the regulatory and acclimation processes were highlighted as potential targets for rice improvement, (Horton, 2000; Horton *et al.*, 2001) in general these aspects have attracted very little attention. Indeed, their existence can be seen a negative factor – the plasticity that results from these processes means that genetic alteration of single enzymes or other components is much less likely to produce the desired changes in total photosynthesis because of compensatory adjustments. Feedback and feed forward regulation of photosynthesis occurs at many levels (Fig. 1a) and enables responses to changing external and internal conditions. In each case, the regulatory mechanism should ideally fulfill a number of criteria: it should react to enable the concentrations of key metabolites such as ATP to be maintained; it should prevent the build-up of potentially damaging species, such as triplet states, or conditions, such as extreme redox states; it should be sufficiently dynamic to produce prompt responses to the altered conditions; and it should have a dynamic range capable of responding to a wide range of changes in the signaling factor. However, fulfilling all of these criteria simultaneously may not be possible and we therefore need away to quantitatively describe the imperfections in these processes. The application of systems approaches promises to provide a means of exploring these features in a way that produces practically useful outcomes. Only when the dynamic structure of the photosynthetic process has been analyzed will it be possible to determine the robust targets for intervention. It is not just plant metabolism that could be targeted in this way: recent data suggest that canopy photosynthesis may be amenable to further manipulation by smart alteration of 3D structure. In warm environments, air temperature and humidity are two factors thought to be critical for determining both stomatal-mediated and chloroplast-level reductions in photosynthetic capacity. In such environments, transpiration plays a key role in lowering leaf temperatures to temperatures that are optimal for C3 photosynthesis. Leaf temperatures in trees are kept remarkably constant across a range of latitudes from subtropical to boreal regions. In cold regions they are even raised above ambient during the periods in which they accumulate most biomass. This homeostasis of leaf temperature is thought to occur by adaptation of canopy architecture and structure, notably the proximity of leaves to each other (Helliker and Richter, 2008). Were this to be applicable to crops it could be exploited in canopy design and in the manipulation of canopy temperature to beneficial manipulation. This may be especially effective in environments where temperature is the overriding limiting factor for photosynthetic rate. However, in a theme central to this review, one may ask: what are the consequences for other processes of such adaptation to temperature.

Canopy photosynthesis

Canopy photosynthesis describes photosynthetic activity per unit ground area and combines genotype efficiency, leaf morphology and canopy architecture (Wells *et al.*, 1986). Once exponential phases of canopy leaf area development are complete, individual leaves

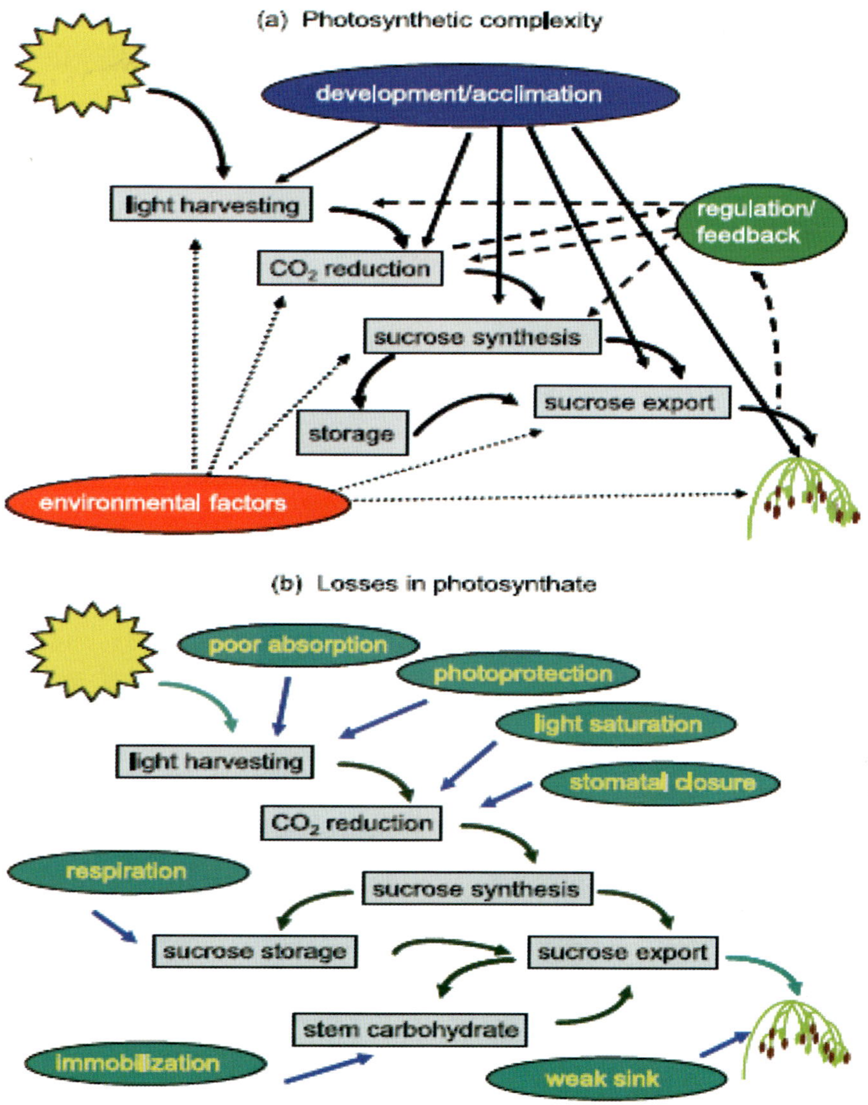

Fig 6.1. Schematic depicting (a) the complexity of photosynthetic processes in the agricultural context with regulation needed at every step from chloroplast level processes through to transport and storage of assimilate (b) The losses which reduce photosynthate production below the potential rate.

begin to senesce, and if not balanced by production of new leaves, contribute to declines in canopy photosynthesis (Kely and Davies, 1988). Contributions of leaf senescence to declining canopy photosynthesis and to C availability for yield have been identified as potential limitations in crop production (Wullschleger and Oosterhuis, 1992).

Net photosynthetic rates of crop plants vary with leaf position, nutritional status, water status, plant species, cultivar within species and growth stage. Crop species are generally classed to have 'C3' (barley, bean, cotton, cowpea, oat, peanut, rice, soybean, sugarbeet and wheat) or 'C4' (maize, millet, sorghum and sugarcane) based on whether their initial sugars produced in photosynthesis consist of three or four C molecules. The two types of photosynthetic pathways differ in chloroplast arrangement, primary photosynthetic enzymes, temperature response, water-use efficiency, light saturation and response to CO_2 concentrations (Table 1).

Table 1: Characteristics of C3 and C4 species

Characteristics	C3	C4
Photosynthetic efficiency	Low	High
Photorespiration	High	Low
Water use efficiency	Low	Light
Optimum temperature for photosynthesis	10-25°C	30-45°C
Response to light intensity	Low	High
Response to CO_2 concentration	Low	High
Response to O_2 concentration	Low	High
Major pathway of photosynthetic CO_2 fixation	Reductive pentose Phosphate cycle	C4 - dicarboxylic acid and reductive pentose phosphate cycle
Transpiration ratios	High	Low
Leaf chlorophyll a to b ratio	Low	High

Source: Fageria *et al.*, (1997)

No clear differences in photosynthetic rates exist among crop species, except for maize and alfalfa. Maize has about twice the photosynthetic rate as many other crops and alfalfa had an intermediate photosynthetic rate (Table 2). Variation in photosynthetic rates is often correlated with concentrations of N compounds in leaves (Hirose and Werger, 1987). This correlation is explained by the fact that some 75% of all N in mesophyll cells of C3 plants is associated with photosynthesis (Lambers, 1987). Further, larger fraction of N (about 25%) becomes components of the enzyme Rubisco (ribulose-bisphosphate carboxylase). Rates of photosynthesis are, therefore, closely correlated with Rubisco activity in leaves (Evans, 1983). Net photosynthesis of individual rice leaves reaches maximum values of about 40 to 60 kilolux (800 to 1200 moles $m^{-2}sec^{-1}$) near half-full sunlight (Yoshida, 1981). However, photosynthesis of well developed canopies increases with increasing light intensity of up to full sunlight, and no indication of light saturation appears to be reached. In early stages of crop growth, the main determinant of photosynthesis is the extent of leaf area development. As LAI increases, so does the extent of light interception, and often exceed 95% for many cereal crops, with LAI values of about 4. Once canopies close from leaf density, further increase in LAI, have little effect on plant photosynthesis, which may be influenced by incident radiation and structures of canopies (Evans and Wardlaw, 1976).

Table 2 : Photosynthetic rate per unit leaf area in various crop plants

Crop plant	Photosynthetic rate *(mg CO_2 $dm^{-2}h^{-1}$)*
Alfalfa	50-55
Common bean	30-40
Maize	60-80
Field pea	25-30
Potato	25-35
Rice (lowland)	40-50
Soybean	30-35
Sugarbeet	30-35
Sunflower	40-42
Wheat	28-29

Source: Tanaka and Osaki (1983)

Net photosynthetic rates of active, healthy single rice (C3-plant) leaves are about 40 to 50 mg CO_2 $dm^{-2}h^{-1}$ under light saturation conditions. Ishii (1988) noted no significant differences in photosynthetic rate of single rice leaves at heading stage of growth for 32 Japanese cultivars bred from 1880 to 1976. However, 25% increase in canopy photosynthesis was observed in new rice cultivars (bred between 1880 to 1913). Improvement in canopy photosynthesis was attributed to higher efficiency of light interception by canopies because of changes in leaf angles. Old cultivars had droopy leaves with high mutual shading, while new cultivars had erect leaves with less mutual shading. Similarly, Conoceno, Egdane and Setter (1998) concluded that much of the large increase in rice yield over the past three decades could be attributed to improvements in canopy structure that enhanced canopy light interception and photosynthesis.

Canopy photosynthesis can be described as the product of canopy light interception and radiation use efficiency (Loomis and Connor, 1992). Positive linear relationships have been noted between crop growth rates and canopy light interception for most field crops. In cotton, a close relationship has been worked out between leaf area index and canopy light interception (Heitholt, 1994; Sharma, 2010) and between light interception and lint yield (Heitholt *et al.*, 1992). Leaf photosynthetic rates depend on incident radiation and leaf absorbency. The latter is in turn affected by leaf external and internal reflectance and by leaf pigment i.e. chlorophyll (Mass and Dunlop, 1989).

C3 and C4 photosynthesis

CO_2 is converted into carbohydrates during photosynthesis by two biochemical processes known as calvin pathways C3 and C4 (Hatch and Slack, 1970). The first stable product from fixation of CO_2 in the calvin cycle is 3-phosphoglyceric acid. In recent years, however, studies revealed that certain species have additional photosynthetic reaction in which the first detectable product resulting from CO_2 fixation is not the 3-C compound (3 PGA), but rather the four carbon compound (OAA), which is quickly transformed into either malic or aspartic acid. To distinguish between plants whose leaves have this adaptation and those that do not, it is customary to refer to species for which the calvin cycle alone functions in photosynthesis as C3 species.

The C4 photosynthetic pathway is present in both monocot and dicot plants, and has reduced energy wasting processes of photorespiration that appears in photosynthetic pathway of C3 plants. In C3 plants, CO_2 is fixed through ribulose bisphosphate (RUBP) carboxylase (RUBPcase) so that RUBPcase mediated carboxylation of RUBP produces the compound 3 PGA. This reaction is inhibited nearly 50% when atmospheric O_2 competes with CO_2 at active sites of RUBPcase. The C4 photosynthetic pathway eliminates photorespiration by splitting photosynthetic reactions between two morphologically distinct cell types : bundle sheaths (BS) and mesophylls (M). RUBPcase is physically separated from atmospheric O_{-2} by being compartmentalized in internal BS cells. This scheme is advantageous in hot, dry conditions but can be energetically wasteful (Nelson and Langdale, 1992). Some studies have indicated that certain C4 plants function as C3 plants when the use of C3 pathway is energetically favorable (Khanna and Sinha, 1973). Plants in C4 group have high photosynthetic efficiency compared with plants in C3 group. C4 plants represent an adaptation to habitats with relatively high temperatures, high irradiance and limited water supply. The C4 plants tend to be at disadvantages in cool climates.

Al-Khalib and Paulsen (1999) compared photosynthetic responses among C3 (wheat and rice) and C4 (millet) plants to note their reactions at high temperature and sensitivity to light reactions. Leaf photosynthesis of millet and rice increased at temperatures from 22 to 32°C before decreasing at 42°C, whereas wheat had highest leaf photosynthesis at 22°C and decreased as temperature increased. These results indicate that leaf photosynthetic response differences at high temperature were associated with light reactions. The extreme sensitivity of wheat to high temperature was attributed to injury of PS II.

The chloroplast as sensor and integrator of photosynthesis

Classically, the chloroplast has been described as the site of photosynthesis, which provides the carbohydrate that fuels plant growth and development. This restricted view of chloroplast function ignores not only the fact that chloroplast metabolism is tightly integrated with the rest of the cell, so that photosynthesis is the result of all of the activity of the cell, but also the fact that the chloroplast is involved in the leaf-level acclimation are unclear, but in dicotyledonous species there is evidence for a role of mature leaves in the cellular differentiation of expanding leaves (Oguchi *et al.*, 2003). When mature leaves of *Chenopodium album* were shaded, a leaf expanding into a high-light environment had a shade anatomy, and vice versa. The signals involved are unknown but may be hormonal or be related to photosynthate supply. It has been proposed that the chloroplast redox state also has a role in long-distance signalling in relation to leaf-level acclimation, and indeed to other aspects of plant growth and development (Wilson *et al.*, 2006). The mechanisms of interaction between chloroplast energy status sensing and other networks such as sugar signalling have yet to be determined. However, progress may be swift: recent work suggests that certain genes may have pivotal roles in directing convergent signals, in particular sugar and energy (e.g. KIN10) (Baena-Gonzalez *et al.*, 2007). These genes seem to confer enhanced starvation tolerance, lifespan and architectural and developmental transitions. It is suggested that they act by linking stress, sugar and developmental signals. Recent work has identified a chloroplast-localized protein that is crucial for cytosolic responses to calcium (Ca^{2+}) fluctuations (Webb, 2008; Weil *et al.*, 2008). This may provide some of the first direct evidence for involvement of the chloroplast in signalling cytosolic events. Chloroplast specialization deserves a mention in this context. There are of course examples of structural and functional differentiation: for example, the differences between mesophyll and bundle sheath chloroplasts in C4 plants are striking and are a reflection of the differing ATP:NADPH requirements of each cell type. However, C4 photosynthesis has been identified in the stems of C3 flowering plants (Hibberd and Quick, 2002), which is considered to be an adaptation to low gaseous diffusion rates in these tissues. More recently, it has been suggested that bundle sheath chloroplasts may have a specialized role in the systemic signalling processes operating in high light and/or high light stress (Fryer *et al.*, 2003; Mullineaux *et al.*, 2006). The chloroplasts exhibit photosynthetic electron transport but have a limited capacity for CO_2 fixation; instead, following specific expression of the ascorbate peroxidase gene APX2, they generate large amounts of H_2O_2, which is exported into the transpiration stream (Chang *et al.*, 2004). This level of specialization of chloroplasts in signalling roles supports the concept of the chloroplast as a sensor and processor of environmental signals (Fig 2). Three types of signal may be initiated: firstly, the redox state of the thylakoid membrane in mesophyll chloroplasts (Huner *et al.*, 1998; Walters, 2005; Wilson *et al.*, 2006) which may act as a local or long-distance signal; secondly, the production and accumulation of specific metabolites in sucrose and starch synthesis, which act as an indicator of the 'energy status' of the plant cell and are sensitive to sink strength; thirdly, specific long-distance signals which are produced by chloroplasts and do not have localized effects, such as H_2O_2. In this way the chloroplast receives, stores and integrates environmental information, and then orchestrates the development of the plant, defining its biochemistry, physiology and morphology (and, in the case of a crop plant, its yield/biomass) by interfacing with the central signaling networks. The roles of these signals and how they integrate with each other should be a focus for future research.

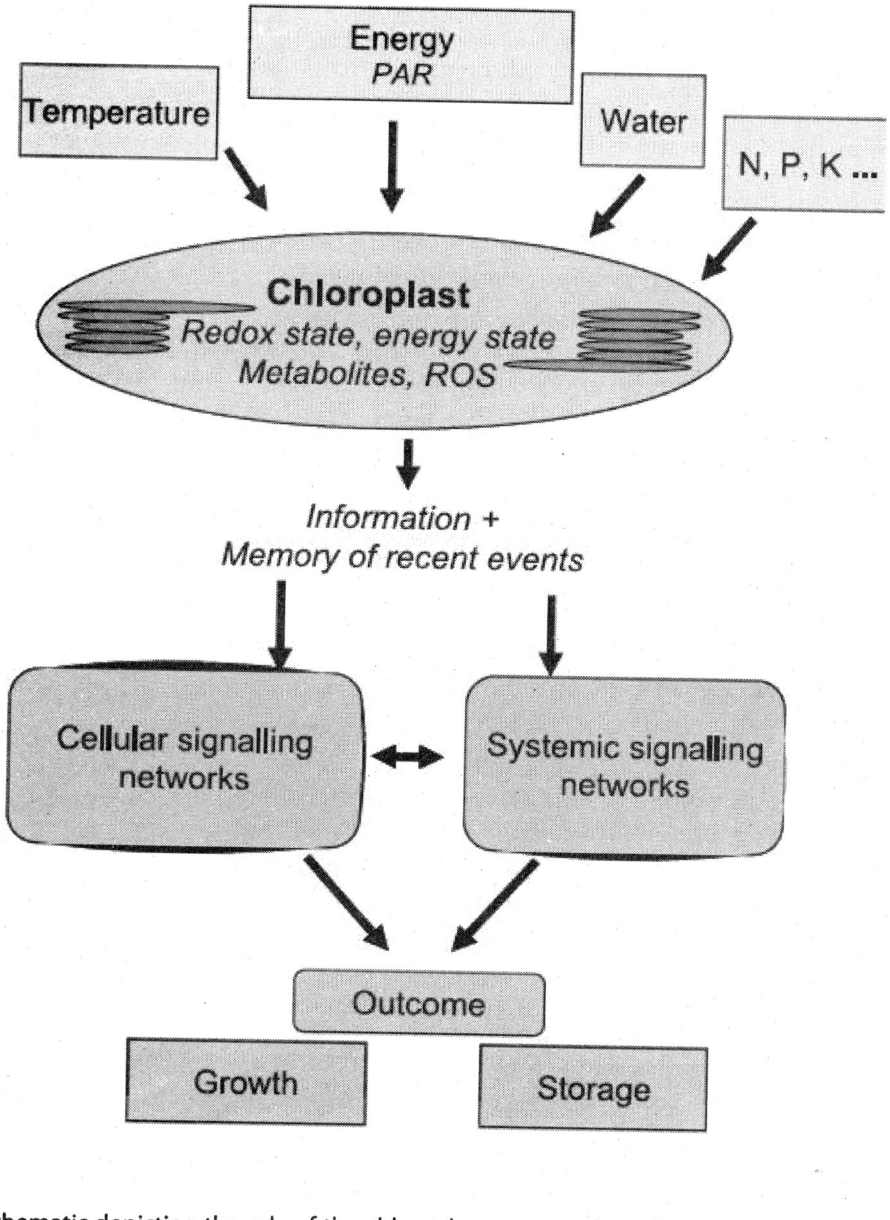

Fig. 2 : Schematic depicting the role of the chloroplast as a supplier of quantitative information that enables the plant to rapidly and accurately match supply of photosynthate with potential growth or storage capacity. This schematic differs from models that largely depict information flowing in the opposite direction only, that is, feedback from sink and feedback from metabolism acting to down-regulate or 'release' energy from the chloroplast. Recent data provide evidence for integration of the chloroplast with Ca2+ signalling in the cytosol (Weinl *et al.*, 2008). PAR, photosynthetically active radiation; ROS, reactive oxygen species.

Photosynthesis: A target for improvement

Basic research in photosynthesis, especially during the past two decades has revealed many possible approaches towards increasing efficiency of CO_2 assimilation by crops. Efficiency is defined as rates of net CO_2 uptake per unit leaf area or per unit ground area (Zelitch, 1975). Total irradiation available for photosynthesis and duration of photosynthesis are undoubtedly important factors determining final plant productivity. Hence, productivity can be increased by planting cultivars with rapid rates of leaf area expansion, using closer plant spacing to capture more sunlight and breeding plants with leaves that have more erect angles of elevation to absorb sunlight more effectively.

Genotypic differences in components of photosynthesis have been found and used as selection criteria for plant improvement (Gifford and Evans, 1981). This means using appropriate genotypes for given environmental conditions can improve photosynthetic rates and consequently crop yields. The genetic variation in chlorophyll content per unit leaf area seems generally to have little impact on variation in CER/productivity, as for example in barley (Ferguson *et al.*, 1973) where even chlorophyll-deficient mutants have near-normal CER (McCathin and Canvin, 1979). The reduction in CER of soybean lines having abnormally low leaf chlorophyll (<35 µg cm^{-2}) may be associated with low specific leaf weight in low chlorophyll genotypes (Lugg and Sinclair, 1979). Similarly, there is no indication that variation in Hill activity/photo-phosphorylation per unit area is reflected as variation in CER (Hanson and Grier, 1973; Treharne, 1972). Determination of Hill activity per unit chlorophyll cannot be interpreted in relation to CER without information on chlorophyll per unit leaf area.

In contrast to light reactions, variations in ribulose bisphosphate carboxylase (RUBPc) in C3 species has been found to correlate with leaf CER, as Randall *et al.* (1977) found when comparing a high CER mutant of tall fescue with normal tall fescues. Since, RUBPc represents 30-50% of the soluble proteins of leaves; there is a risk that breeding for its high activity might be at the expense of other important enzymes. Selection for high specific activity forms of the enzyme might be better, but none have been found yet, although a ryegrass tetraploid with a lower than normal K_m (CO_2) for RUBPc has been described (Garrett, 1978; Rathnam and Chollet, 1980). Also, the oxygenase activity of the carboxylase protein yields a substrate for photorespiration for which a function has not yet been established. The substantial effort to breed out photorespiration by several methods including screening seedlings for low CO_2 compensation point (Apel, 1979) has proved fruitless so far. There is no unequivocal evidence of C3 plant genotypes with low oxygenase and normal carboxylase activity.

A morphological character which often (Dornhoff and Shibles, 1976) but not always (Brinkman and Frey, 1978), correlates with CER is specific leaf weight or more simply, leaf thickness (Charles Edwards, 1978). Intergenotypic variation in SLW at a chosen ontogenetic stage can show stability of ranking from season to season and is heritable. Whether or not, it is a good breeding strategy to select for high SLW depends on its relationship to leaf area development. Expansion of leaf area and thickening of leaves can be inversely related (Motto *et al.*, 1979). A sound strategy might be to produce plants which expand large thin leaves early in the season and then thick leaves after the canopy intercepts all the light.

Selection for stomatal frequency and length has not met with success in changing CER through stomatal conductance, since frequency and length are inversely related and other countering compensations in leaf area and longevity can also occur. Further, stomatal aperture is a much powerful determinant of stomatal conductance and behaves as if it were continuously controlled to resolve optimally the conflict between the carbon needs of growth and the environmental demand on plant water through transpiration. So, even if stomatal frequency were increased, it is likely that the water loss/carbon gain optimization process would lead to the same resolution in terms of stomatal conductance.

The complexity of the internal controls over CER, together with the little success at improving, by selection for component processes, suggests that direct selection for CER itself might be more effective. Evidences exist for heritable variation in leaf CER for high and low lines which have been selected in several species. Unfortunately the relationship between CER, growth and yield is tennous; so much so that it was possible to select all tall fescues with high growth rate but low CER and vice-versa (Wilhelm and Nelson, 1978). The fact that CO_2 enrichment of the atmosphere will enhance growth and yield of C3 crops shows that growth is photosynthetically limited in some environments. So, selection for high CER might be failing because of counter productive associations.

Ontogenetic changes in the partitioning of leaf dry weight between leaf area, leaf thickness together with sink demand effects, can result in poor correlation between CER at one ontogenetic stage and at another. In wheat, the persistence of flag leaf photosynthesis was greatest for the modern hexaploids, so that towards grain maturity they have higher CER than the primitive types despite a reverse ranking early in grain filling (Evans and Dunstone, 1970). An inverse relationship between area per leaf and maximum CER among wheat species and within triticales (Planchon, 1979) and maize (Hanson and Grier, 1973) alfa-alfa (Delancy and Dobrenz, 1974) and soybean (Buttia *et al.*, 1973), is partly a reflection of the inverse association between leaf thickness and leaf area. It also reflects a negative relation between mesophyll cell size and CER.

Nitrogen supply can affect plant growth and productivity by altering both leaf area and photosynthetic capacity (Novoa and Loomis, 1981). Strong relationship between leaf N concentration and single leaf photosynthetic rates occur (Bondada *et al.*, 1996), which appears to be associated with large fractions of leaf N compounded in photosynthetic enzymes (Shiraiwa and Sinclair, 1993). High correlation between light saturated leaf CO_2 assimilation and leaf N per unit leaf area have been reported and hyperbolic relationships between CO_2 assimilation rates and N per unit leaf area differed markedly among species (Sinclair and Horie, 1989). Thus, use of adequate N is an important strategy for improving photosynthesis in crop plants. High percentages of N contribute in maintaining photosynthetic integrity e.g. 50 to 70% of the total N in leaves was directly associated with chloroplasts in maize (Hageman, 1986). Strong correlation has been demonstrated between net photosynthetic rates and leaf N or protein contents (Sinclair and Horie, 1989). Pn rates increased linearly with leaf N contents in maize until critical leaf N values have been reached, after which Pn rates remained constant above critical N contents despite further increase in N (Wong *et al.*, 1985). Since, N can limit crop growth in many environments, high correlations of Pn with leaf soluble protein indicates that leaf-N may be a good criterion for estimating photosynthetic capacity in some environmental conditions (Edwards, 1986). Plant deficiencies in N will have lower Pn, accumulate less dry matter and produce lower yields (Dwyer *et al.*, 1991).

Numerous physiological processes associated with wheat growth are influenced by N fertility. In many wheat producing areas, increasing rates of N fertility results in maximum LAI, vegetative dry weights and leaf N concentration (Frederick and Camberato, 1994). Flag leaf CO_2 exchange rates during grain fill are usually positively correlated with N concentration (Hunt and Poorten, 1985) except when drought induced stomatal closure occurs. Maintenance of green leaf areas late into a growing season often results in longer grain filling and higher yields in wheat (Fredrick and Camberato, 1994). Gwathmey and Howard (1998) reported that soil applications of K may also increase canopy photosynthetic photon flux density in cotton.

Growth is affected by planting date and planting before or after optimal dates results in reduced LAI, leaf area duration, total dry matter production and grain yield. Appropriate row spacing, plant densities, growing crops under optimum temperatures, are other management strategies for improving photosynthetic efficiency in crop plants and consequently yields. Development of crop cultivars that produce leaf areas quickly and tolerate high plant densities could enhance yields. A breeding strategy for increasing leaf areas per plant is to incorporate leafy traits into inbred lines. Plants bearing leafy traits are characterized by extra leaves above ears, low ear placement, highly lignified stalks and leaf parts, early maturities and high yield potentials (Shaver, 1983). Increasing plant population is a management tool for increasing capture of solar radiation within canopies. However, efficiency or conversion of intercepted light into economic yields can decrease, due to mutual shading of plants. We need more standardized system for measuring photosynthetic variables, more important at the whole plant level. A lack of correlation between photosynthetic rates and economic yield may result from the chemical partitioning of carbon between starch and sucrose. As this relationship between CER and yield is weak, this area of research does not appear to be over promising.

With the advent of transportable infrared CO_2 analyzers, opened the opportunity for selecting crcp genotypes on the basis of leaf photosynthetic rates (Long *et al.*, 1996). Many studies indicate that leaf photosynthesis limits crop productions. The modern bread wheat cultivars have lower leaf photosynthetic rates than their wild ancestors (Evans and Dunstone, 1970). Lack of correlation between crop yield and leaf photosynthetic rate is noted frequently in other studies. The lack of correlation indicates these plants differ genetically in many respects beyond photosynthesis. While it is implicit in equation that photosynthetic efficiency of the whole crop averaged over time. Many surveys of leaf photosynthesis are based on the light -saturated rate of a single leaf at a single stage in crop development (Long, 1988). The relationship between single leaf measurements and the whole crop will be complex, and net intuitive. Firstly, as much as 50% of crop carbon may be assimilated by leaves under light limiting conditions in which different biochemical and biophysical properties determine photosynthetic rate (Long, 1993). Secondly, increase in leaf area may often be achieved by decreased investment per unit leaf are, thus, light-saturated photosynthetic rate is commonly lower in species with thinner leaves, quite simply because the apparatus is spread more thinly (Beadle *et al.*, 1985).

Photosynthesis can be limited by sink capacity (i.e. ability to use photosynthate). After flowering, the major sink in the grain crops is the number and potential size of the seed formed. Decreased sink capacity, as may be induced by removing filling grains can feedback to decrease photosynthetic capacity (Preet and Kramer, 1980). It may be expected, however that breeding selects for the cultivar that are able to make maximum use of photosynthetic capacity. For example, if weather favours increased photosynthesis, an effectively selected cultivar should have sufficient capacity for formation of grain to use the additional photosynthate. However, a recent detailed analysis that reviewed the magnitude of seed dry weight changes in response to manipulations in assimilate availability during seed filling for wheat, maize and soybean has concluded that in all three crops, yield is usually more limited by sink than by source i.e. photosynthesis (Borras *et al.*, 2004).

Contrary to the finding of Evans and Dunstone (1970), Watanake *et al.* (1994) show a strong positive correlation between leaf photosynthetic rate and date of release of Australian bread wheat cultivars. The difference may be explained by the fact that the latter study is limited not only to a single species, but to a narrow range of germplasm within that species. Here, variability in leaf area per plant and in distribution would be smaller and variation in leaf photosynthetic rate is not confounded with large difference in total or specific leaf area. The potential of leaf photosynthetic rate in improving potential crop yield can only be evaluated when other factors, in particular leaf canopy size and architecture, are constant. Sinclair *et al.* (2004) stated that even in these circumstances, a 33% increase in leaf photosynthesis may translate into an 18% increase in biomass and only a 5% increase in grain yield or a 6% change in grain yield in the absence of additional nitrogen. Leaf photosynthesis has little potential in increasing crop yields are based on comparisons of different genotypes in which differences in leaf photosynthesis are confounded with many other genetic differences (Borras *et al.*, 2004) or are limited to untested theoretical analysis (Sinclair *et al.*, 2004).

Lack of correlation between leaf photosynthesis and yield, coupled with evidence that yield is sink rather than source limited have led to a view that crop yields cannot be improved by increasing leaf photosynthetic rates. Fortunately, the focus on atmospheric CO_2-concentration increase has provided tests in abundance.

Energy transduction

During photosynthesis oxygen is evolved, only a limited portion of the solar spectrum can be used. Photons in the wavelength 350-740 nm may be used, below 400 nm and above 700 nm; photons can only be used at low efficiency, if at all. Photosynthetically active radiation (PAR) for practical purposes ranges from 400-700 nm, representing 48.7% of the total incident solar energy i.e. 51.3% is lost at this point. The weaker absorbance of chlorophyll in the green band, vegetation is not a perfect absorber of PAR, which limits maximum interception of 400 nm to 700 nm in healthy leaves to approximately 90%. A blue photon (400 nm) has 75% more energy than a red photon (700 nm), higher.

In the last decade, increase in yield for some major crops such as rice has shown little improvement (Peng *et al.*, 2009). The slow pace of yield increase is occurring in context of increasing world population, climate change, the diversion of an increasing proportion of the grain harvest to meat production, and the emergence of bioenergy production. This is in addition to losses of agricultural land to urbanization and soil degradation. In 2008, the world saw the lowest wheat stockpiles of the past 30 years (Wiebe *et al.*, 2008) and fears of a rice shortage incited riots in some countries. Coupled with the rapid growth in the Chinese and Indian economics has resulted in never before seen demands on grain supplies. Increasing grain crop productivity is the foremost challenge facing agricultural research. Although photosynthesis is the ultimate basis of yield, improving photosynthetic efficiency has played only a minor role in improving yields till-date. However, the yield traits that drove the remarkable yield increases during the green revolution appear to have little remaining potential for further increases. After rapid increases in the yield over the latter half of the twentieth century, further yield increases appear harder to obtain. While realized yields have improved in part through better fertilization and improved protection, they have also improved very substantially as a result of increased genetic yield potential (Y). This is defined as the yield that crop can attain under optimal management practices and in the absence of biotic and abiotic stresses. Adapting the equation of Monteith (1977).

$$Y = 0.487 \times S_t \times E_i \times E_c \times E_p$$

Where S_t is the total incident solar radiation across the growing season. Leaves of healthy crops typically absorb approximately 90% of the photosynthetically active radiation (400-700 nm) but transmit most of the near infrared radiation (>700 nm), approximately half of the energy of sunlight. To limit the analysis to photosynthetically active radiation, S_t is multiplied by 0.487. Light interception efficiency (E_i) of photosynthetically active radiation is determined by the speed of canopy development and closure, leaf absorbance, canopy longevity, size and architecture. Conversion efficiency (E_c) is the combined gross photosynthesis of all leaves within the canopy, less all plant respiratory losses. Partitioning efficiency (E_p) also termed as harvest index, is the amount of the total biomass energy partitioned into the harvested portion of the crop.

This equation gives the harvestable yields in MJm^{-2} converting it to mass depends on the energy content of the harvested material. For non-oil grains, it is 18 MJg^{-1} but can rise to 35-40 MJg^{-1} for oil-rich seeds. According to this equation, increase in potential yield over the past 50 years has resulted largely from increase in E_p and E_i. Increased E_p has resulted in large part through dwarfing of stem and increase in potential number of seed set. Increased E_i has resulted through the development of large leafed cultivars and more rapid coverage of the ground after germination. Dwarfing has also indirectly improved realized E_i by improving the standing power of the crop to adverse weather conditions, such as wind, rain/hail (i.e. decreasing lodging) (Beadle *et al.*, 1985, Evans 1993, Hay, 1995).

Healthy crops of modern cultivars at optimal spacing intercepts most of the available radiation within their growing season, limiting prospects for any further improvement of E_i. One caveat is that most crops do not currently use the full potential growing season viz. period when temperature and water are adequate for plant growth. The effects on biomass production of extending the growing season can be seen by comparing biomass production of the unusually cold tolerant perennial C4 gross *Miscanthus* x *giganteus,* with its relative maize. Although its E_c was almost identical to maize, it produced 60% more biomass in the Midwest, where recorded yields of maize are among the highest in the world. The higher productivity of Miscanthus x giganteus was simply due to a closed canopy, with an E_{i-}>0.9, four weeks before maize and having maintained it for a further four weeks after the maize had senesced (Dohleman and Long, 2009). Extending the growing season increased the cumulative intercepted radiation by approximately 60% (Baker *et al.*, 1983; Beale *et al.*, 1999).

Soybean, the most important dicotyledonous crop, in terms of total global grain production and further most important grain crop, after maize, rice and wheat. Modern soybean cultivar developed for Midwest conditions, grown under normal conditions and at current atmospheric CO_2 intercepted almost 90% (E_i=0.89) of photo synthetically active radiation across the growing season. Further, 60% of the biomass energy was partitioned into the harvested seed (E_p=0.60). Thus, breeding has succeeded in maximizing both E_i and E_p in soybean. The crop will inevitably fail to intercept some radiation between sowing and canopy closure and then the cell wall material cannot be recycled to the seed from leaves, roots and stems there is little or no prospect of further improvement in E_i or E_p. Analysis of the other major grain crops i.e. maize, wheat and rice, provide similar findings. In context to the equation, only two prospects which remain are extending the growing season to increase S_t or increasing E_c. The reluctance to invest for increasing photosynthesis and hence E_c are due to two reasons "firstly there is no correlation between the yield of abroad range of crops and photosynthesis and secondly the yield is limited by sinks for photosynthate and not by photosynthetic capacity. Elevated CO_2 increased leaf photosynthesis in the soybean cv.93815. Corresponding in turn to an 18.8% increase in E_c and an 18.2% increase in total above ground biomass (Benarcehi *et al.*, 2002). The increase in yield of 15% as compared to a 23% increase in photosynthesis reflects an increase in respiration associated with greater biomass and yield (Leakey *et al.*, 2009) and may also indicate a lack of adequate sink capacity to fully utilize the increased supply of photosynthate but it nevertheless results in a large increase in yield. This review examine the opportunities for further improvement in yield via photosynthesis, the only major trait available for any further increase in yield, increasing leaf photosynthesis will increase yield when other factors are constant, analysis of theoretical limits to the efficiency of photosynthetic processes can reveal the key targets for improvement and further specific options for engineering improved leaf photosynthesis and crop yield might be realized on a relatively short time scale excited states of chlorophyll very rapidly relax, and all photochemistry is driven in the photosynthetic reaction. Centres with the energy of a red photon regardless of the wavelength that was originally absorbed accounting for a 6.6% energy loss as heat, the "photochemical inefficiency". It is assumed that in non-cyclic electron transport, the partitioning of photon between photosystem I (PSI) and photosystem II (PSII) is equal.

At the reaction centres, thermodynamics limit the amount of energy available to do photosynthetic work in terms of charge separation. Thermodynamics limit the energy available for work to 63% of the total energy in red photon (685 nm), resulting in an energy loss of 37%.

Energy expenditure associated with electron and proton transport and in the reduction of CO_2 to carbohydrate in C3 cycle, with the additional losses in the C4 photosynthesis. In C3 photosynthesis, a minimum of 3ATP and 2 NADPH is required to assimilate one molecule of CO_2 into carbohydrate and to regenerate 1 RUBP to complete the C3 cycle. In whole chain linear electron transport the absorption of a minimum of 4 photons is needed to reduce one molecule of NADPH while translocating a maximum of 6 photons into the thylakoid lumen 2 from water oxidation and 4 from plastoquinol oxidation by the cytochrome b_6/f complex via the Q cycle (Kramer *et al.*, 2003). Since 2 NADPH are required for assimilation of one CO_{-2} into carbohydrate, the absorption of 8 photons results in a maximum of 12 protons transported into the lumen. As 4 protons are needed for synthesis of I ATP (Ferguson, 2000, Richter, 2004, Turina *et al.*, 2003), these 12 protons transported are just sufficient to support the synthesis of 3 ATP required to balance 2 NADPH in the assimilation of one CO_2. The 8 moles of photons, the minimum required to convert 1 mole of CO_2 to carbohydrate, represents 874 KJ energy available for work. One sixth of a mole of glucose viz. 1 C carbohydrate unit, contains 477 KJ of energy. Therefore, the minimum energy expenditure in "Carbohydrate biosynthesis" is 1 or 10.78% of the original incident. In turn, the maximal energy conversion efficiency (E_c) of C3 photosynthesis, prior to photorespiration and respiration is 12.6%.

All the major C4-crops – maize, sorghum, sugarcane as well as the emerging bioenergy crop *Miscanthus* belong to the most efficient C4 subtype (NADP-malic enzyme) (Ehleringer and Pearcy, 1983). This subtype requires an additional 2 ATP relative to C3 photosynthesis for the phosphorylation of pyruvate to phosphoenol pyruvate (PEP) i.e. 5 ATP and 2 NADPH are required to assimilate 1 CO_2. Increased demand for ATP is underlined by the fact that bundle sheath chloroplasts in the C4 subtype are often deficient in grana and photosystem II, thus, increasing cyclic electron transport. Here, electrons from photosystem I are required to the Cyt b6/f complex, resulting in the translocation of 2 protons per photon into the thylakoid lumen (Crammer *et al.*, 2006; Zhu *et al.*, 2004). Therefore, the translocation of the 8 protons needed to produce the 2 additional ATP requires the absorption of an additional 4 photons by PSI, raising the minimum total quantum requirement for CO_2 assimilated in C4 photosynthesis to 12. Following earlier calculations for C3 photosynthesis (Zhu *et al.*, 2008), the maximal energy conversion efficiency (E_c) of C4-photosynthesis, prior to respiration is 8.5%.

With the investment of 2 extra ATP, CO_2 is concentrated at the site of carboxylation by Rubisco in bundle sheath cells to a sufficient extent to competitively inhibit oxygenation (Furbank and Hatch, 1987), under most conditions. However, in C3 species, oxygenation and the ensuring photorespiratory metabolism represents a significant energy loss, essentially halving the maximum energy conversion efficiently from 12.6% to 6.1%. The 'quantum requirement' penalty for each oxygenation event is ~ 9 photons (Amthor, 2007). The actual extent of this penalty in raising the quantum requirement for CO_2 fixation in C3 leaf depends on the Rubisco specificity factor, temperature and CO_2. At 25°C under current atmospheric CO_2 (387 ppm) for a typical C3 crop, Rubisco specificity factor, photorespiration raises the minimum quantum requirement of a C3 plant from 8 to 13 photons per CO_2 assimilated.

Mitochondrial respiration is another necessary expenditure of energy that must be subtracted in estimating the theoretical maximal E_c. Mitochondrial respiration has been phenomenological subdivided into maintenance respiration and growth respiration (Penning *et al.*, 1974). Growth respiration is the portion invested in biosynthesis whereas maintenance respiration amounts for the energy expenditure to maintain plant cells in dynamic environments viz. replacement of proteins, metabolite transport and repair of cell damage.

There is no known quantitative mechanistic link between photosynthetic and respiration rates. Negative correlations between the respiration of mature leaves and production have been reported for maize (Earl and Tollenaar, 1998) and ryegrass (Wilson and Jones, 1982), implying the selection for lines with lower respiratory rates, while maintaining photosynthetic rate, may be an approach to improve E_c. Ratios of respiratory CO_2 loss as a fraction of photosynthetic CO_2 uptake for major crops vary from 30% to 60% (Amthor, 1989). 30% is the assumed to be the minimum respiratory expenditure that might be achieved without otherwise adversely affecting plant growth. This represents a loss of the original incident solar energy of 1.9% (C3) and 2.5% (C4) with the result the maximum conversion efficiencies of solar radiation into biomass are 4.6% (C3) and 6% (C4) at 30°C or 9.4% and 12.3% of PAR.

Energy conversion efficiency (E_c) and yield

Commonly measured as the mean slope of the relationship between the accumulations of biomass energy in the crop versus the cumulative amount of intercepted radiation. It can be measured over any portion or the entire growing season, by combining sequential harvests of biomass with measurements of intercepted radiation. The amount of intercepted radiation is determined by continuously measuring the amount of radiation incident above the crop and subtracting the amount that penetrates below the canopy. Light levels are highly variable both spatially and temporally at the base of canopy. Line radiation sensors provide a means to average across the heterogeneity. At any point of time, the fraction of light intercepted by the canopy (α_c) is:

$$\alpha_c = 1 - Z_c$$

Where, Z_c is the fraction of incident radiation transmitted by the canopy. α_c varies with sun angle and day of year, cumulative absorbed radiation is typically calculated by summing measurements made at short intervals (i) viz. every hour across the growing season.

$$I^1 = \text{“}\ I_i\ \alpha_c i$$

Radiation interception might be overestimated at the end of growing season due to presence of necrotic shoot tissue in the upper canopy and senescing floral parts, which can be overcome by estimating the proportion of dead/senescing parts using imaging methods (Beale *et al.*, 1999). Radiation capture can also be estimated if leaf area index and leaf angular distribution are known. To get a true measure of E_c for the full growing season, the total biomass consisting of leaves, stem, root and seeds and also including those shed before crop matures, needs to be taken into account (Dermody *et al.*, 2008; Eberhard *et al.*, 2008). The total energy content is calculated based on the biomass quantity and the energy content of each biomass component (Dermody *et al.*, 2008). Despite the simplicity and importance of this measurement in providing a link between crop production and photosynthesis such complete data sets are rare. Accumulation of biomass (W) versus cumulative intercepted radiation (I^1) describes relationship (Fig.3), implying that E_c is relatively constant. Crops respond to stress by altering their canopy size i.e. E_i rather than E_c (Monteith, 1977). This indicates that success in genetic improvement in E_c may be as valuable under suboptimal growth conditions as under optimal.

Monteith (1977) reassessed maximum growth rates for C3 and C4 crops, found maximum short term E_c of 0.029 and 0.042 respectively on the basis of photosynthetically active radiation. The advantage of C4 photosynthesis diminishes as temperature declines but still theoretical advantage to the simulated daily integral of canopy CO_2 uptake even down to 5°C (Long *et al.*, 2006) at current CO_2 although other physiological and biochemical factors conspire to limit this advantage to temperatures above ~14°C in maize (Ortiz *et al.*, 1990) and other C4 grain crops. However, certain C4 species have shown to maintain their advantage at lower temperatures (Wang *et al.*, 2008). Increased nitrogen fertilizers dramatically increases E_c of major crops such as barley, oat, rice and wheat (Muurinen *et al.*, 2006) as does irrigation (Ercoli *et al.*, 1999). E_c can differ with developmental stage. For example, the E_c of oat is higher before heading compared to post heading in contrast, barley and wheat showed higher E_c after heading (Muurinen *et al.*, 2006). Compared to C4 species, C3 species usually have smaller E_c across the growing season. The maximum E_c that a genotype, may achieve under optimal conditions may fall short of the theoretical maximum for two seasons. Firstly, leaves become light saturated, energy is wasted and efficiency drops. Secondly, the theoretical E_c may be increased by decreasing photorespiration. The inherent mechanisms within photosynthesis that underlie the lower E_c achieved in field crops relative to the theoretical E_c.

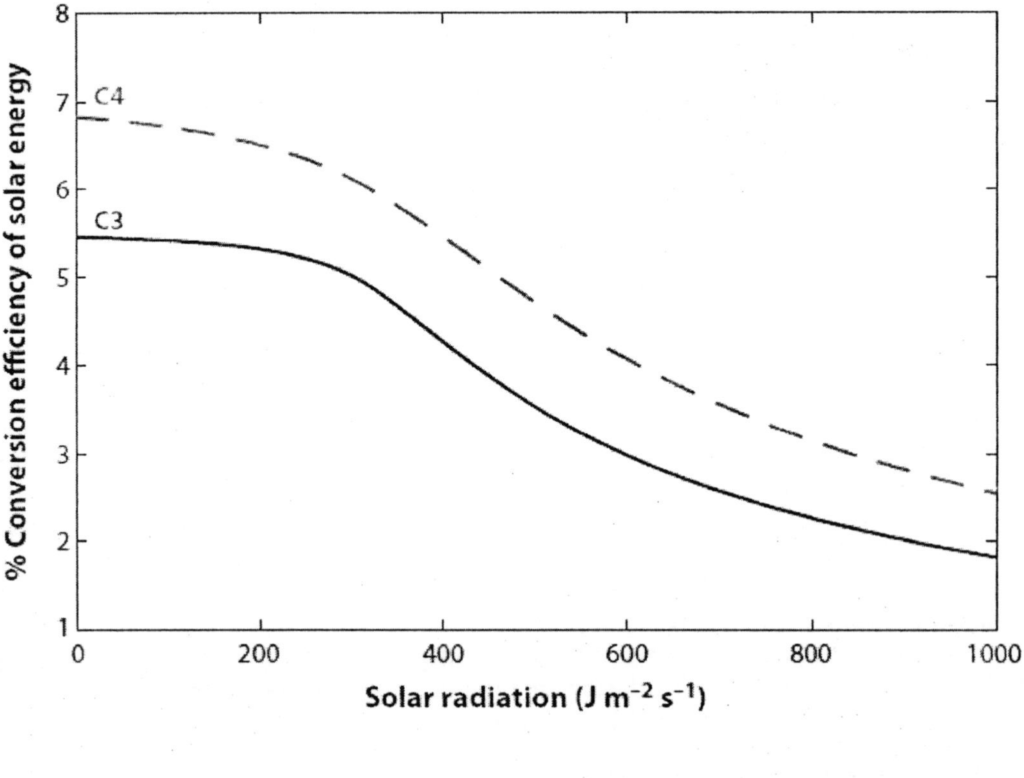

Fig 3. Decline in conversion efficiency of photosynthesis at the leaf level with increasing incident solar radiation .The lines are calculated using maximum observed quantum efficiencies and maximum leaf photosynthetic rates for C4 and C3 species (Long *et al.*, 2006)

Light saturation

The response of the rate of leaf photosynthetic CO_2 uptake per unit leaf area (A) to sunlight intensity is commonly referred in terms of the response to the number of photons rather than energy; because the response is largely independent of wavelength and so independent of energy content of the photons within the photosynthetically active wavelengths (400-700 nm). This response has been vividly described by (Falk *et al.*, 1992; Leverenz *et al.*, 1990; Marshall and Biscoe, 1980).

At low light, more than 80% of the absorbed photosynthetically active quanta can be used (Bjorkman and Demming, 1987), but at ½ of the full sunlight (~ 1000 μmol $m^{-2}s^{-1}$), as little as 25% of the absorbed quanta are used; at full sunlight this value falls to <10% (Long *et al.*, 1994). Full spectrum sunlight at Earth's surface will typically contain ~ 2μmol of photons (400-700 nm) per J. Based on this conversion and assuming that every g of CH_2O synthesized represents 17.5 KJ of stored energy. Fig.3 shows how, at the leaf level, efficiency of radiation use declines with increase in radiation received by the leaf. Under optimal conditions efficiency is high and close to theory in low light. Extensive measurements of the actual efficiency of photosynthesis in low light have shown that for unstressed leaves (a) while C3 and C4 are distinct within these groups, there is little variation even between young and old leaves, and (b) the value is often close to the theoretical maximum (Long *et al.*, 1993). For healthy leaves acclimated to high sunlight, the high efficiency may be maintained approximately $1/10^{th}$ of full sunlight. Beyond this point efficiency declines. A higher E_c could, therefore, be achieved by selecting canopy structures or photosynthetic pigment concentrations that spread light within the canopy to minimize occurrence of light levels above $1/10^{th}$ full sunlight or by increasing capacity for photosynthesis at light saturation.

C3 photosynthesis has been summarized in a simple and widely validated steady state biochemical model developed by Farquhar *et al.*, (1980) with subsequent minor modification (Harly and Sharkey, 1991; Von Caemmerer, 2000). According to these models, the steady state light saturated leaf photosynthetic CO_2 uptake rate (A_{Sat}) is determined by three processes. (1) Rubisco catalyzed RUBP carboxylation; (2) RUBP regeneration and (3) Triose phosphate utilization.

Under given light, CO_2 and O_2 conditions, A is determined by the slowest of these three processes. Under optimal conditions which determine maximum yield potential for triose-phosphate utilization which typically reflects inability to utilize photosynthate, would not be expected to be limiting. At low CO_2, photosynthesis is limited by capacity for Rubisco catalyzed carboxylation and at high CO_2, by the capacity for regeneration of RUBP. The latter may be limited both by the rate of whole claim electron transport and/or by the activity of enzymes involved in the regeneration of RUBP within the C3 carbon reduction cycle. In well fertilized C3 crops under current ambient atmospheric CO_2 control appears to be shared between the Rubisco and regeneration of RUBP. Light saturated photosynthesis could, therefore, be increased by increasing Rubisco, RUBP regeneration or/and CO_2 at the site of carboxylation.

Photoinhibition lowers the maximum efficiency

The CO_2 concentration at the site of carboxylation (C_c) is determined by both the ambient CO_2 concentration (C_a) and the conductance of the different path from the atmosphere to the chloroplast stroma. The diffusion path includes the leaf boundary layer, stomatal aperture, substomatal cavity, cell wall, cell membrane and the cytosol of the mesophyll (Ball *et al.*, 1987; Farquhar *et al.*, 1980). At light saturation, CO_2 in the intercellular space (C_i) is 0.7 of C_a in C3 plants. This fraction appears remarkably constant across species, even when C_a is elevated (Ainsworth and Long, 2005; Leakey *et al.*, 2006). The decline between C_i and C_c is similar to that between C_a and C_i. At the current atmospheric CO_2 concentration of 387 ppm (Meehl *et al.*, 2007), the typical CO_2 of Rubisco at light saturation would, therefore, be ~194ppm. The remarkably constant ratio of C_i/C_a appears to result from coordination between the rate of CO_2 assimilation and stomatal conductance. A higher conductance and C_i/C_a would deliver a higher photosynthetic CO_2 uptake rate (A). Due to the response of A to C_i is nonlinear, an increase in stomatal conductance (gs) results in diminishing returns but a linearly proportional increase in transpiration. Higher A through increased gs, would, therefore, come at the expense of decreased efficiency of water use and a disproportionate increase in transpiration. Diffusion of CO_2 from the intercellular air space through the liquid phase to the site of carboxylation is governed by the mesophyll conductance (gm). Factors controlling gm are poorly understood and have been associated with the mesophyll surface area exposed to the intercellular air space, carbonic anhydrase and aquaporins. Higher gs could be an important approach to increase CO_2 at the site of carboxylation and in turn photosynthetic rate. Unlike increase gs, there is no evidence of a penalty for increased gm (Niyogi *et al.*, 2005). Light-saturated photosynthesis in C3 species has been shown to be closely related to the amount of Rubisco in a leaf. However, there is substantial evidence that in well fertilized C3 grain crops; there may be no physical capacity for more Rubisco and other soluble proteins in the mesophyll.

In C4 species, an analogous steady-state model of photosynthesis has been developed. Light-saturated photosynthesis is limited by the activity of the primary carboxylase of C4 photosynthesis, phosphoenol pyruvate carboxylase (PEPcase) at low C_i. At high C_i photosynthesis is co limited by activity of Rubisco, in C4 plants is limited to bundle sheath and by pyruvate P_i dikinase (PPDK) which regenerates PEP in the mesophyll (Van Caemmerer, 2000) unlike C3 photosynthesis, light saturated C4 photosynthesis in healthy leaves is generally CO_2-saturated, because C4 pathway is a mechanism for concentrating CO_2 and elevating C_i in the bundle sheath so that Rubisco is CO_2 saturated leading to increase in gs or gm will not increase photosynthesis.

Rubisco and Elevated CO_2 Concentration

Atmospheric CO_2 over the past several million years has been averaged 220 ppm. With the beginning of Industrial Revolution, CO_2 has begin to rise, providing little time in which plants could adopt to this increase and in the absence of natural adaptation, only engineered adaptation can suffice.

Rubisco activity, particularly carboxylase is sensitive to variation in CO_2. Rubisco catalyses the competitive reactions of RUBP carboxylation and RUBP oxygenation. It is well recognized that genetic modification of Rubisco to enhance its specificity for CO_2 relative to O_2, would decrease photorespiration and potentially increase C3 photosynthesis and thus, crop productivity. Analysis of the natural genetic variation in the kinetic properties of Rubisco from divergent photosynthetic organisms reveals that forms with higher oxygenation have lower maximum catalytic rates of carboxylation per active site (K_c) and vice-versa (Bainbridge *et al.*, 1995; Zhu *et al.*, 2004). The inverse relationship implies that higher oxygenation, would increase light limited photosynthesis, while the associated decrease in K_c would lower the light saturated rate of photosynthesis. The daily integral of CO_2 uptake by a crop canopy is determined by a dynamic combination of light-limited and light saturated photosynthesis, so the benefit of increasing oxygenation at the expense of K_c is not intuitive. By using the model of canopy photosynthesis that determined the daily cause of light level on both the sunlit and shaded leaves in the canopy, Zhu *et al.* (2004) showed the average oxygenation found in present day C3 crops exceeds the level that would be optimal for the present atmospheric CO_2 of >380 ppm but would be optimal for ~220 ppm i.e. close to be average prior to Industrial Revolution (Barnola *et al.*, 1999).

According to simulation model, for the same amount of Rubisco, 10% more carbon could be assimilated if oxygenation were optimized for the current atmospheric CO_2. Greater improvement could be achieved for the same total quantity of Rubisco, if a low oxygenation and high K_c could be expressed in the upper canopy and then be replaced by current C3 Rubisco as these leaves become shaded by new leaves, as the canopy grows. The possibility of increased CO_2 favour evolutionary selection of form of Rubisco with increased K_c and decreased oxygenation is consistent with the observation that Rubisco from C4 plants, where enzyme that originated in C3 ancestors, now functions in a high CO_2. Some C4 Rubiscos have the predicted higher K_c and lower oxygenation than that of C3 land plants (Sage, 2002, Shikanoi, 2007).

Variations in Rubisco catalytic rate and specificity do exist in nature e.g. Rubisco from red algae has a specificity 3 times higher than C3 crop species (Tabita, 1999).

In higher plants, Rubisco with higher oxygenation values has been reported in plants adapted to drier environments and in semi deciduous or evergreen species (Galmes *et al.*, 2005). There is a trade off between specificity and catalytic rate i.e. the tighter binding is required for specificity results in slower catalytic turnover rate. So, improving catalytic rate could be at the expense of specificity (Zhu *et al.*, 2004). If Rubisco from the red alga (*Griffithsia monilis*) could be expressed in place of present C3 crop. Rubisco, then daily canopy carbon gain would be predicted to increase by 27%. Large gains could also be achieved by expressing Rubisco from C4 dicot (*Amaranthus*). Rubisco in the upper canopy leaves exposed to full sunlight and a high oxygenation than the Rubisco in the shaded lower canopy leaves results in even greater gains (Long *et al.*, 2006).

Eukaryotic Rubisco has eight large chloroplast encoded and eight small nuclear encoded protein subunits. The large submit (LSU) contains the structure needed for the catalysis, so the current Rubisco genetic screening and mutagenesis research focuses on the LSU, with the aim of improving the catalytic efficiency and specificity of Rubisco to CO_2 versus O_2. Though, amino acid substitutions to different areas of LSU have been attempted with the help from holoenzyme crystal structure, no more efficient enzymes have been produced so far; in fact only less efficient enzymes have been produced (Parry *et al.*, 2003; Spretizer and Salvucci, 2002, Matsumura *et al.*, 2005). Comparison of the three dimentional X-ray structure of Rubisco from multiple prokaryotic and eukaryotic sources suggest a notable difference in one region of small subunit, indicating that small subunit engineering holds same potential to increase Rubisco efficiency and specificity (Anderson and Taylor, 2003; Karkehabadi *et al.*, 2005).

Progress is still needed in order to efficiently transform foreign Rubisco into crop species. Replacing Rubisco in one plant species with Rubisco from a different species is challenging because of the different coding locations of the subunits of Rubisco and the intricate mechanism coordinating the expression, post-translational modifications and assembly of the subunits into the functional hexadecamer (L_8S_8) enzyme within the chloroplast stroma (Houtz and Portis, 2003; Whitney and Andrews, 2001) and says nothing about silencing the native genes. Replacing Rubisco in tobacco with the simple homodimeric form of the enzyme from the á-proteobacterium *Rhodospirillum rubrum* which has no small subunits and no special assembly requirements, produced plants that were autotrophic and reproductive although they required CO_2 supplementation, thereby establishing that Rubisco from a very different phylogeny can be integrated into chloroplast photosynthetic metabolism without prohibitive obstacles (Whitney and Sharwood, 2007). The tobacco chloroplast genome transformed with plastid DNA containing the Rubisco large subunit (rbcL) gene from both sunflower and cynobacterium PCC6301, although the catalytic activities of the recombinant enzymes were only 25% of the native tobacco enzyme (Kanevski *et al.*, 1999). Compared to the genetic manipulation of the large subunit, engineering the small subunit is difficult because a gene family of multiple closely related genes codes for the small subunit, making targeted mutagenic or placement strategies problematic. A potential alternative route is simultaneous expression of both the large and small subunits as a fusion protein as demonstrated recently by the success of linking the subunits of Synebacoccus PCC6301 Rubisco to generate correctly assembled Rubisco in *E. coli* with catalytic capacity similar to wild type *Synebacoccus* (Whitney and Andrews, 2001). Improved E_c of introducing Rubisco better adapted for current and future conditions are fully evident, technical obstacles are preventing implementation. Acclimation and adaptation have been insufficient to ensure maximal E_c in the face of rapid environmental change is the inability of plants to optimally deploy nitrogen resources within the photosynthetic apparatus (Zhu *et al.*, 2010a).

Potential approach for manipulation in carbon metabolism

Rubisco has been the primary focus of research to improve photosynthetic efficiency, other enzymes in the C3 cycle have been manipulated in different plants and their impacts on photosynthesis evaluated (Raines, 2003, 2006). The experimental results clearly demonstrated that metabolic control of CO_2 fixation rate is shared among different enzymes (Raines, 2003). More than 60 + metabolic reaction in photosynthetic carbon metabolism and associated cellular metabolism in sucrose synthesis and photorespiration, there are many potential permutations of change which could not be addressed practically without some means of directing the efforts, Zhu *et al.* (2007) extended existing dynamic metabolic models of the C3 cycle by including the photorespiratory pathway and cellular metabolism to starch and sucrose to develop a complete dynamic model of photosynthetic carbon metabolism. The model consisted of a series of linked differential equations, with each differential equation representing the concentration change of one metabolite. Initial concentration of metabolites and maximal activities of enzymes were extracted from literature. The dynamics of CO_2 fixation and metabolite concentrations were simulated by numerical integration, such that the model could mimic well established physiological phenomena. Using an evolutionary optimization algorithm in which partitioning of a fixed quantity of protein-nitrogen enzymes was allowed to vary, and selecting on photosynthetic rate, resulted after several generations in individuals with a light saturated photosynthetic rate that was 60% higher. This suggests that the typical partitioning of resources among enzymes of photosynthetic carbon metabolism in C3 crop leaves is not optimal for maximumizing the light saturated rate of photosynthesis under current or future conditions. There appears to be an overinvestment in enzymes of the photorespiratory pathway and marked underinvestment in ADP glucose pyrophosphorylase and SBPase (Sedoheptulose 1, 7 bisphosphtase) enzymes which occupy key control points in carbon metabolism. Under the elevated CO_2 conditions predicted for the near future, this pattern of under and over investment is amplified, suggesting that manipulation of partitioning of resources among enzymes could greatly increase carbon gain without any increase in the total protein-nitrogen investment in the apparatus for photosynthetic carbon metabolism. Direct support for this prediction comes from the overexpression of SBPase, which increases photosynthesis and biomass of tobacco (Lefebvre *et al.*, 2005) whereas small decrease in SBPase decreased photosynthesis and biomass (Harrison *et al.*, 2001, 1998).

Rubisco-Related Bottlenecks

Three approaches to improve the C3 cycle have targeted Rubisco-associated bottlenecks. Expression of the cyan bacterial ictB protein in higher plants to reduce the oxygenation reaction of Rubisco resulted in the production of transgenic tobacco (*Nicotiana tabacum*) with higher photosynthetic rates under limiting, but not saturating, CO_2 levels. Similar results were obtained in Arabidopsis (*Arabidopsis thaliana*) plants expressing the ictB from Anabaena PCC7120; in low humidity conditions, the growth of transgenic Arabidopsis plants was faster and the final dry weight was higher than that of wild-type plants (Lieman-Hurwitz *et al.*, 2003). A second approach introduced the enzymes of the bacterial glycolate pathway into plants, creating a photorespiratory bypass in the chloroplast (Peterhansel and Maurino, 2011). In Arabidopsis, this resulted in reduced photorespiration and increased CO_2 assimilation and biomass yield (Kebeish *et al.*, 2007; Leegood, 2007). These two strategies for overcoming the oxygenase reaction of Rubisco offer more immediate alternatives to improve crop photosynthesis in the near future, compared to the more ambitious CO_2-concentrating strategies.

The activation state of Rubisco is dependent on the enzyme Rubisco activase, which at temperatures above 30° C has been shown to be unstable in vitro. This led to the suggestion that Rubisco activase is responsible for the reversible temperature-sensitive reduction in the activation state of Rubisco, leading toinhibition of photosynthesis when plants are subjected to mild heat stress (Salvucci and Crafts-Brandner, 2004). The introduction of a thermostable version of the Rubisco activase enzyme into an Arabidopsis activase deletion mutant resulted in transgenic lines with higher photosynthetic rates and increased biomass and seed yield when compared with control plants expressing the wild-type activase (Kurek *et al.*, 2007). All three of these approaches have demonstrated the potential to stimulate photosynthesis and growth in model species, and the next step will be to test this in crop plants under natural environments.

Increasing Regenerative Capacity of the C3 Cycle/Target enzymes

Transgenic antisense studies identified SBPase, plastid aldolase, and TK as enzymes that might be targets for improving C3 cycle flux. Indeed, over expression of either a bifunctional

SBPase/FBPase or plant SBPase in tobacco plants resulted in increased photosynthetic CO_2 fixation and growth (Miyagawa *et al.*, 2001; Lefebvre *et al.*, 2005). Analysis of CO_2 response curves revealed that this increase in photosynthesis could be attributed to an increase in the capacity to regenerate the CO_2 acceptor molecule RuBP. Measurements of ambient photosynthesis over the daily cycle revealed an increase in photosynthetic rates of between 6% and 12%. Of particular interest for crop improvement is that these plants also displayed an increase in leaf area and biomass of up to 30%. Young seedlings (4/5 leaves) with increased SBPase activity were shown to have an increase in leaf area compared to wild-type plants. Chlorophyll fluorescence imaging of these young plants revealed a higher photosynthetic capacity at the whole plant level. These plants also showed an increased growth rate during the early phase of development (Lefebvre *et al.*, 2005). This result is of significance in an agricultural context as rapid, early seedling establishment is known to contribute to yield increases at maturity. In contrast to the results with tobacco, increasing SBPase activity in rice (*Oryza sativa*) did not lead to increases in photosynthesis or growth. However, if the plants were subjected to heat or salt stress conditions, photosynthesis rates in the transgenic plants with increased SBPase plant were higher than in wild-type controls (Feng *et al.*, 2007a, 2007b). The rice studies highlight the fact that increasing photosynthesis and yield by manipulation of SBPase is likely to be dependent not only on species but also growth conditions. Plastid TK was also identified as a potential target to increase the regenerative capacity of the C3 cycle (Henkes *et al.*, 2001). To explore this possibility, transgenic plants have been produced over expressing TK, either on its own or together with over expression of SBPase. Analysis of these transgenic plants has shown that photosynthetic rates were similar to wild type but that the plants have reduced growth and display a chlorotic phenotype. The basis of this phenotype has not been fully resolved, but preliminary analysis would suggest that increasing TK activity has perturbed the balance of export of carbon from the C3 cycle (M.Khozaei, S.C. Lefebvre, T.Lawson, and C.A. Raines, unpublished data). This is in accordance with the antisense studies that revealed a reduction in carbon flow from the C3 cycle to the shikimate pathway that correlated with reductions in TK activity (Henkes *et al.*, 2001). Thus there is need for modeling approaches that not only allow identification of potential targets for improving CO_2 assimilation but that will also allow flux from the C3 cycle to be determined.

Photorespiration

At 25°C and current CO_2, about 30% of the carbohydrate formed in C3 photosynthesis is lost through photorespiration (Monteith, 1977). The loss increases with temperature so that photorespiration is a particularly significant inefficiency for C3 crops in tropical climates and during hot summer weather in temperate climate. Photorespiration results from the apparently unavoidable oxygenation of RUBP by Rubisco (Giordano *et al.*, 2005). Beyond this point, the purpose of photorespiratory metabolism is to recover the carbon diverted into this pathway. Blocking photorespiratory (C2) metabolism downstream of Rubisco simply results in the carbon entering a dead end metabolic pathway. Indeed, mutations that lack any one of the photorespiratory enzymes die unless the plants are rescued at low oxygen or at very high CO_2 to inhibit oxygenation of RUBP. The only remaining prospect for decreasing photorespiration then, is decreased oxygenation.

Photorespiratory metabolism can dissipate excess excitation energy at high PPFD, involves the synthesis of serine and glutamate and transfers reductive power from the chloroplast to the mitochondria. However, photorespiration is essential for normal plant function (Barber, 1998; Evans, 1998). However, xanthophylls provide a far more effective means of dissipating excess energy. Unlike photorespiration, this dissipation mechanism is not a significant drain on the ATP and NADPH produced by the light reactions. Further, dissipation of energy as heat through xanthophylls is induced by excess light and is reversed when light is no longer in excess. So, unlike photorespiration, it does not continue to divert energy from photosynthesis when light is no longer in excess. In addition, the photosynthetic cell has pathways besides photorespiration for amino acid synthesis and transfer of reductive energy to the cytosol (Long, 1988), which suggests the supposed beneficial functions of photorespiration are reductant within the cell. Further, this process can be eliminated without detriment to the plant by growing plants in a very high concentration of CO_2, a competitive inhibitor of the oxygenase activity of Rubisco. Commercial growers of some greenhouse crops increase CO_2 to 3 or 4 times the normal atmospheric concentration (Chalabi *et al.*, 2002). This inhibits the oxygenation reaction of Rubisco increasing photosynthetic efficiency and final yield.

Synthetic biology, however, is now opening new opportunities of altering C_2 metabolism downstream of oxygenation (Foyer *et al.*, 2009). Kebeish *et al.* (2007) produced plants in which chloroplastic glycolate can be converted directly to glycerate in the chloroplast by introducing five genes of the *E. coli* glycolate catabolic pathway into *Arabidopsis thaliana* chloroplasts. This created a bypass of the energy-intensive conversion that otherwise involves the cytosol, peroxisomes and mitochondria. The bypass decreased the energy required to recycle glycolate back to the C3 pathway as glycerate and correspondingly increased photosynthesis and biomass production (Kebeish *et al.*, 2007). The increase in photosynthetic rate is attributed to the increase in CO_2 around. Rubisco, since CO_2 is released in the chloroplast rather than the mitochondria and because of the bypass decreased the ATP required by avoiding ammonium refixation. If engineering could completely bypass the normal photorespiratory pathway then it would raise maximum efficiency in C3 photosynthesis at 25°C and current atmospheric CO_2- by 13% (Zhu *et al.*, 2008).

Another approach to decrease photorespiratory loss is to engineer the C4 photosynthetic processes into C3 plants. C4 photosynthesis has significantly higher E_c under current atmospheric CO_2 than C3 photosynthesis, although this efficiency advantage will decline as atmospheric CO_2 continues to rise, reaching parity by the end of this century, except at very high leaf temperatures.

Conversion of C3 species to C4 photosynthetic metabolism

The polyphyletic evolution of the C4 pathway (Sage, 2004), characteristics of the C4 pathway in some cell types of C3 species (Hibbered and Quick, 2002), the C3- pattern of cell differentiation in some tissues of C4 species (Langdale *et al.*, 1988) and the switch between C3 and C4 photosynthesis in some plants (Cheng *et al.*, 1989). All suggests that the transition from C3 to C4 species may be controlled by relatively few genes and mechanisms controlling the C3 and C4 photosynthesis differentiation are flexible. Efforts to transfer C3 plants to express the C4 pathway enzymes to create C4 photosynthesis in a single cell (Miyao, 2003) have had very little success so far (Matsuoko *et al.*, 2001). A single cell type C4 might not be able to support a high E_c, even though single C4 photosynthesis in multicellular plants exists in nature with PEP_{case} and Rubisco spatially separated by distance elongated cells (Voznesenskaya *et al.*, 2002, 2001). These plants are slow growing and usually grow in hot, semiarid environments consistent with the theoretical predictions that a single cell C4- system would allow a positive carbon balance only under high light and drought conditions, but now with high efficiency of light use due to increased ATP demand for CO_2 fixation due to CO_2 leakage and refixation (Von Caemmerer, 2003).

For all the domesticated and high yielding C4 crop, Kranz anatomy and the compartmentation of photosynthetic enzymes are closely linked, conversion of C3 to a C4 crop will inevitably require the elucidation of the interaction between leaf morphology and photosynthesis. Understanding of the factors controlling the development of mesophyll and bundle sheath cells in C4 leaves will be crucial and critical.

Modifications for enhancing photosynthesis

1. Plant architecture

Dwarf stature in the cereal crops has lead to improvement in HI. Beyond maximizing HI, ideal crop plant architecture should minimize the highest photon flux density at an individual leaf level; at the same time maximize the total solar energy absorbed by the canopy per unit ground area. Ideally, the plant architecture and leaf biochemical properties should be designed so that the light levels match the photosynthetic capacity at different layers within the canopy (Hikosaka and Terashima, 1995). It is not well understood how nitrogen distribution within a canopy tracts light distribution, although the photosynthetic apparatus show clear differentiation under different light conditions (Tazoe *et al.*, 2006).

When leaf area index (LAI) is lower than 2, canopies with horizontal leaves will enable the greatest interception of daily incident solar irradiance (Loomis *et al.*, 1967). For a canopy with a higher LAI, however, an ideal plant architecture will have a more vertical leaf angle at the top of the canopy that gradually dcreases with the depth into the canopy (Long *et al.*, 2006). This will enable the light to spread more evenly through the canopy. Theoretical analysis suggested that compared to a canopy with horizontal leaves, a canopy with a gradual decreased leaf angle can increase the daily integral of carbon uptake up to 40% on a sunny day at mid latitude. A season long improvement of E_c of ~20% could result from the avoidance of severe light saturation at the top of canopy and severe light limitation within the canopy due to improved canopy architecture.

Substantial progress has been made in elucidating the genetic basis of plant architecture determinants (Morinako *et al.*, 2006). In rice dwarf stature is due to loss of function of brassinosteroid insensitive ortholog (OSBRII). One allele of OSBRII, d61-7, confers important agronomic traits like semi dwarf stature, erect leaves and in turn leads to 30% more grain yield than wild type at high plant densities. Genes for the erect leaves likely exist in most of the current crops, so if searching for d61-7 like alleles may help for improving E_c. Further selection of plants with gradually decreasing leaf angles at different layers of canopy has the potential to further increase E_c. Hypothetically, optimal architecture in plant monocultures will differ among species that vary in plant stature, leaf chlorophyll content, canopy albedo, and other species specific characters. Further, geographic location and time of the year matters because canopies with higher LAI and more erect leaves show the greatest advantage with higher solar elevation e.g. during summer/at low latitude (Duncan, 1971).

2. *Photosynthetic unit/Light harvesting molecules*

Engineering a smaller antenna size is another opportunity to optimize light energy distribution within a canopy to improve E_c (Melis, 2009). Glick and Melis (1998) estimated the minimum number of chlorophyll molecules needed for the assembly of the photosystem core complexes to be 37 "chlorophyll" a molecules for PSII and 95 "chlorophyll" a molecules for PSI, which is approximately 25% of the number of chlorophylls normally associated with a typical plant photosystem (Nelson and Yocum, 2006).

In bright light, high chlorophyll content results in over excitation, the induction of nonphotochemical quenching and greater potential for photo damage (Meles, 1999). A high chlorophyll concentration also directly deprives cell at lower layers of a canopy or even lower cells within the leaf of light, which lowers E_c. Therefore, a smaller antenna size would not only mitigate efficiency losses associated with NPQ, but also allow a greater transmittance of light into lower layers of the crop canopy/cells towards the lower surface of the leaf (Neidhardt *et al.*, 1998). Correspondingly increasing E_c. The advantage of a smaller antenna size and the scarcity of nitrogen in the field, then why lower chlorophyll content has not shown a selective advantage. This could be because low chlorophyll may be a disadvantage in a competitive natural habitat with other species, the circumstances to which most plant evolutionary selection has occurred. If a plant intercepts light that it cannot itself use, it is still disadvantages a competitor that might otherwise have received that light. The ideal for a crop plant would be a minimum antenna size in upper canopy leaves, which increases as leaves become progressively more shaded. Early in the season i.e. at the germination stage and early vegetative stage, when canopy density is insufficient to absorb nearly all of the PAR, a reduced antenna will be a disadvantage due to lower E_s. However, this hypothesis has not been rigorously tested in crops but some proof of the concept exists in mutants of soybean and rice. Mutants of soybean cultivar clark Y_9 and Y_{11} contain about half of the chlorophyll of the wild type, yet mature canopies of these low chlorophyll mutants shows higher daily integral of photosynthesis than the wild types (Sage, 2002). In rice mutant (*Oryza sativa* L. var. *Zhenhui* 249Y) with a low content of chlorophyll b and a high chla/b-ratio of 4-7 (Chen *et al.*, 2007) was reported to have slightly decreased A, but with improved resistance to photoinhibition, perhaps indicating that when canopy light penetration is improved by more erect leaf deployment, the benefits of reduced antennae size are less. However, the method by which the antenna is reduced is also important to determine the extent to which E_c is improved. Lowering chlorophyll content by drastically reducing chlorophyll b synthesis, as in the above cited mutants, might imbalance the antennae size of PSI and PSII create a respiratory drag i.e. as LHCII will continue to be synthesized, but cannot be stabilized in the absence of chloropbhyll b, and, therefore, reduce excit on transfer among PSII centres – all the factors that would be expected to constrain improvements to E_c.

Down regulating chlorophyll synthesis early in the pathway might be better option. The complicated interactions of lowering leaf chlorophyll on canopy light dynamics, which will vary with location and time of the year; suggests an important role for modeling in optimizing chlorophyll content to improve E_c.

3. Photoprotection

When there is light in excess than used by photosynthesis, the normally efficient light harvesting system of PSII switches to a photo protective state, where there is thermal dissipation of the potentially harmful excess energy (Wakao *et al.*, 2009). This photoprotective heat dissipation is measured and is often called 'non-photo-chemical quenching' (NPQ). This thermal dissipation of chlorophyll – excited states competes with PSII fluorescence emission as well with photochemistry. Dissipating more energy as heat instead of driving primary charge separation decreases the quantum yield of PSII (Niyogi, 1999). The down regulation of efficiency in PSII drives a commensurate quenching in PSI, in this case due to quenching by elevated amounts of P_{700}^{+} (Ort, 2001).

In excess light cannot be dissipated safely, photodamage can occur and lead to oxidative damage to PSII, especially to D_1 protein (Long *et al.*, 1999), which would lower photosynthetic efficiency and requires repair and replacement of the protein before efficiency could be restored. The detailed energetic cost of photo damage, repair, and protective mechanisms has not been analyzed thoroughly; such costs are possibly smaller than the lost carbon uptake due to photo protection because not many proteins are involved (Long *et al.*, 1994). This suggests that plants with increased capacity of photo protection and repair will gain competitive advantage in high light stress conditions.

Over expressing beta-carotene hydrolase in *Arabidopsis thaliana*, which controls the biosynthesis of the xanthophylls cycle carotenoids, changed the rate of formation and relaxation of NPQ, though the final amplitude of NPQ was unaltered (Johnson *et al.*, 2008). Further, the kinetics of NPQ correlate with the deep oxidation state of the xanthophylls cycle pool and not the amount of zeaxanthin, thus, suggesting that zeaxanthin and violaxanthin antagonistically regulate the switch between the light harvesting and photoprotective modes of light harvesting system. This further suggests that fine tuning of the xanthophyll cycle pool size might be a feasible approach to engineer optimal NPQ kinetics. Since, the crystal structure of LHCII shows a single xanthophyll cycle carotenoid per monomer, it is puzzling how the over synthesis of xanthophyll cycle carotenoids acts to affect NPQ. Similarly, the overexpression of P_sbS induces enhanced NPQ saturating at $5P_s bS/D_1$ (Niyogi *et al.*, 2005), whereas the wild-type titer is a single $P_s bS/D_1$, again raising the question of how the extra copies interact to impact NPQ. Genetic variations within a single species or among species susceptibile to photoinhibition either by different decrease of maximum quantum efficiency CO_2 ($ØCO_2$) or by different rates of recovery of $ØCO_2$ after photoinhibition (Wang *et al.*, 2008) are another resource that can be used to identify and then engineer optimal NPQ kinetics.

4 Improving Known Bottlenecks

The oxygenase reaction of Rubisco and subsequent photorespiratory pathway are clear targets for improvement of the C3 cycle (Whitney *et al.*, 2011). Over the last 20 years, considerable effort has been focused on reengineering the Rubisco protein, using site directed mutagenesis to increase the carboxylase relative to the oxygenase reaction. However, this approach has yet to produce an improved Rubisco (Parry *et al.*, 2007). Recently, it has been shown that it is possible to form an active Rubisco *in vitro* using unfolded large subunits with the chaperone proteins GroEL/ES, RbcX, and ATP (Liu *et al.*, 2010). This paves the way for *in vitro* analysis of mutants and has the potential to allow screening for improvements in the Rubisco catalytic parameters of genetically engineered mutants. A new opportunity has also arisen from interspecies comparison of Rubisco catalytic properties that has revealed variation in the specificity, kcat, and temperature response of Rubisco from natural vegetation in the Mediterranean (Sage, 2002; Galmes *et al.*, 2005). The variation in the properties would indicate that transferring Rubisco from these species into crop plants could substantially increase the rate of CO_2 assimilation. The recent demonstration of the successful expression of Rubisco large and small subunits as a single fusion protein in the chloroplast that assembled into an active holoenzyme provides the opportunity to use expression of foreign Rubisco as a strategy to improve the C3 cycle (Whitney and Sharwood, 2007; Whitney *et al.*, 2009). CO_2-concentrating mechanisms (CCMs) have been found in all plant groups, in algal species, and in the cyanobacteria. These adaptations present the possibility to exploit existing solutions that have evolved in these organisms to improve C3 plants (Gowik and Westhoff, 2011; Price *et al.*, 2011). To realize the potential of these strategies will likely be in the long term; nevertheless, it is essential to begin the research now to determine the feasibility as a realistic option for future improvements. Transferring C4 photosynthesis to C3 species is one strategy to reduce photorespiration. But engineering C3 plants to carry out C4 metabolism will be a challenge as not only is there a requirement for additional metabolism but there are also changes in anatomy needed (Hibberd *et al.*, 2008; Wang *et al.*, 2009; Zhu *et al.*, 2010b). Fundamental research aimed at understanding the molecular basis of the anatomical specialization and the regulatory processes determining the location and the level of expression of the C4-specific enzymes is under way (Hibberd and Covshoff, 2010). Although this has not yet led to major breakthroughs, the combination of new sequencing and proteomic technologies together with a large international program give optimism for future success in this area. The prokaryotic carboxysome provides a micro compartment in the cell that transports and fixes CO_2, acting as a CCM (Badger *et al.*, 2006; Moroney and Ynalvez, 2007; Price *et al.*, 2008). Eukaryotic algae (and some primitive land plants) possess a structure, the pyrenoid that may act to concentrate CO_2 at Rubisco (Moroney and Ynalvez, 2007). Both of these CCM provide components that could be transferred to higher plants to reduce photorespiration and increase water and nitrogen use efficiency. However, it is likely that engineering the complete CCM from either of these groups of organisms into higher plants will be a complex feat of gene engineering, and given that the components of the pyrenoid remain to be elucidated, these approaches are likely to be successful only in the longer term. However, both the carboxysome and pyrenoid CCMs involve more than just the micro compartmentation of Rubisco. It might be more attainable in the shorter term to exploit the CO_2 transport systems in these organisms to boost CO_2 levels in C3 plants and thereby have a positive effect on atmosphere CO_2 assimilation by Rubisco (Price *et al.*, 2011; Raines, 2011).

5. *Identifying Novel Targets*

Redistribution of enzymes of carbon fixation

Although antisense studies described above revealed enzymes in the C3 cycle as targets for manipulation, it would be impossible to analyze every potential manipulation in this way. It is even less attainable to select multiple targets that may impact synergistically to improve the cycle. In order that significant advances can be made in a sensible time frame, it will be essential to have a dynamic kinetic model that is predictive. A model of photosynthetic carbon metabolism has been developed, and its application, in conjunction with an evolutionary algorithm, indicated that the distribution of resources between enzymes of carbon metabolism (that is the amount of each) is not optimal (Zhu *et al.*, 2007). An overinvestment in enzymes of the photorespiratory pathway, namely Glydecarboxylase, and an under investment in ADP-Glcpyrophosphorylase, SBPase, and plastid aldolase enzymes were identified. This suggests that the rate of light-saturated photosynthesis in existing C3 plants could be increased significantly by redistribution of the amounts of these enzymes in carbon assimilation, photorespiration, and starch biosynthesis. The hypotheses generated from this modeling can easily be tested using current transgenic technology using a model plant system which has the potential to provide clear strategies for improvement of photosynthesis in a time frame of 2 to 3 years.

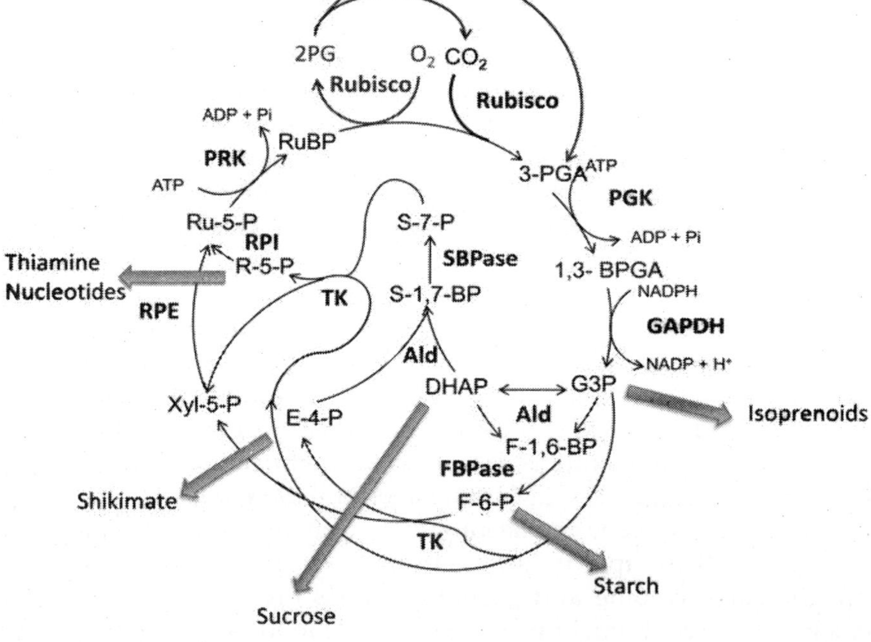

Fig 4 The C3 cycle. The carboxylation reaction catalyzed by Rubisco fixes CO_2 into the acceptor molecule RuBP, forming 3-PGA.The reductive phase of the cycle follows with two reactions catalyzed by 3-PGA kinase (PGK) and GAPDH, producing G-3-P. The G-3-P enters the regenerative phase catalyzed by aldolase (Ald) and either FBPase or SBPase, producing Fru-6-P (F-6-P) and sedoheptulose-7-P (S-7-P). Fru-6-P and sedoheptulose-7-P are then utilized in reactions catalyzed by TK, R-5-P isomerase (RPI), and ribulose-5-P (Ru-5-P) epimerase (RPE), producing Ru-5-P. The final step converts Ru-5-P to RuBP, catalyzed by PRK. The oxygenation reaction of Rubisco fixes O_2 into the acceptor molecule RuBP, forming PGA and 2-phosphoglycolate (2PG), and the process of photorespiration (shown in red) releases CO2 and PGA. The five export points from the pathway are shown with blue arrows (source Raines, 2011)

A neglected aspect of the C3 cycle is the control of flux from the cycle to the output pathways of starch, Suc, isoprenoids, shikimate, and nucleotides (Fig 4). Although detailed analysis of the regulation of Suc, starch, and shikimate biosynthetic pathways have been undertaken, there is no complete model that can simulate the way the relative flux to these pathways changes during development, in response to stress or availability of light. This lack of knowledge places limitations on the ability to exploit the enormous potential for manipulation of the cycle for the synthesis of high value chemicals such as pharmaceuticals, plant protectants, nutraceuticals, and flavor/color compounds that are products of the shikimate and isoprenoid output pathways of the C3 cycle.

Introduction of designed synthetic carbon fixation pathway

Although the majority of photosynthetic organisms fix atmospheric CO_2 through the C3 cycle, in nature there are five other CO_2 fixation pathways that are found in the Archaea (organisms that have structure and metabolism similar to bacteria and transcriptional regulation more closely aligned with the eukaryotes (Berg *et al.*, 2010)). Although none of these organisms use either Rubisco or phosphoenolpyruvatecarboxylase as the carboxylation enzyme, they do reveal a diversity of metabolism that has evolved in response to a range of environmental niches. This raises the possibility of employing a combination of modeling together with synthetic biology, to construct completely novel CO_2 fixation pathways that maybe more efficient than the C3 cycle. In addition, this approach may offer a solution to the problem of Rubisco without having to redesign plant development, required for introduction of the C4 pathway. Using a constraint-based modeling approach that included approximately 5,000 enzymes known to occur naturally, a number of hypothetical alternative carbon fixation pathways were identified (Bar-Even *et al.*, 2010). This study showed that some of these hypothetical pathways may be more efficient than the naturally occurring C3 cycle. Interestingly, all of the pathways predicted to be 2 to 3 times faster than the C3 cycle included the enzyme phosphoenolpyruvate carboxyiase, which is the carboxylating enzyme found in C4 species, and not any of the carboxylating enzymes found in the Archaea. These model-generated, hypothetical pathways also share a core set of reactions and have been termed the malonyl-CoA-oxaloacetateglyoxylate (MOG) pathways (Fig 5.). Although the glyoxylate produced could be converted to G-3-P by introducing enzymes from Escherichia coli, there is still a requirement to synthesize hexose phosphates, Fru-6-P, E-4-P and R-5-P, needed to supply the essential carbon compounds for growth and development of the plant. If a MOG-type synthetic pathway was to be engineered into it may require coupling to a partial Calvin cycle, creating Rubisco-independent carbon assimilation. To maintain the MOG pathway, it would also be necessary to use some of the G-3-P to regenerate the phosphoenolpyruvate needed for CO_2 assimilation (Fig. 5). The technical challenges to engineer such large changes in the core CO2 pathway(s) in higher plants using synthetic biology will be significant. But redesigning metabolism based on the modeling strategy of Bar-Even *et al.* (2010) offers a novel and exciting approach that warrants experimental study.

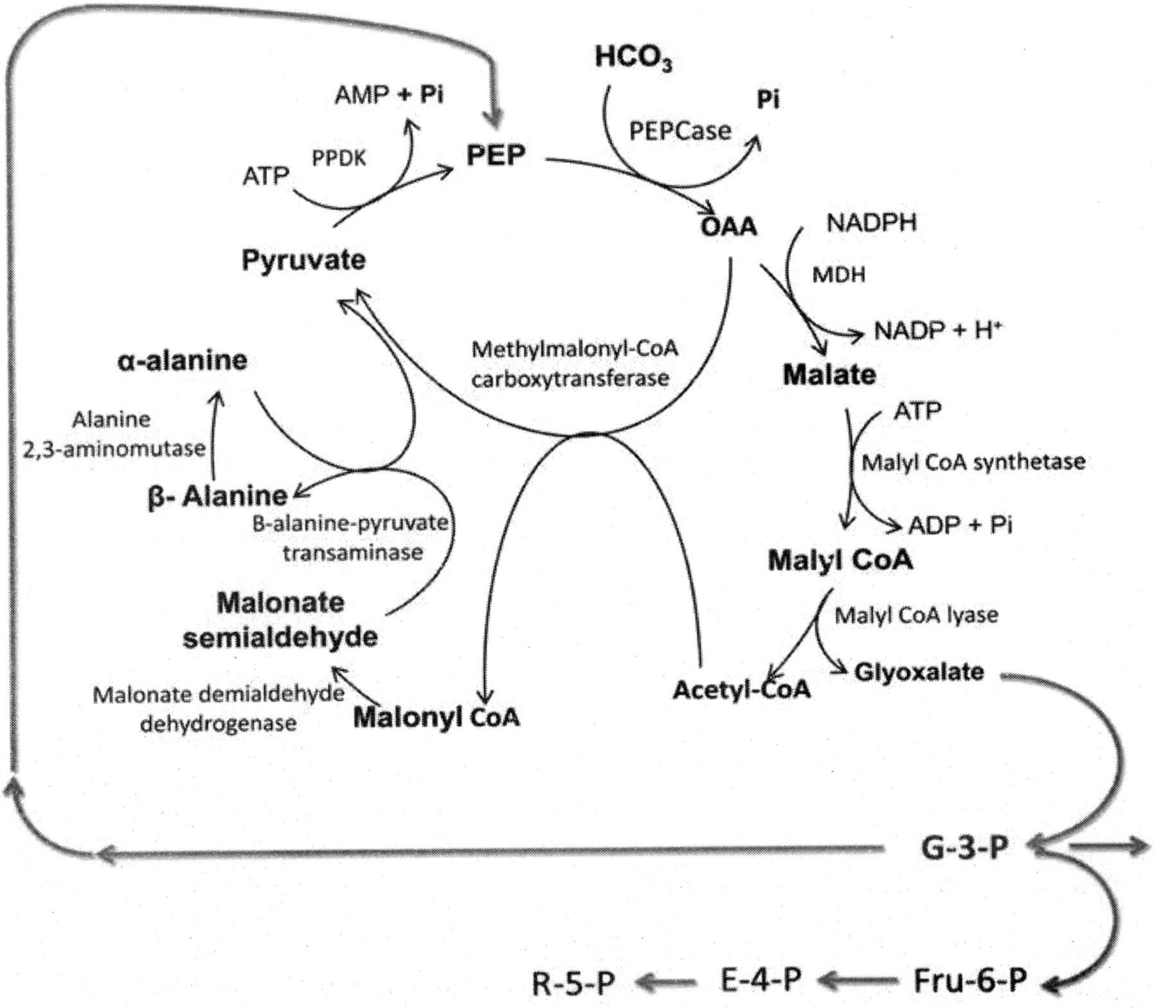

Fig 5 Rubisco-independent carbon assimilation. C4-glyoxalate (MOG) cycle—a model-developed hypothetical carbon fixation pathway. This pathway uses the enzyme phosphoenolpyruvate carboxylase (PEPCase) to fix atmospheric CO_2 into oxaloacetate and the only output product of this cycle is glyoxalate. PPDK, Pyruvate phosphate dikinase; MDH, malate dehydrogenase. The arrows in blue are speculative routes for carbon that may be need to be engineered to maintain the MOG cycle in addition to linking with a partial C3 cycle to provide the intermediates for biosynthesis of isoprenoids (G-3-P), shikimate (E-4-P), Suc (G-3-P), starch (Fru-6-P), and nucleotides (R-5-P). Adapted from Bar-Even *et al.*, (2010).

Deciphering the Underlying mechanisms for Natural Variation in Photosynthesis

There is considerable variation in photosynthetic rates per unit leaf area between C3 species. This is an overlooked and untapped resource yet has huge potential to identify natural mechanisms that have evolved to allow plants to survive in different environments. Systematic analysis of plants from different geographic regions, using high-throughput in vivo photosynthesis tools, has the potential to identify novel targets for manipulation in crop plants.

Development of photosynthetic capacity also differs in the same species when grown under different environmental conditions; which results in changes in morphology and anatomy and is termed developmental acclimation. In addition, some plant species are able to alter photosynthetic capacity in fully mature leaves, displaying a dynamic acclamatory response. Developmental and dynamic acclimation is distinct processes and dynamic acclimation may have a role in increasing the fitness of plants in natural environments (Athanasius *et al.*, 2010). However, most studies focus on plants grown in constant conditions or after a shift to differenconditions, e.g. low to high light. Only few studies are there where plants have been exposed to fluctuating conditions. Elucidation of the mechanisms that are employed by plants during environmental fluctuations can provide novel targets for improving crop yield. This is particularly relevant given that many crop species lack any dynamic range to respond to this variation (Murchie *et al.*, 2009).

Another aspect where little is known about the relative importance of regulatory mechanisms in determining carbon flux in the C3 cycle. It has been known for several decades that a number of enzymes of the C3 cycle, FBPase, SBPase, PRKase, and GAPDH, are subjected to light activation via the ferredoxin/thioredoxin system. Importantly, there is a lag in this response when plants experience a dark to light shift, resulting in a delay of minutes in photosynthetic rates reaching a steady-state maximum (Sassenrath-Coleand Pearcy, 1994; Sassenrath-Cole *et al.*, 1994). It has also been shown recently that the enzymes PRK and GAPDH are subject to rapid inactivation/activation in response to changes in light availability, mediated by a small protein, CP12, together with Trxf (Howard *et al.*, 2008; Marri *et al.*, 2009). These regulatory mechanisms are believed to match the rate of carbon fixation with the availability of energy from the light reactions. However, the relative importance of this regulation in determining photosynthetic capacity is unknown but may be important in the acclamatory response of plants dealing with fluctuating conditions. Furthermore, the extent to which this regulation varies between species is also relatively unexplored goal.

Thrust areas

For improving photosynthetic efficiency, the in-depth knowledge *i.e.* how alterations made to the photosynthetic process at the chloroplast level will scale, because it is the impact on the integral of canopy photosynthesis, not the instantaneous rate of chloroplast or single leaf photosynthesis which is related to biomass production and ultimately yield.

- Photosynthesis is strongly affected by external environmental factors as well as internal processes such as respiration, nitrogen metabolism and water transport.

This consideration enables to select the changes in the photosynthetic process intended to improve biomass production and crop yield must take care of a complex matrix of interacting elements.

- Overall, the world has limited capacity to sustain the expansion of crop land in which it is shrinking in many developed countries. For meeting the future increase in demand will have to come from a near doubling of productivity on a land area basis. Important gains in productivity is possible through reducing stress-induced and post-harvest losses and further improvement in interception efficiency, partitioning efficiency may be possible in less developed crop species. However, a major break through will come from improved photosynthetic conversion efficiency (E_c). The task of improving E_c, will take next 10-20 years to bring such innovation to farms in commercial cultivars at adequate scale.
- Transferring C4 Rubisco into the chloroplasts of C3 plants faces technical obstacles of transforming both genomes i.e. ensuring proper import, post-translational processing, assembly, silencing of native genes and efficient interaction with regulating partner i.e. Rubisco activase, needs to be overcome. Further, the molecular features of Rubisco holoenzyme that control the discrimination between oxygen and carbondioxide are unknown and the reaction mechanism of Rubisco precludes the possibility of engineering any significant decrease in oxygenation activity. Reduction of the Rubisco oxygenase reduction remains a target for future improvement photosynthesis. However, the current strategies that might be exploited to achieve this goal are conceptually straightforward, and all of these approaches will be technically demanding requiring fundamental research to identify genes involved.
- The goal of converting C3 crops to C4 photosynthetic metabolism remains critical e.g. discovering the genetic basis of Kranz anatomy and the developmental compartmentation of the processes of C4 photosynthesis are still unknown. Another long-term goal to be achieved is to engineer increase in mesophyll conductance to CO_2, since the critical information about the physiological and physical factors controlling the mesophyll conductance's are lacking/meager.
- Improvement of the C3 cycle is not just about increasing CO_2 fixation but should aim to increase both nitrogen use efficiency and water use efficiency while maintaining high productivity.
- The range of genetic and molecular techniques that are currently available, together with the development and application of rapid in vivo techniques to allow in - field analysis of wider range of species in their natural environment, will facilitate the wider analysis of natural variation in photosynthetic carbon assimilation. This approach has enormous and unexplored potential for future exploitation to improve yield.

Conclusion

Several changes in the photosynthetic process have been identified that are well supported by theory to increase canopy photosynthesis and production. For some implementation is limited by technical issues that can be overcome by sufficient investment, whereas, in other cases, too little of the science has been undertaken to identify what needs to be altered to effect an increase in yield. Improving photosynthetic conversion efficiency will require a systematic approach coupled with the informative models, to correlate a change made in chloroplast to yield and that implements a full suite of tools specially breeding, gene transfer, synthetic biology to bring about the designed alteration in photosynthesis.

REFERENCES

Ainsworth EA, Long S P (2005) What have we learned from 15 years of free-air CO_2 enrichment (FACE) ?A meta-analytic review of the responses of photosynthesis, canopy. *New Phytol.* 165:351–71

Ainsworth EA, Rogers A, Nelson R, Long SP (2004) Testing the 'sourcesink' hypothesis of down-regulation of photosynthesis in elevated [CO_2] in the field with single gene substitutions in *Glycine max*. *Agricultural and Forest Meteorology* 122: 85–94

Al-Khatib K, Paulsen GM (1999) High temperature effects on photosynthetic processes in temperate and tropical cereals. *Crop Sci* 39 : 119-123

Amthor JS (2007) Improving photosynthesis and yield potential. In Improvements of Crop Plants for Industrial End Uses, ed. P Ranalli, pp. 27–58. Dordrecht, Neherlands Springer

Amthor JS (1989) Respiration and Crop Productivity. New York: Springer-Verlag

Andersson I, Taylor TC (2003) Structural framework for catalysis and regulation in ribulose-15-bisphosphate carboxylase/oxygenase. *Arch Biochem Biophys* 414 :130-40

Apel P (1979) Dark and photorespiration. In Crop Physiology and Cereal Breeding : Procedings of a Eucarph Workshop ed. J.H.J. Spertz. Th Kramer, pp. 102-05, Wageningen. Centre Agric. Publ. Dec. 186 pp

Athanasiou K, Dyson BC, Webster RE, Johnson GN (2010) Dynamic acclimation of photosynthesis increases plant fitness in changing environments. *Plant Physiol* 152: 366–373

Austin RB, Bingham J, Blackwell RD, Evans LT, Ford MA, Morgan CL, Taylor M (1980) Genetic improvements in winter wheat yields since 1900 and associated physiological changes. *Journal of Agricultural Science* 94: 675–689

Austin RB, Morgan CL, Ford MA (1982) Flag leaf photosynthesis of *Triticum aestivum* and related diploid and tetraploid species. *Annals of Botany* 49: 177–189

Badger MR, Price GD, Long BM, Woodger FJ (2006) The environmental plasticity and ecological genomics of the cyanobacterial CO2 concentrating mechanism. *J Exp Bot* 57: 249–265

Baena-Gonzalez E, Rolland F, Thevelein JM, Sheen J (2007) A central integrator of transcription networks in plant stress and energy signalling. *Nature* 448: 938–U10

BainbridgeG, Madgwick P, Parmar S, Mitchell R, Paul M (1995) Engineering Rubisco to change its catalytic properties. *J Exp Bot* 46: 1269–76

Baker NR, East T M, Long SP (1983) Chilling damage to photosynthesis in young *Zea mays*. *J Exp Bot* 34 :189-97

Ball J T, Woodrow IE, Berry JA (1987) A model predicting stomatal conductance and its contribution to the control of photosynthesis under different environmental conditions. In Progress in Photosynthesis Research, ed. J Biggens, vol. 4, pp. 221–24. Dordrecht, Netherlands: Martinus Nijhoff

Barber J (1998) What limits the efficiency of photosynthesis and can there be beneficial improvements?. In Feeding a World Population of More Than Eight Billion People-A Challenge to Science (eds J.C. Water low, D.G. Armstrong, L. Fowden &R. Riley) , pp. 112–123. Oxford University Press, Cary, NC

Bar-Even A, Noor E, Lewis NE, Milo R (2010) Design and analysis of synthetic carbon fixation pathways. *Proc Natl Acad Sci USA* 107: 8889–8894

Barnola JM, Raynaud D, Lorius C, Barkov N (1999) Historical CO2 record from the Vostok ice core.In Trends: A Compendium of Data on Global Change, Carbon Dioxide Information Analysis Center, U.S.Dept. of Energy, Oak Ridge National Laboratory, Oak Ridge, TN

Beadle CL, Long SP, Imbamba S K, Hall DO, Olembo RJ (1985) Photosynthesis in relation to plant production in terrestrial environments. UN Environ. Programme (UNEP) , Oxford, UK: Tycooly Int.156 pp

Beale CV, Morrison JIL, Long SP (1999) Water use efficiency of C4 perennial grasses in a temperate climate. *Agric For Meteorol* 96:103-5

Berg IA, Kockelkorn D, Ramos-Vera WH, Say RF, Zarzycki J, Hu¨ gler M, Alber BE, Fuchs G (2010) Autotrophic carbon fixation in archaea. *Nat Rev Microbiol* **8**: 447-460

Bernacchi CJ, Portis AR, Nakano H, Von Caemmerer S, Long SP (2002) Temperature response of mesophyll conductance. Implications for the determination of Rubisco enzyme kinetics and for limitations to photosynthesis *in vivo*. *Plant Physiol* 130: 1992-98

Bj¨orkmanO , Demmig B (1987) Photon yield of O_2 evolution and chlorophyll fluorescence characteristics at 77K among vascular plants of diverse origins. *Planta* 170: 489-504

Black CC, Tu ZP, Counce PA, Yao PF, Angelov MN (1995) An integration of photosynthetic traits and mechanisms that can increase crop photosynthesis and grain production. *Photosynthesis Research* 46: 169–175

Bohnert HJ, Gong Q, Li P, Ma S (2006) Unraveling abiotic stress tolerance mechanisms – getting genomics going. *Current Opinion in Plant Biology* 9: 180–188

Bondada BR, Oosterhuis DM, Norman RJ, Baker WH (1996) Canopy photosynthesis, growth, yield and boll ^{15}N accumulation under nitrogen stress in cotton. *Crop Sci.* 36 : 127-133

Brinkman MA, Frey KJ (1978) Flag leaf physiological analysis of Oat isolines that differ in grain yield from their recurrent parents. *Crop Sci.* 18 : 67-73

Buttia SS, Edje OI, Wahah AR (1973) Effects of seed size on seedling performance in soybean II. Seedling growth photosynthesis and field performance. *Crop Sci.* 13 : 206-10

Calderini DF, Dreccer MF, Slafer GA (1995) Genetic improvement in wheat yield and associated traits. A re-examination of previous results and the latest trends. *Plant Breeding* 114: 108–112

Chang CCC, Ball L, Fryer MJ, Baker NR, Karpinski S, Mullineaux PM (2004) Induction of ASCORBATE PEROXIDASE 2 expression in wounded Arabidopsis leaves does not involve known wound-signalling pathways but is associated with changes in photosynthesis. *Plant Journal* 38: 499–511

Charles-Edwards DA (1978) An analysis of the photosynthesis and productivity of vegetative crops in U.K. *Am Bot.* 42 : 717-32

Cheeseman JM, Clough BF, Carter DR, Lovelock CE, Eong OJ, Sim RG (1991) The analysis of photosynthetic performance in leaves under field conditions – a case-study using bruguiera mangroves. *Photosynthesis Research* 29: 11–22

Chen X, Zhang W, Xie Y, Lu W, Zhang R (2007) Comparative proteomics of thylakoid membrane from a chlorophyll b–less rice mutant and its wild type. *Plant Sci.* 173 : 397

Cheng SH, Demoore B, Wu JR, Edwards GE, Ku MSB (1989) Photosynthetic plasticity in flaveriabrownii-growth irradiance and the expression of C4 photosynthesis. *Plant Physiol* 89: 1129-35

Conocono EA, Eefdane JA, Setter TI (1998) Estimation of canopy photosynthesis in rice by means of daily increases in leaf carbohydrate concentration. *Crop Sci* 38 : 987-995

Cook MG, Evans LT (1983) Some physiological-aspects of the domestication and improvement of rice (*Oryza* spp.) . *Field Crops Research* 6: 219–238

Cooper CS, MacDonald PW (1970) Energetic of early seedlings growth in corn. *Crop Sci* 10: 136-139

Cramer WA, Zhang HM, Yan JS, Kurisu G, Smith JL (2006) Transmembrane traffic in the cytochromeb6f complex. *Annu Rev Biochem* 75 : 769-90

De-Datta SK (1981) Principles and Practices of Rice Productions. New York John Wiley Sons

Delaney RH, Dobrenz AK (1974) Morphological and anatomical features of alfalfa leaves as related to CO_2 exchange. *Crop Sci.* 14 : 444-47

Dermody O, Long SP, Mc Connaughay K, DeLucia EH (2008) How do elevated CO_2 and O3 affect the interception and utilization of radiation by a soybean canopy? *Glob. Change Biol.* 14 : 556-64

Dohleman FG, Long SP (2009) More productive than maize in the Midwest-How does Miscanthus do it? *Plant Physiol.* 150 : 2104-15

Duncan WG (1971) Leaf angle, leaf area and crop photosynthesis. *Crop Sci.* **11**: 482-85

Duncan WG, Hesketh JD (1968) Net photosynthetic rates, relative growth rates and leaf numbers of 22 races of maize grown at eight temperatures. *Crop Sci.* **8** : 670-74

Dwyer LA, Tallenaar M, Stewart DW (1991) Changes in plant density dependence of leaf photosynthesis of maize (*Zea mays* L.) hybrid. 1959 to 1988. *Can J Plant Sci.* 71 : 1-11

Earl HJ, Tollenaar M (1998) Difference among commercial maize (*Zea mays* L.) hybrids in respiration rates of mature leaves. *Field Crops Res.* **59** : 9-19

Eberhard S, Finazzi G, Wollman FA (2008) The dynamics of photosynthesis. *Annu Rev Genet.* 42 :463-515

Ehleringer J, Pearcy RW (1983) Variation in quantum yield for CO_2 uptake among C3 and C4 plants. *Plant Physiol.* 73 : 555-59

Ercoli L, Mariotti M, Masoni A, Bonari E (1999) Effect of irrigation and nitrogen fertilization on biomass yield and efficiency of energy use in crop production of Miscanthus. *Field Crops Res.* 63 : 3-11

Evans LT (1998) Greater crop production: whence and whither? In Feeding a World Population of More Than Eight Billion People - A Challenge to Science (eds. J.C .Waterlow, D.G. Armstrong, L. Fowdenand and R. Riley) , pp. 89-97.OxfordUniversity Press, Cary, NC, USA

Evans LT (1993) Crop Evolution, Adaptation and Yield. Cambridge: Cambridge Univ. Press

Evans LT , Dunstone RL (1970) Some physiological aspects of evolution in wheat. *Australian Journal of Biology Sciences* 23 :725-741

Evans LT, Wardlow IF (1976) Aspects of the comparative physiology of the grain yield in cereals. *Adv Agron* 28 : 301-359

Evans JR (1983) Nitrogen and photosynthesis in the flag leaf of wheat. *Plant Physiol* 72 : 297-302

Evans LT (1993) *Crop evolution, adaptation and yield.* Cambridge, UK: Cambridge University Press

Fageria NK, Baligar VC, Jones CA (1997) Growth and mineral nutrition of field crops. Second Edition, New York : Marcel Dekker

Falk S, Leverenz JW, Samuelsson G, Oquist G (1992) Changes in photosystem II fluorescence in Chlamydomonas reinhardtii exposed to increasing levels of irradiance in relationship to the photosynthetic response to light. *Photosynth Res* 31:31–40

Farquhar GD, Von Caemmerer S, Berry J A (1980) A biochemical model of photosynthetic CO_2 assimilation in leaves of C3 species. *Planta* 149 : 78-90

Feng LL, Han YJ, An BG, Yang J, Yang GH, Li YS, Zhu YG (2007a) Over expression of sedoheptulose-1,7-bisphosphatase enhances photosynthesis and growth under salt stress in transgenic rice plants. *Funct Plant Biol* 34: 822–834

Feng LL, Wang K, Li Y, Tan YP, Kong J, Li H, Li Y Zhu YG (2007b) Overexpression of SBPase enhances photosynthesis against high temperature stress in transgenic rice plants. *Plant Cell Rep* 26: 1635–1646

Ferguson H, Eslick RF, Aase JK (1973) Canopy temperature of barley as influenced by morphological characteristics. *Agron J* 65 : 425-28

Ferguson SJ (2000) ATP synthase: What dictates the size of a ring? *Curr Biol* 10 : 804-8

Fischer RA, Rees D, Sayre KD, Lu ZM, Condon AG, Saavedra AL (1998) Wheat yield progress associated with higher stomatal conductance and photosynthetic rate, and cooler canopies. *Crop Science* 38: 1467–1475

Foyer H, Bloom AJ, Queval G, Noctor G (2009) Photorespiratory metabolism: genes, mutants, energetics ,and redox signaling. *Annu Rev Plant Biol* 61 : 455-84

Frederick JR, Camberato JJ (1994) Leaf net CO_2-exchange rate and associated leaf traits of winter wheat grown with various spring nitrogen fertilization rates. *Crop Sci.* **34** : 432-439

Fryer MJ, Ball L, Oxborough K, Karpinski S, Mullineaux PM, Baker NR (2003) Control of ascorbate peroxidase 2 expression by hydrogen peroxide and leaf water status during excess light stress reveals a functional organisation of Arabidopsis leaves. *Plant Journal* 33: 691–705

Furbank RT, Hatch MD (1987) Mechanism of C4 photosynthesis-the size and composition of the inorganic carbon pool in bundle sheath cells. *Plant Physiol.* 85 : 958-64

Galmes J, Flexas J, Keys A J, Cifre J , Mitchell RAC (2005) Rubisco specificity factor tends to be larger in plant species from drier habitats and in species with persistent leaves. *Plant Cell Environ.* 28 : 571-79

Galmes J, Flexas J, Keys AJ, Cifre J, Mitchell RAC, Madgwick PJ, Haslam RP, Medrano H, Parry MAJ (2005) Rubisco specificity factor tends to be larger in plant species from drier habitats and in species with persistent leaves. *Plant Cell Environ* 28: 571–579

Gan S, Amasino RM (1995) Inhibition of leaf senescence by autoregulated production of cytokinin. *Science* 270: 1986–1988

Garrett MK (1978) Control of photorespiration at RUBP carboxylase oxygenase level in ryegrass cultivars. *Nature* 274 : 913-15

Gifford RM, Evans LT (1981) Photosynthesis, carbon partitioning and yield. *Annu Rev Plant Physiol* 32 : 485-509

Giordano M, Beardall J, Raven J A (2005) CO_2 concentrating mechanisms in algae: mechanisms, environmental modulation and evolution. *Annual Review of Plant Biology* 56 : 99-131

Glick RE, Melis A (1988) Minimum photosynthetic unit size in system-I and system-II of barley chloroplasts. *Biochim Biophys Acta* 934 : 151-55

Gowik U, Westhoff P (2011) The path from C3 to C4 photosynthesis. *Plant Physiol* 155: 56–63

Gutierrez-Rodriguez M, Reynolds MP, Larque-Saavedra A (2000) Photosynthesis of wheat in a warm, irrigated environment – II. Traits associated with genetic gains in yield. *Field Crops Research* 66: 51–62

Gwathmey CO, Howard DD (1998) Potassium effects on canopy light interception and earliness of no-tillage cotton. *Agron J* 90 : 144-149

Hageman RH (1986) Nitrate metabolism in roots and leaves. pp. 105-116. in JC Shannon, DP Knievel and CD Boayer (eds.) . Regulation of carbon and nitrogen reduction and utilization in maize. Rockville M.D. *Am Soc Plant Physiol*

Hanson WD, Grier RE (1973) Rates of electron transfer and of non-cyclic photophosphorylation for chloroplasts isolated from maize population selected for juvenile productivity and in leaf width. *Genetics* 75 : 247-57

Harley PC, Sharkey TD (1991) An improved model of C3 photosynthesis at high CO_2: reversed O_2 sensitivity explained by lack of glycerate reentry into the chloroplast. *Photosynth Res.* 27 :169-78

Harrison EP, Olcer H, Lloyd JC, Long SP, Raines CA (2001) Small decreases in SBPase cause a linear decline in the apparent RuBP regeneration rate, but do not affect Rubisco carboxylation capacity. *J Exp Bot* 52 : 1779-84

Harrison E P, Willingham NM, Lloyd JC, Raines CA (1998) Reduced sedoheptulose-1,7-bisphosphatase levels in transgenic tobacco lead to decreased photosynthetic capacity and altered carbohydrate accumulation. *Planta* 204 : 27-36

Hay RKM (1995) Harvest index-a review of its use in plant-breeding and crop physiology. *Ann Appl Biol* 126 :197-216

Hay R, Porter J (2006) The physiology of crop yield. Oxford, UK: Blackwell Publishing Ltd

Heitholt JJ (1994) Canopy characteristics associated with deficient and excessive cotton plant population densities. *Crop Sci.* 34 : 1291-1297

Heitholt JJ, Pettigrew WT, Meredith Jr WR (1992) Light interception and lint yield of narrow-row cotton. *Crop Sci.* 32 : 728-33

Helliker BR, Richter SL (2008) Subtropical to boreal convergence of tree-leaf temperatures. *Nature* 454: 511–514

Henkes S, Sonnewald U, Badur R, Flachmann R, Stitt M (2001) A small decrease of plastid transketolase activity in antisense tobacco transformants has dramatic effects on photosynthesis and phenylpropanoid metabolism. *Plant Cell* 13: 535–551

Hibberd JM, Quick WP (2002) Characteristics of C4 photosynthesis in stems and petioles of C3 flowering plants. *Nature* 415 : 451-54

Hibberd JM, Covshoff S (2010) The regulation of gene expression required for C4 photosynthesis. *Annu Rev Plant Biol* 61: 181–207

Hibberd JM, Sheehy JE, Langdale JA (2008) Using C4 photosynthesis to increase the yield of rice-rationale and feasibility. *Curr Opin Plant Biol* 11: 228–231

Hikosaka K, Terashima I (1995) A model of the acclimation of photosynthesis in the leaves of C3 plants to sun and shade with respect to nitrogen use. *Plant Cell Environ.* 18 : 605-18

Hirose T, Werger MJA (1987) Nitrogen use efficiency in instantaneous and daily photosynthesis of leaves in the canopy of a *Solidago altissmastand. Physiol. Plant* 70 : 215-222

Horton P (2000) Prospects for crop improvement through the genetic manipulation of photosynthesis: morphological and biochemical aspects of light capture. *Journal of Experimental Botany* 51: 475–485

Horton P, Johnson MP, Perez-Bueno ML, Kiss AZ, Ruban AV (2008) Photosynthetic acclimation: does the dynamic structure and macroorganisation of photosystem II in higher plant grana membranes regulate light harvesting states? *FEBS Journal* 275: 1069–1079

Horton P, Murchie EH, Ruban AV, Walters RG (2001) Increasing rice photosynthesis by manipulation of the acclimation and adaptation to light. In: Goode JA, Chadwick D, eds. *Rice biotechnology: improving yield, stress tolerance and grain quality. Novartis Foundation Symposium 236.* Chichester, UK: John Wiley & Sons, 117–130

Houtz RL, Portis AR (2003) The facts and mysteries. Arch. *Biochem. Biophys.* 414 :150-58

Howard TP, Metodiev M, Lloyd JC, Raines CA (2008) Thioredoxinmediated reversible dissociation of a stromal multiprotein complex inresponse to changes in light availability. *Proc Natl Acad Sci USA* 105: 4056–4061

Hubbart S, Peng S, Horton P, Chen Y, Murchie EH (2007) Trends in leaf photosynthesis in historical rice varieties developed in the Philippines since 1966. *Journal of Experimental Botany* 58: 3429–3438

Huner NPA, Oquist G, Sarhan F (1998) Energy balance and acclimation to light and cold. *Trends in Plant Science* 3: 224–230

Hunt IA, Poorten GV (1985) Carbon dioxide exchange rates and leaf nitrogen contents during aging of the flag and penultimate leaves of five spring wheat cultivars. *Can J Bot.* 63 : 1605-1609

Ishii R (1988) Varietal differences of photosynthesis and grain yield in rice. *Korean J Crop Sci.* 33 : 315-321

Johnson MP, Davison PA, Ruban AV, Horton P (2008) The xanthophyll cycle pool size controls the kinetics of nonphotochemical quenching in *Arabidopsis thaliana. FEBS Lett.* 582 : 262-66

Kanevski I, MalRhoade DF, Gutteridge S (1999) Bisphosphate carboxylase/ oxygenase in tobacco to form a sunflower large subunit and tobacco small subunit hybrid. *Plant Physiol.* 119 : 133-41

Karkehabadi S, Peddi SR, Anwaruzzaman M, Taylor TC, Cederlund A (2005) Chimeric small subunits influence catalysis without causing global conformational changes in the crystal structure of ribulose-1,5-bisphosphate carboxylase/oxygenase. *Biochemistry* 44 : 9851-61

Kebeish R , Niessen M, Thiruveedhi K, Bari R, Hirsch H (2007) Chloroplastic photorespiratory bypass increases photosynthesis and biomass production in Arabidopsis thaliana. *Nat Biotechnol.* 25 : 593-99

Kelly MO, Davies PJ (1988) The control of whole plant senescence CRC. *Crit Rev Plant Sci* 6: 611-616

Khan MA, Tsundo S (1970) Growth analysis of cultivated wheat species and their wild relatives with special reference to dry matter distribution among different plant organs and to leaf area expansion. *Tohoka J Agri Res* 21 : 47-59

Khanna R, Sinha SK (1973) Changes in the predominance from C4 to C3 pathway following anthesis in sorghum. *Biochem Biophys Res Commun* 52 : 121-124

Kramer DM, Cruz JA, Kanazawa A (2003) Balancing the central roles of the thylakoid proton gradient. *Trends Plant Sci.* 8 : 27-32

Kudla J (2008) A plastid protein crucial for Ca^{2+}-regulated stomatal responses. *New Phytologist* 179: 675–686

Kurek I, Chang TK, Bertain SM, Madrigal A, Liu L, LassnerMW, Zhu GH (2007) Enhanced thermostability of Arabidopsis Rubisco activase improves photosynthesis and growth rates under moderate heat stress. *Plant Cell* 19: 3230–3241

Lambers H (1987) Does variation in photosynthetic rate explain variation in growth rate and yield? *Neth. J Agric Sci* 35 : 505-519

Langdale JA, Zelitch I , Miller E, Nelson T (1988) Cell position and light influence C4 versus C3 patterns of photosynthetic gene-expression in maize. *EMBO J* 7 : 3643-51

Leakey ADB, Bernacchi CJ, Ort DR, Long SP (2006) Growth of soybean under free-air [CO_2] enrichment (FACE) does not cause stomatal acclimation. *Plant Cell Environ.* 29 : 1794-1800

Leakey ADB, Xu F, Gillespie KM, McGrath JM, Ainsworth EA, Ort DR (2009) Genomic basis for stimulated respiration by plants growing under elevated carbondioxide. *Proc. Natl. Acad. Sci. USA* 106 : 3597-602

Leegood RC (2007) A welcome diversion from photorespiration. *Nat Biotechnol* 25: 539–540

Lefebvre S, Lawson T, Zakhleniuk, OV, Lloyd JC, Raines CA (2005) Increased sedoheptulose-1,7-bisphosphatase activity in transgenic tobacco plants stimulates photosynthesis and growth from an early stage in development. *Plant Physiol.* 138 : 451-60

Leverenz JW, Falk S, Pilstrom CM, Samuelsson G (1990) The effects of photoinhibition on the photosynthetic light-response curve of green plant-cells (chlamydomonas-reinhardtii) . *Planta* 182 :161-68

Li Z, Wakao S, Fischer BB, Niyogi KK (2009) Sensing and responding to excess light. *Annu. Rev. Plant Biol.* 60 : 239-60

Lieman-Hurwitz J, Rachmilevitch S, Mittler R, Marcus Y, Kaplan A (2003) Enhanced photosynthesis and growth of transgenic plants that express ictB, a gene involved in HCO3- accumulation in cyanobacteria. *Plant Biotechnol J* 1: 43–50

Liu CM, Young AL, Starling-Windhof A, Bracher A, Saschenbrecker S, Rao BV, Rao KV, Berninghausen O, Mielke T, Hartl FU, et al., (2010) Coupled chaperone action in folding and assembly of hexadecameric Rubisco. *Nature* 463: 197–202

Long SP, Humphries SW, Falkowski PG (1994) Photoinhibition of photosynthesis in nature. *Annu Rev Plant Physiol Plant Molec Biol.* 45 : 633-62

Long SP, Postl WF, Bolharnordenkampf HR (1993) Quantum yields for uptake of carbon-dioxide in C3 vascular plants of contrasting habitats and taxonomic groupings. *Planta* 189 : 226-34

Long SP, Zhu XG, Naidu SL, Ort DR (2006) Can improvement in photosynthesis increase yield? *Plant Cell and Environment* 29 : 315-330

Loomis RS, Connar DJ (1992) Crop Ecology, Productivity and Management in Agricultural Systems, New York, Cambridge University Press

Loomis RS, Williams WA, Duncan WG (1967) Community architecture and the productivity of terrestrial plant communities. In Harvesting the Sun: Photosynthesis in Plant Life, ed. A San Pietro, FA Greer, TJ Army, pp. 291–308. New York: Academic

Lugg DG, Sinclair TR (1979) A survey of soybean cultivars for variability in specific leaf weight. *Crop Sci.* 19 : 887-92

Lush WM, Wien HC (1980) The importance of seed size in early growth of wild and domesticated cowpea. *J Agric Sci.* 94 : 177-82

Mae T (1997) Physiological nitrogen efficiency in rice : Nitrogen utilization, photosynthesis and yield potential. pp. 51-60. In : Ando K Fujita, T. Mae H Tsumoto, S. Mori and J Sekiya (eds.). Plant Nutrition for Sustainable Food Production and Environment. Dordrecht, the Netherlands, Kluwer Acd. Publ.

Marri L, Zaffagnini M, Collin V, Issakidis-Bourguet E, Lemaire SD, Pupillo P, Sparla F, Miginiac-Maslow M, Trost P (2009) Prompt and easy activation by specific thioredoxins of calvin cycle enzymes of Arabidopsis thaliana associated in the GAPDH/CP12/PRK supramolecular complex. *Mol Plant* 2: 259–269

Marshall B, Biscoe PV (1980) A model for C3 leaves describing the dependence of net photosynthesis on irradiance 0.1. derivation. *J Exp Bot* 31 : 29-39

Mass SJ, Dunlop JR (1989) Reflectance, transmittance and absorbance of light by normal, etiolated and albino corn leaves. *Agron J* 81 : 105-110

Matsumura I, Patel M, Greene D (2005) Directed evolution of Rubisco through genetic selections of metabolically engineered Escherichia coli. *FASEB J* 19 : A292

Matsuoka M, Furbank RT, Fukayama H, Miyao M (2001) Molecular engineering of C4 photosynthesis. *Annu. Rev. Plant Physiol. Plant Molec. Biol.* 52 : 297-314

McCashin BG, Canvin DT (1979) Photosynthetic and photorespiratory characteristics of mutants of *Hordeum vulgare* L. *Plant Physiol.* 64 : 354-60

Meehl GA, Stocker TF, Collins WD, Friedlingstein P, Gaye AT (2007) Global climate projections. In Climate Change 2007: The Physical Science Basis. Contribution of Working Group I to the Fourth Assessment Report of the Intergovernmental Panel on Climate Change, ed. S Solomon, D Qin, M Manning, Z Chen, M Marquis, et al., pp. 119–234. Cambridge: Cambridge Univ. Press

Melis A (1999) Photosystem-II damage and repair cycle in chloroplasts: What modulates the rate of photo damage in vivo? *Trends Plant Sci.* 4 :130-35

Melis A (2009) Solar energy conversion efficiencies in photosynthesis: Minimizing the chlorophyll antennae to maximize efficiency. *Plant Sci.* 17 : 272-80

Miralles DJ, Slafer GA (1997) Radiation interception and radiation use efficiency of near-isogenic wheat lines with different height. *Euphytica* 97: 201–208

Mitchell PL, Sheehy JE, Woodward FI (1998) *Potential yields and the efficiency of radiation use in rice.* IRRI Discussion Paper Series No. 32. Manila, the Philippines: International Rice

Research Institute

Miyagawa Y, Tamoi M, Shigeoka S (2001) Overexpression of a cyanobacterial fructose-1,6-/sedoheptulose-1,7-bisphosphatase in tobacco enhances photosynthesis and growth. *Nat Biotechnol* 19: 965–969

Miyao M (2003) Molecular evolution and genetic engineering of C4 photosynthetic enzymes. *J Exp Bot.* 54 :179-89

Monteith JL (1977) Climate and the efficiency of crop production in Britain. *Philos Trans R Soc Lond. Ser. B 281* : 277-94

Morinaka Y, Sakamoto T, Inukai Y, Agetsuma M, Kitano H (2006) Morphological alteration caused by brassinosteroid insensitivity increases the biomass and grain production of rice. *Plant Physiol.* 141 : 924-31

Moroney JV, Ynalvez RA (2007) Proposed carbon dioxide concentrating mechanism in Chlamydomonas Reinhardt. *Eukaryotic Cell* 6: 1251–1259

Motto M, Soressi G P, Salamini F (1979) Growth analysis in a reduced leaf mutant of common bean (*Phaseolus vulgaris* L.) . *Euphytica* 28 : 593-600

Mullineaux PM, Karpinski S, Baker NR (2006) **Spatial** dependence for hydrogen peroxide-directed signaling in light-stressed plants. *Plant Physiology* 141: 346–350

Murchie EH, Horton P (1997) Acclimation of photosynthesis to irradiance and spectral quality in British plant species: chlorophyll content, photosynthetic capacity and habitat preference. *Plant, Cell & Environment* 20: 438–448

Murchie E, Horton P (2007) Toward C4 rice: learning from acclimation of photosynthesis in the C3 leaf. In: Sheehy JE, Mitchell PL, Hardy B, eds.Charting pathways to C4 rice. Los Banos, the Philippines: International Rice Research Institute, 333–350

Murchie EH, PintoM, Horton P (2009) Agriculture and the new challenges for photosynthesis research. *New Phytol* 181: 532–552

Murchie EH, Chen YZ, Hubbart S, Peng SB, Horton P (1999) Interactions between senescence and leaf orientation determine in situ patterns of photosynthesis and photoinhibition in field-grown rice. *Plant Physiology* 119: 553–563

Muurinen S, Peltonen-Sainio P (2006) Radiation-use efficiency of modern and old spring cereal cultivars and its response to nitrogen in northern growing conditions. *Field Crop Res* 96 : 363-73

Neidhardt J, Benemann JR, Zhang LP, Melis A (1998) Photosystem-II repair and chloroplast recovery from irradiance stress: relationship between chronic photoinhibition, light-harvesting chlorophyll antenna size and photosynthetic productivity in *Dunaliella salina* (green algae). *Photosynth Res.* 56 :175-84

Nelson CJ, Larson KL (1984) Seedling growth pp.93-129. In : MB Tesar (ed.) Physiological Basis of Crop Growth and Development. Madison WI, *Am Soc Agron* and *Crop Soc Am.*

Nelson N, Yocum CF (2006) Structure and function of photosystems I and II. *Annu Rev Plant Biol.* 57 : 521-65

Nelson T, Langdale JA (1992) Developmental genetics of C4 photosynthesis. *Annu Rev*

Plant Physiol. Plant Mol. Biol. 43 : 25-47

Niinemets U (2007) Photosynthesis and resource distribution through plant canopies. *Plant Cell Environ.* 30 :1052-71

Niyogi KK (1999) Photo protection revisited: genetic and molecular approaches. *Annu Rev Plant Phy siol Plant Mol Biol* 50 : 333-59

Niyogi KK , Li XP, Rosenberg V, Jung HS (2005) Is PsbS the site of non-photochemical quenching in photosynthesis? *J Exp Bot* 56 : 375-82

Novoa R, Loomis RS (1981) Nitrogen and plant production. *Plant Soil* 58 : 177-204

Oguchi R, Hikosaka K, Hirose T (2003) Does the photosynthetic light acclimation need change in leaf anatomy? *Plant, Cell & Environment* 26: 505–512

Ort D R (2001) When there is too much light. *Plant Physiol.* 125 : 29-32

Ortiz-Lopez A, Nie GY, Ort DR, Baker NE (1990) The involvement of the photoinhibition of photosystem II and impaired membrane energization in the reduced quantum yield of carbon assimilation in chilled maize. *Planta* 181 : 78-84

Parry MAJ, Andralojc PJ, Mitchell RAC, Madgwick PJ, Keys AJ (2003) Manipulation of Rubisco: The amount, activity, function and regulation. *J Exp Bot* 54 :1321-33

Parry MAJ, Madgwick PJ, Carvalho JFC, Andralojc PJ (2007) Prospects for increasing photosynthesis by overcoming the limitations of Rubisco. *J Agric Sci* 145: 31-43

Peng S B, Tang Q, Zou Y (2009) Current status and challenges of rice production in china. *Plant Prod Sci.* 12 : 3-8

Peng S, Laza RC, Visperas RM, Sanico AL, Cassman KG, Khush GS (2000) Grain yield of rice cultivars and lines developed in the Philippines since 1966. *Crop Science* 40: 307–314

Penning de Vries FWT, Brunsting, AHM, Van Laar HH (1974) Products, requirement and efficiency of biosynthesis: A quantitative approach. *J Theor Biol.* 45 : 339-77

Peterhansel C, Maurino VG (2011) Photorespiration redesigned. *Plant Physiol* 155: 49–55

Planchon C (1979) Photosynthesis, transpiration, resistance to CO_2 transfer and water efficiency of flag leaf of bread wheat, durum wheat and Triticale. *Euphytica* 28 : 403-08

Price GD, Badger MR, Woodger FJ, Long BM (2008) Advances in understanding the cyanobacterial CO2-concentrating-mechanism (CCM) : functional components, Ci transporters, diversity, genetic regulation and prospects for engineering into plants. *J Exp Bot* 59: 1441–1461

Price GD, Badger MR, von Caemmerer S (2011) The prospect of using cyanobacterial bicarbonate transporters to improve leaf photosynthesis in C3 crop plants. *Plant Physiol* 155: 20–26

Raines CA (2003) The Calvin cycle re-visited. *Photosynth Res.* 75 :1-10

Raines CA (2006) Transgenic approaches to manipulate the environmental responses of the C3 carbon fixation cycle. *Plant Cell Environ.* 29:331–39

Raines CA (2011) Increasing photosynthetic carbon assimilation in C3 plants to improve crop yield: Current and future strategies. *Plant Physiol* 155: 36-42

Randall DD, Nelson CJ, Asay KH (1977) Ribulose-bisphosphate Carboxylase altered genetic expression in tall fescue. *Plant Physiol.* 59 : 38-41

Rathnam CKM, Chollet R (1980) Photosynthetic and photorespiratory carbon metabolism in mesophyll protoplasts and chloroplasts isolated from isogenic diploid and tetraploid cultivars of ryegrass. *Plant Physiol.* 65 : 489-94

Richter ML (2004) Gamma-epsilon interactions regulate the chloroplast ATP synthase. *Photosynth Res* 79 : 319-29

Sage RF (2002) Variation in the K (cat) of Rubisco in C3 and C4 plants and some implications for photosynthetic performance at high and low temperature. *J Exp Bot* 53: 609-20

Sage RF (2004) The evolution of C4 photosynthesis. *New Phytologist* 161: 341-70

Salvucci ME, Crafts-Brandner SJ (2004) Mechanism for deactivation of Rubisco under moderate heat stress. *Physiol Plant* 122: 513–519

Sassenrath-Cole GF, Pearcy RW (1994) Regulation of photosynthetic induction state by the magnitude and duration of low-light exposure. *Plant Physiol* 105: 1115–1123

Sassenrath-Cole GF, Pearcy RW, Steinmaus S (1994) The role of enzyme activation state in limiting carbon assimilation under variable light conditions. *Photosynth Res* 41: 295–302

Sharma P (2010) Translocation of photoassimilates, carbon partitioning and crop productivity (communicated).

Shaver DL (1983) Genetics and breeding of maize with extra leaves above the ear. *Proc Annu Corn Sorghum Res Conf* 38 : 161-180

Shearman VJ, Sylvester-Bradley R, Scott RK, Foulkes MJ (2005) Physiological processes associated with wheat yield progress in the UK. *Crop Science* 45: 175–185

Shikanai T (2007) Cyclic electron transport around photosystem I: genetic approaches. *Annu Rev Plant Biol* 58 : 199-217

Shiraiwa T, Sinclair TR (1993) Distribution of nitrogen among leaves in soybean canopies. *Crop Sci* 33 : 804-808

Sinclair TR, Horie T (1989) Leaf nitrogen, photosynthesis and crop radiation use efficiency review. *Crop Sci* 29 : 90-98

Sinclair TR, Purcell LC, Sneller CH (2004) Crop transformation and the challenge to increase yield potential. *Trends Plant Sci* 9: 70–75

Spielmeyer W, Ellis MH, Chandler PM (2002) Semidwarf (sd-1) , 'green revolution' rice, contains a defective gibberellin 20-oxidase gene. *Proceedings of the National Academy of Sciences, USA* 99: 9043–9048

Spreitzer RJ, Salvucci ME (2002) RUBISCO: Structure, regulatory interactions, and possibilities for a better enzyme. *Annu Rev Plant Biol.* 53 : 449-75

Stitt M, Quick WP, Schurr U, Schulze ED, Rodermel SR, Bogorad L (1991) Decreased ribulose-1,5-bisphosphate carboxylase- oxygenase in transgenic tobacco transformed with antisense rbcS. II. Flux control coefficients for photosynthesis in varying light, CO_2, and air humidity. *Planta* 183 : 555-66

Tabita FR (1999) Microbial ribulose 1,5-bisphosphate carboxylase/oxygenase: a different perspective. *Photosynth Res* 60 : 1-28

Tanaka A, Osaki M (1983) Growth and behaviour of photosynthesized ^{14}C in various crops in relation to productivity. *Soil Sci Plant Nutr.* 29 : 147-158

Taylor M (1980) Genetic improvements in winter wheat yields since 1900 and associated physiological changes. *Journal of Agricultural Science* 94: 675–689

Tazoe Y, Noguchi KO, Terashima I (2006) Effects of growth light and nitrogen nutrition on the organization of the photosynthetic apparatus in leaves of a C4 plant, *Amaranthus cruentus. Plant Cell Environ.* 29 : 691-700

Treharne KJ (1972) Biochemical limitation to photosynthetic rates. In Crop Proccesses in Controlled Environments. Ed. A.R. Rees, pp. 285-303. New York Academic, 391 pp

Turina P, Samoray D, Graber P (2003) H^+/ATP ratio of proton transport–coupled ATP synthesis and hydrolysis catalysed by CF0F1-liposomes. *EMBO J.* 22 : 418-26

Uemura K, Anwaruzzaman M, Miyachi S, Yokota A (1997) Ribulose-1,5-bisphosphate carboxylase/oxygenase from thermophilic red algae with a strong specificity for CO_2 fixation. *Biochem Biophys Res Commun.* 233 : 568-71

Ueno O (1998) . Induction of Kranz anatomy and C4 like biochemical characteristics in a submerged amphibious plant by abscisic acid. *Plant Cell.* 10 : 571-83

Valliyodan B, Nguyen HT (2006) Understanding regulatory networks and engineering for enhanced drought tolerance in plants. *Current Opinion in Plant Biology* 9: 189–195

Von Caemmerer S (2000) Biochemical models of leaf photosynthesis. Collingwood, Aust: CSIRO

Von Caemmerer S (2003) C-4 photosynthesis in a single C-3 cell is theoretically inefficient but may ameliorate internal CO2 diffusion limitations of C-3 leaves. *Plant Cell Environ.* 26:1191–97

Voznesenskaya EV, Franceschi, VR, Kiirats O, Artyusheva EG, Freitag H, Edwards GE (2002) Proof of C4 photosynthesis without kranz anatomy in *Bienertia cycloptera* (Chenopodiaceae) . *Plant J.* 31 : 649-62

Voznesenskaya EV, Franceschi VR, Kiirats O, Freitag H, Edwards GE (2001) Kranz anatomy is not essential for terrestrial C4 plant photosynthesis. *Nature* 414 : 543-46

Walters RG (2005) Towards an understanding of photosynthetic acclimation. *Journal of Experimental Botany* 56: 435–447

Wang DF, Portis AR, Moose SP, Long SP (2008) Cool C4 photosynthesis: pyruvate P-i dikinase expression and activity corresponds to the exceptional cold tolerance of carbon assimilation in *Miscanthus giganteus. Plant Physiol.* 148 : 557-67

Wang XY, Gowik U, Tang HB, Bowers JE, Westhoff P, Paterson AH (2009) Comparative genomic analysis of C4 photosynthetic pathway evolution in grasses. *Genome Biol* 10: R68

Watanabe Y, Nakamura Y, Ishii R (1997) Relationship between starch accumulation and activities of the related enzymes in the leaf sheath as a temporary sink organ in rice (*Oryza sativa*) . *Functional Plant Biology* 24: 563–569

Webb AR (2008) The chloroplast as a regulator of Ca^{2+} signalling. *New Phytologist* 179: 568–570

Weinl S, Held K, Schlücking K, Steinhorst L, Kuhlgert S, Hippler M, Wells R , Meredith Jr. WR , Williford JR (1986) Canopy photosynthesis and its relationship to plant productivity in near isogenic cotton lines differing in leaf morphology. *Plant Physiol.* 82 : 635-640

Weinl S, Held K, Schlücking K, Steinhorst L, Kuhlgert S, Hippler M, Kudla J (2008) A plastid protein crucial for Ca^{2+}-regulated stomatal responses. *New Phytologist* 179: 675–686

Whitney SM, Andrews TJ (2001) The gene for the ribulose-1,5-bisphosphate carboxylase/ oxygenase (Rubisco) small subunit relocated to the plastid genome of tobacco directs the synthesis of small subunits that assemble into Rubisco. *Plant Cell.* 13 :193-205

Whitney SM, Andrews TJ (2001) Plastome-encoded bacterial ribulose-1,5-bisphosphate carboxylase/oxygenase (Rubisco) supports photosynthesis and growth in tobacco. *Proc Natl Acad Sci. USA* 98 : 14738-43

Whitney SM, Sharwood RE (2007) Linked Rubisco subunits can assemble into functional oligomers without impeding catalytic performance. *J Biol Chem.* 282 : 3809-18

Whitney SM, Houtz RL, Alonso H (2011) Advancing our understanding and capacity to engineer nature's CO2-sequestering enzyme, Rubisco. *Plant Physiol* 155: 27–35

Whitney SM, Kane HJ, Houtz RL, Sharwood RE (2009) Rubisco oligomers composed of linked small and large subunits assemble in tobacco plastids and have higher affinities for CO_2 and O_2. *Plant Physiol* 149: 1887–1895

Whitney SM, Sharwood RE (2007) Linked Rubisco subunits can assemble into functional oligomers without impeding catalytic performance. *J Biol Chem* 282: 3809–3818

Wiebe K (2008) The State of Food and Agriculture 2008. Biofuels: Prospects, Risks and Opportunities. Food and Agriculture Organization of the United Nations, Rome, Italy

Wilhelm WW, Nelson CJ (1978) Irradiance response of tall fescue genotypes with contrasting levels of photosynthesis and yield. *Crop Sci.* 18 : 405-08

Wilson D, Jones JG (1982) Effect of selection for dark respiration rate of mature leaves on crop yields of *Lolium perenne* cv. S23. *Ann Bot.* 49 : 313-20

Wilson KE, Ivanov AG, Oquist G, Grodzinski B, Sarhan F, Huner NPA (2006) Energy balance, organellar redox status, and acclimation to environmental stress. *Canadian Journal of Botany–Revue Canadienne De Botanique* 84: 1355–1370

Wong SC, Cowan IR , Farquhar GD (1985) Leaf conductance in relation to rate of CO_2 assimilation.I. Influence of nitrogen nutrition, phosphorus nutrition, photon flux density and ambient partial pressure of CO_2 during Ontogeny. *Plant Physiol.* 78 : 821-825

Wullschleger SD, Oosterhuis DM (1992) Canopy leaf area development and age class dynamics in cotton. *Crop Sci.* 32 : 451-456

Yeo ME, Yeo AR, Flowers TJ (1994) Photosynthesis and photorespiration in the genus *Oryza. Journal of Experimental Botany* 45: 553–560

Yoshida S (1981) Fundamentals of Rice Crop Science. Los Bairas, Philippines. *Int Rice Res. Inst.*

Zelanski MI, Mogilena G, Shitova I, Fattakhova F (1978) Hill reaction of chloroplasts from some species varieties and cultivars of wheat. *Photosynthetica* 12 : 428-35

Zelitch J (1975) Improving the efficiency of photosynthesis. *Sci* 188 : 626-633

Zhang WH, Kokubun M (2004) Historical changes in grain yield and photosynthetic rate of rice cultivars released in the 20th century in Tohoku region. *Plant Production Science* 7: 36–44

Zhu XG, de Sturler E, Long SP (2007) Optimizing the distribution of resources between enzymes of carbon metabolism can dramatically increase photosynthetic rate: a numerical simulation using an evolutionary algorithm. *Plant Physiol.* 145 : 513-26

Zhu XG, Ort DR, Whitmarsh J, Long SP (2004) The slow reversibility of photosystem II thermal energy dissipation on transfer from high to low light may cause large losses in carbon gain by crop canopies: a theoretical analysis. *J Exp Bot* 55 : 1167

Zhu X, Portis AR Jr, Long SP (2004) Would transformation of C3 crop plants with foreign Rubisco increase productivity? A computational analysis extrapolating from kinetic properties to canopy photosynthesis. *Plant Cell Environ.* 27 : 155-65

Zhu XG, Long SP, Ort DR (2008) What is the maximum efficiency with which photosynthesis can convert solar energy into biomass? *Curr Opin Biotechnol.* 19 :153-59

Zhu XG, de Sturler E, Long SP (2007) Optimizing the distribution of resources between enzymes of carbon metabolism can dramatically increase photosynthetic rate: a numerical simulation using an evolutionary algorithm. *Plant Physiol* 145: 513–526

Zhu XG, Long SP, Ort DR (2010a) Improving photosynthetic efficiency for greater yield. *Annu Rev Plant Biol* 61: 235–261

Zhu XG, Shan LL, Wang Y, Quick WP (2010b) C4 rice—an ideal arena for systems biology research. *J Integr Plant Biol* 52: 762–770

7

Utilization of Nanotechnology in Crop Improvement

Himakshi Bhati-Kushwaha and *C P Malik*

With the help of nanotechnology various nano-devices have been devised having positive impact on agricultural practices. This technology holds the promise of controlled release of agrochemicals and site targeted delivery of various macromolecules which are needed for improved disease resistance plant, efficient nutrient utilization and enhanced plant growth. They are considered to be the safer system for the detection and treatment of plant diseases and were sure to help reduce the pathogen risk in the near future. Nanotechnology is basically a hybrid of chemistry and engineering that has opened up a whole new world of possibilities in the field of sciences. Nano-encapsulation is considered to be the more efficient and safer means of handling pesticides, with less exposure to the environment that guarantees ecoprotection. Nanoparticle mediated plant transformation has the potential to raise genetic modification in plants for further improvement. Specifically, application of nanoparticle technology helps to resolve the ill effects of plant–pathogen interactions by specifically targetting agricultural problems and provide new ways for crop protection. Unarguably, this will raise the crop productivity and testify the improvement in quality of produce. If it is possible to have a distribution of properly functionalized nanoparticles throughout the plant vascular system and guide them to targeted sites; then these nanoparticles can be successfully used to unload chemicals (fungicides, insecticides, etc.), or other substances (plant hormones, elicitors, nucleic acids) into localized areas of plant tissues which ultimately leads to increase in crop yield with desired traits. This paper reviews the potential applications of synthetic biology which involves designing and constructing new biological entities and nanotechnology, a recent science that deals with the design and manufacture of small and complex molecules, electronic circuits and mechanical devices in crop improvement efforts, and food security.

Introduction

Crop improvement is an essential continual process meant to increase food security and crop production by improving yields and product quality. With the growing concerns of adverse environmental effects, technologies that will enhance the development and utilisation of new crop varieties to perform more efficiently under adverse soil, water and climate conditions are sought. Synthetic biology in this concern has wide applicability that involves designing and construction of new biological entities and nanotechnology. Nanotechnology is a recent science that deals with the design and manufacture of small and complex molecules, electronic circuits and mechanical devices in crop improvement efforts, and food security (Lokko *et al.*, 2011). Nanotechnology can be defined as any technology which exploites phenomena and structure that can only occur at a nanoscale. Nanotechnology was first introduced in 1959, by the Noble Prize- winning physicist Richard Feyman, in "There's Plenty of Room at the Bottom". The term nanotechnology was introduced by Eric Dexler in 1986. The ability to understand and manipulate both matter and life at a nano-scale level is closely related to the ability to understand and manipulate both matter and life at most basic level (Alivisator, 1996). All the living organism from microorganism to human is powered by highly endowed molecular and cellular mechanism at operates at a nano scale. Nanotechnology is the manipulation or self-assembly of individual atoms, molecules, or molecular clusters into structures to create materials and devices with new or vastly different properties. Nanotechnology can work from the top down (which means reducing the size of the smallest structures to the nano-scale e.g. photonics applications in nano-electronics and nano-engineering) or the bottom up (which involves manipulating individual atoms and molecules into nanostructures and more closely resembles chemistry or biology).The potential of nanotechnology to revolutionise the health care, textile materials, information and communication technology and energy sectors has been well publicised. In fact several products enabled by nanotechnology are already in the market, such as antibacterial dressings, transparent sunscreen lotions, stain-resistant fabrics, scratch free paints for cars, and self cleaning windows. Changes that had occurred in agricultural technology have been a major factor in shaping modern agriculture. Today the latest innovations, nanotechnology has occupied a prominent position in transforming agriculture and food production. The development of nano-devices and nano-materials could open up novel applications in plant biotechnology and agriculture (Scrinis and Lyons, 2007). Currently, the main thrust of research in nanotechnology focuses on applications in the field of electronics (Feiner, 2006), energy (Hu and Chen, 2007), medicine and life sciences (Caruthers *et al.*, 2007; Patolsky *et al.*, 2006). The successful application of various nanoplatforms in medicine under in vitro conditions leads to the development of genetically modified crops, plant protecting chemicals and precision farming techniques. Some of the nano-devices are CNT's, nano-biosensors, nanoparticles etc. These nano-devices have contributed in crop improvement, disease detection, disease treatment, pathogen detection, food and packaging etc. All these nano-devices have helped to increase the crop yield with high quality and also for food safety and biosecurity (Kushwaha and Malik, 2012; 2013). Nanotechnology technology also helps in the controlled release of agrochemicals and site targeted delivery of various macromolecules which are needed for improved disease resistance plant, efficient nutrient utilization and also leads to enhanced plant growth. Processes such as nano-encapsulation show the benefit of more efficient use and safer handling of pesticides with less exposure to the environment that guarantees ecoprotection.

The uptake efficiency and effects of various nanoparticles on the growth and metabolic functions vary differently among plants. Nanoparticle mediated plant transformation has the potential for genetic modification of plants for further improvement. Specifically, application of nanoparticle technology in plant pathology targets specific agricultural problems in plant-pathogen interactions and provide new ways for crop protection. In this review we tried to explain the utility of delivering nanoparticulate materials to plants and their ultimate effects which could provide some insights for the safe use of this novel technology for the improvement of crops. Nanotechnology ultimately brings a new revolution in the field of agriculture and food industry. Nano-devices with novel properties make the agriculture system more strong and smart. These smart systems have the capability to deliver specific chemicals in a controlled and targeted manner which was very much similar to the nano-drug delivery in humans (Roco, 2003; Lu *et al.*, 2008). "Smart Delivery Systems" in agriculture should possess extensive properties of combinations of time controlled, specifically targeted, remotely regulated/ pre-programmed/self-regulated and multifunctional characteristics to avoid biological barriers for successful targeting. Nano-technology brings a new revolution in the field of agriculture and food industry. With the help of nanotechnology new tools for the molecular treatment of diseases, rapid disease detection and speedy uptake of nutrients by plants have been devised. Nanotechnology shows greater applicability for exploring life process, analyzing and determining products quality, monitoring plant and animal health, developing novel techniques for pollution detection and reduction (air, soil and water) (Karnakarn and Jayes, 2005). Nanobioprocessing emerges to be a boon to agriculture and food processing industry. This review presents some of the recent developments in plant science and agriculture which states the application of nanoparticles for more effective and safe use of chemicals for plants, plants as source of "green synthesis" for nanoparticle formulation, the effects of nanoparticles on the growth and metabolic functions of different plants and nanoparticle mediated plant genetic transformation.

History of Nanotechnology

Nanotechnology which we thought to be a modern science has its history dating to as back as 9th century. Paper "Experimental relations of Gold (and other metals) to light" gave the first scientific description about the properties of nanoparticles. After this a describing talk was given by Richad Feymon in his lecture "There's plenty of space at bottom" in 1959. After this nanotechnology attain a new height and became the foremost area of research. In late 1960's Peter Paul and his research group investigates first polyacrylamide beads for the oral administration. This is followed by the synthesis of first nanoparticles for drug delivery and vaccines. This states that nanorevolution starts from 1950's -1960's. In Japan an experiment was performed for binding 5-flourouracil to the albumin nanoparticles in 1977 and temperature dependent denaturation is also studied when it is injected intravenously in a mice tail vein (Sugibayashi, 1977). This also reports the increase in the life span, when nanoparticles were injected intraperitonial into Charlich Ascites Carcinoma bearing mice (Kreutter, 2007). The first paper on nanoparticles was published in 1981 by K. Eric Dexler entitled "An Approach to the development of general capabilities of molecular manipulations" which means that the amount of gap available for us for information storage (or other uses) is abundant. He also proposed the term "Nanotechnology" for the first time in 1981. He suggested the possibilities of the nano sized objects that would be self replicating robots, or nano-machines. This nano sized object would roam around the human body killing the cancer cells. The major step in the direction of nanotechnology was the invention of the Scanning Tunnelling Microscopy during 1081-1982 and Atomic Force Microscopy in 1986 by Gerd Benning and Heinrich Rocher. These tools use nano scale probe to image a surface at the atomic resolution. The two strategies which are used to synthesize nanoparticles were given by Don Eigles and Echead Schewieger in 1990's. These two approaches are "Top- down" approach and "Bottom –up" approach. Top-down approach involves starting with the larger block of material and then creating the smaller size and desired shape material from them.In Bottom –up approach assembly of various smaller size sub units, unites to make a larger structure material. Both the methods faced certain challenges. In Top- down approach the challenge is to create a smaller structure with sufficient accuracy whereas in Bottom- up approach is the creating of the structure larger enough is very difficult. In 1996 Prof Richarg Smalley discovered a new and improved form of carbon and nano sized product which he named as "super- carbon". For the first time potentials of nanotechnology in medical field was realized by Prof. Mauro Ferrari. He devised therapeutic micro- tubes devices which were similar to the manufacturing of the silicon microchips. In 1985 researchers reported the discovery of the "bulk ball" a lovely round molecule consisting of 60 carbon atoms.This in turn leads to discovery of 'carbon nano-tubes' in 1991. These were discovered by the Japenese Electrom Microscopist Sumio Iijima. In 1992 Dexler proposed "molecular nanotechnology" or "molecular manufacturing" to distinguish his manufacturing ideas from simpler products to the focused research. With the greater advances in nanotechnology, its contribution in chemistry, biology, engineering and applied physics was remarkable (Fig. 1). Nanotechnology thus is a projected technology based on general ability to built objects of complex atomic specification but with better functionality. Nanotechnology will be based on programmable machine tools called as "assemblers" which will enable the construction of a wide range of molecular structures.

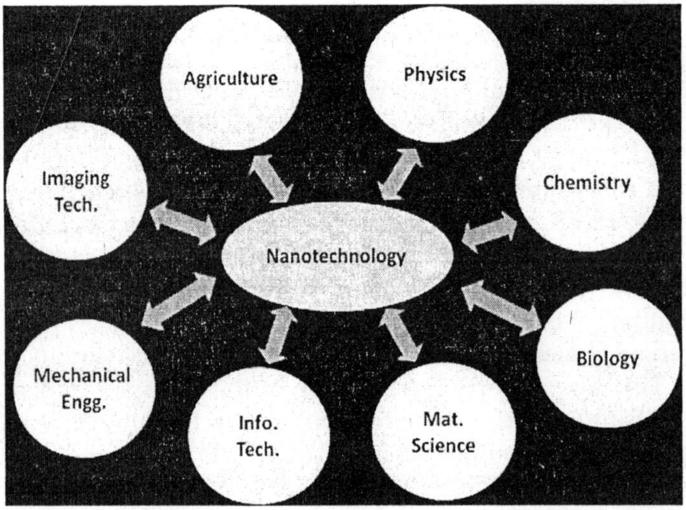

Fig 1 Nanotechnology as multidisciplinary field

Nanotechnology: Current status

Nanotechnology being multi-directional and fast-paced is still at its early stages of development. Many universities have taken initiative for its exploration by forming nanotechnology centres and the number of papers and patent applications in the concern area. Many tools have been developed that might help in its viable application and could end-up on the 'technology shelf' in the future but offcorse there are definite benefits. Nanobiotechnology is an interdisciplinary filed that brings together life scientists and engineers. This aspect of interconnection can fuels further growth of ideas and innovations.

What makes Nanotechnology a unique field?

1) Small size (High surface to volume ratio), therefore requires self assemblers.
2) Significantly higher hardness, breaking strength and toughness at low temperatures and super plasticity at high temperatures, additional electronic states, high chemical selectivity of surface active sites which significantly increased surface energy.
3) New entry ways (high mobility in human body, plants and environment).

Applications of Nanotechnology in Agriculture

1) Crop improvement
2) Nanobiotechnology
3) Analysis of gene expression and Regulation
4) Soil management
5) Plant disease diagnostics
6) Efficient pesticides and fertilizers
7) Water management
8) Bioprocessing
9) Post Harvest Technology
10) Monitoring the identity and quality of agricultural produce
11) Precision agriculture

Scope of Nanotechnology in Agriculture

Nanotechnology is considered as one of the possible solutions to problems in food and agriculture (see Kushwaha and Malik, 2012; 2013). Nanoparticles can serve as 'magic bullets', containing herbicides, chemicals, or genes, which target particular plant parts to release their content. Nano-capsules can enable effective penetration of herbicides through cuticles and tissues, allowing slow and constant release of the active substances (Perea-de-Lugue and Rubiales, 2009). In the last four years USA has invested 3.7 billion USD through NNI (National Nanotechnology Initiative). It is followed by Japan and Europe that have invested around 750 and 1.2 billion per year. According to reports available China has contributed nearly 7.5% till 1995 which has now gone up to 18.3% in 2004. China come to occupy second position in the world (Ranking the Nation).Recent years have seen great interest in food technology especially in India, South Korea, Iran and Thailand. It is estimated that the nano-food market has surged from 2.6 billion USD to 20.4 billion USD in the past two years. Asia constitutes the largest market for the nano-food led by China (Kaiser, 2004). Number of companies working in the area of nanotechnology is 400 in number which will be expected to increase up to 1000 in the next 10 years. According to the research and industrial analysis, made by Business Communication Company, the market for nanotechnology was 7.6 billion USD in 2004 and has arisen up to 1 trillion USD in 2006.

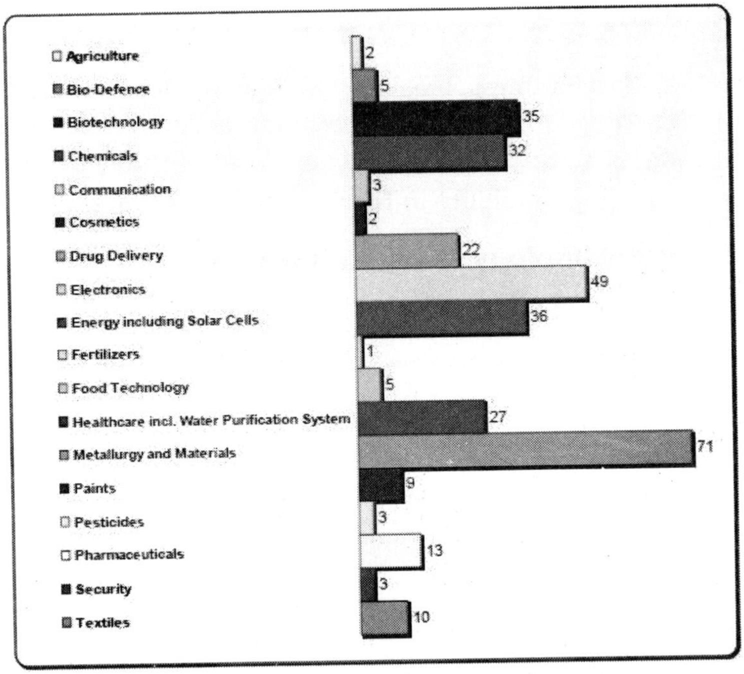

Fig. 2 Present areas of activities in the field of Nanotechnology in India

Nanotechnology: An Effective Delivery System for Crop Improvement

Nanoformulations

Nanotechnology has proved to be very important in the field of agriculture. Today agrochemicals has been a major concern in the field of agriculture. It is so because when the chemicals are sprayed on the plants in a specified quantity, the appropriate quantity of the sprayed chemicals does not reaches to the plant due to the leaching of chemicals, photolytic degradation, hydrolysis and microbial degradation. To overcome these problems repeated spraying of chemicals are required which ultimately leads to the soil as well as environment pollution. Nanoencapsulation is one of the most important processes in which certain chemicals, which can be effectively used against pathogens are slowly and efficiently released in a particular host plant for insect control. Through this process pesticides are effectively absorbed through diffusion, dissolution, biodegradation and osmotic pressure (Scrinis and Lyons 2007; Vidhyalakshmi *et al.*, 2009; Ding and Shah, 2009). DNA and other desired chemicals are delivered efficiently to the plant tissue through this method which protects the plant against pests (Torney, 2009). Through this process potentialized drugs/PGRs that significantly increase plant growth, chlorophyll, protein and water contents are efficiently delivered to the leaves (Sukul *et al.*, 2009).These potential drugs interfere with the integral membrane proteins of the leaves and modulate the cell physiology towards rapid growth (Kushwaha and Malik, 2012; 2013). On the other hand nano-encapsulated agrochemicals should be designed in such a way that they possess all necessary properties such as effective concentration (with high solubility, stability and effectiveness), time controlled release in response to certain stimuli, enhanced targeted activity and less exotoxicity with safe and easy mode of delivery and thus are helpful in avoiding repeated application (Green and Beestman, 2007; Wang *et al.*, 2007; Boehm *et al.*, 2003, Tsuji, 2001). Properly functionalized nanocapsules also provide better penetration through cuticle and allow slow and controlled release of active ingredients on reaching the target weed. Hydrophobic nano-silica has been successfully used to against a range of agricultural insect pests (Barik *et al.*, 2008 and Rahman *et al.*, 2009). Properly functionalised lipophilic nanosilica gets absorbed into the cuticular lipids of insects by the process of physiosorption and leads to the damage of protective wax layer [made of various fatty acids and lipids that acts as an effective barrier for the evaporation of water] and finally induces death by desiccation (Athanassiou *et al.*, 2007; Mewis and Ulrichs, 2001). The use of porous hollow silica nanoparticles (PHSN), which are characterized by a thick shell of nearly 15 nm and a pore diameter of 4–5 nm, providing shielding protection to pesticides from degradation by UV light was reported (Li *et al.*, 2007). These carriers improve the photostability of the pesticide, avermectin, which was loaded into the inner core and avermectin showed a typical sustained-release pattern from the carrier. Such carriers have a promising future in the sustained-release of various photosensitive components. The effects of slow and controlled-release of fertilizers which were cemented and coated by nanomaterials; clay-polyester, humus-polyester and plastic starch, on crops were studied with wheat (Liu *et al.*, 2006; Zhang *et al.*, 2006). It was found that these nano-composites prove to be safer for wheat seed germination (over 99% germination), emergence and growth of seedlings. However it was also observed that a very high concentration of nanosilica-silver induces some chemical injuries on the tested plants (cucumber leaves and pansy flowers).

Silver nanoparticles (Ag NPs) exhibiting an important property of controlling phytopathogens are successfully adopted in diverse medical streams. These silver nanoparticles act as antifungal and antibacterial agents (Panacek *et al.*, 2009; Singh *et al.*, 2008). Ag NPs possessing broad spectrum of antimicrobial activity reduce various plant diseases caused by spore producing fungal pathogens (Jo and Kim, 2009). The effectiveness of Ag NPs can be improved by applying them before the penetration and colonization of fungal spores within the plant tissues. The small size of the active ingredient (diameter of 1–5 nm) can effectively control the fungal diseases like powdery mildew (Park *et al.*, 2006). Ag NPs can be used as an alternative to pesticides for the control of sclerotium forming phytopathogenic fungi was already reported (Min *et al.*, 2009). When Ag NPs were exposed to fungal hyphae it causes severe damages, by the separating the layers of hyphal wall and leading to the collapsing of hyphae. Detrimental effects of Ag NPs on unidentified fungal species of the genus *Raffaelea* causing mortality of oak trees was also investigated and studies showed harmful effects of Ag NPs on conidial germination (Kim *et al.*, 2009). Reports were there that also proves the efficacy of Ag NPs in extending the vase life of gerbera flowers and also showed inhibited microbial growth and reduction in vascular blockage which increased the water uptake and maintained the turgidity of gerbera flowers (Solgi *et al.*, 2009; Liu *et al.*, 2009). With all these benefits, improved nanoparticle delivery systems need to be developed for specifically targeting the infected tissues alone. Studies have also reported the use of biocide-containing polymeric nanoparticles for introducing organic wood preservatives and fungicides into wood products thereby reducing wood decay (Liu *et al.*, 2001; 2002a; 2002b; 2002c and 2003). Nano Alumino Silicates are the most active pesticides used as they are more actively picked up by the hairs of the insects and insect consumes it easily. On the other hand they are most eco-friendly. Mesoporous silica nanoparticles are actively used for the delivery of DNA and other chemicals into the plant tissue and thus cause targeted death of the pathogens (Wang *et al.*, 2002). Though nanoparticles have positive effect on plant growth but the nanoparticle of inorganic nature as well as those of larger size (more than 100 nm), show negative effect on plant growth. The nanoparticles made up of cupric oxide as well as larger size nanoparticles are responsible for causing lesions in DNA of the plant cell (Donald *et al.*, 2012).

Role of Nanoparticles in Plant System

Plants have proved to be useful source where nanoparticles can be successfully applied. Nanoparticles have shown their valuable contribution in the field of biological sciences since the 1990's (Kausch and Bruce, 1994; Salata, 2004) and now nanomaterials has been a most advanced level of nanotechnology which are being employed in plant research. Some of the areas of interest are fluorescent labels (Ravindran *et al.*, 2005), gene delivery (Roy *et al.*, 2005), bio detection of pathogens (Li *et al.*, 2005), detection of proteins (Nam *et al.*, 2003), probing of DNA (Cao and Mirkin, 2002; Storhoff *et al.*, 2004), bio-imaging (Lewis, 2006) and separation and purification of biological molecules and cells (Werner *et al.*, 2006). Various studies involving the application and effects of different nanoparticles on plant growth and metabolic functions have been reported. The most potent applications of nanotechnology in agricultural and food systems involves: pathogen and contaminant detection, identity preservation and tracking, smart treatment delivery, nanodevices are formulated for molecular and cellular biology and agricultural waste treatment and processing (Kuzma, 2006). Nanoparticles are also effectively used for bioremediation of the contaminated environment, biocides and antifungal on textiles (Barik *et al.*, 2008). Colloidal nano silver solution which is prepared by chemical reaction of silver ions by physical method using reducing agents and stabilizers shows their potential against antifungal diseases like rose powdery mildew. This disease is caused by the fungus *Sphaerotheca pannosa*. Silver nanoparticles of average size 1.5 nm are found to be effective against this disease. These colloidal solutions of Ag nanoparticles also show affectivity against fungi and can be, therefore, used as fungicides. Silver is also found to be excellent plant growth stimulator, powerful, non-toxic and healing agent. Nano Silica- Silver Composite consist of silicon which is essential for plant to increase disease resistively as well as stress resistance (Brecht *et al.*, 2003). Silicate aqueous solution shows great affinity towards powdery mildew or downy mildew in plants. Silver on the other hand exhibit antimicrobial activity in ionic state (Thomas and Mc Cubin, 2003). Here silver is combined with silica at a nano scale. This composition exhibits antifungal effect on powdery mildew in pumpkin at 0.3 ppm concentration. This composite has shown results with greater efficiency in fungus *Phythium ultimum, Magnoporthe grisea, Colletotrichum gloesporioides* and *Rhyzoctonia solani* at 10 ppm concentration (Park *et al.*, 2006).This also acts against *Bacillus subtilis, Azotobacter chrococcum, Rhizobium tropici, Pseudomonas syringae* and *Xanthomonas comperthis* pv. at 100 ppm concentration. Nano Alumino Silicates are the most active pesticides used as they are more actively picked up by the hairs of the insects and insect consumes it easily. On the other hand they are most eco-friendly. Mesoporous silica nanoparticles are actively used for the delivery of DNA and other chemicals into the plant tissue and thus cause targeted death of the pathogens (Wang *et al.*, 2002). Nanoparticles also prove to be a promising source that can easily enter the cell wall of the plant. Plant cell wall acts as a barrier by inhibiting the entry of external agent including nanoparticles into it. This sieving property of the plant cell wall is determined by their pore diameter which ranges from 5 to 20 nm (Fleischer, 1999). Hence, only nanoparticles or nanoparticle aggregates having their diameter less than the pore diameter of the cell wall could easily pass through and reach the plasmamembrane (Navarro *et al.*, 2008; Moore, 2006). The problem of interference by the plant cell wall pore can be easily solved by using the engineered nanoparticles, which increases the rapid uptake of of nanoparticles by increasing the pore size of the cell wall. Nanoparticles can be readily up taken by the leaves by means of stomatal opening or through trichomes and from there it can be easily translocated to various tissues (Uzu *et al.*, 2010; Fernandez and Eichert, 2009; Eichert *et al.*, 2008). But the problem faced with this mechanism is the accumulation of nanoparticles on photosynthetic surface which leads to foliar heating causing alterations in gas exchange due to stomatal obstruction that produce changes in various physiological and cellular functions of plants (Da Silva *et al.*, 2006). Experiments on effect of Gold-nanoparticles were determined on the growth profile and yield of *Brassica juncea*, under field conditions. During the experiment five different concentrations (0, 10, 25, 50 and 100 ppm) of gold-nanoparticles were applied through foliar spray. Atomic absorption spectroscopy was used to detect the presence of Gold-nanoparticles in the leaf tissues. During the study it was found that various growth and yield related parameters, including plant height, stem diameter, number of branches, number of pods, seed yield etc. were positively affected by the nanoparticle treatment. The treatment increases the number of leaves per plant; by least affecting the average leaf area. Optimal increase in seed yield was recorded at 10 ppm of Gold nanoparticle treatment. Reducing sugars as well as total sugar contents were found to be increased by 25 ppm of Gold nanoparticle treatment (Sandeep *et al.*, 2012).

Plants as Potent Source for Nanoparticle Synthesis

It was evident in many reports and studies where plant also proves to be potential source of synthesis of nanoparticles. This has evolved as a process of "green synthesis" where biological material has replaced the physical and chemical methods of nanoparticles synthesis. The process is more often adopted, being eco-friendly, cost effective, rapid means of synthesis and more controlled methods of synthesis. Chemical methods are also used for the production of nanoparticles but some of them were toxic as these methods sometimes cannot avoid the use of toxic chemicals (Jain *et al.*, 2007). On the other hand biosynthesis of nanoparticles synthesized using the physical and chemical processes were costly. To overcome these problems safe methods have been utilized which involve the synthesis of silver nanoparticles from biological organism (Vdayasoorian *et al.*, 2009). Biological methods of nanoparticles synthesis using micro organism, enzyme (Willner *et al.*, 2006) and plant or plant extract have been suggested as possible eco-friendly alternatives to chemical and physical methods. The formation and growth of gold nanoparticles [Au NPs] inside live alfalfa plants and *Sesbania* seedlings has already been reported (Garnea-Torresdey *et al.*, 2002; Sharma *et al.*, 2007). Silver nanoparticles were synthesized using the *Capsicum annum* L. extract (Li *et al.*, 2007). *C. annum* L. extract is known to contain a number of biomolecules such as proteins, enzymes, polysaccharides, amino acids and vitamins. These biomolecules could be used as bio-reductants to react with metal ions and they could also be used as scaffolds to direct the formation of nanoparticles in solution. The bioreduction behaviour of various plant leaf extract has been demonstrated. These are *Helianthus annus, Baselia alba, Oryza sativa, Saccharum officinarum, Sorghum bicolour* and *Zea mays* (Arangasamy and Munusamy, 2008). Recently, works in India have reported the green synthesis of silver nanoparticles using the leaves of the noxious weed, *Parthenium hysterophorus*. Particles in the size range of 30- 80 nm were obtained after 10 min of reaction. The use of this noxious weed has an added advantage in that it can be used by nanotechnology processing industries (Parashar *et al.*, 2009). *Mentha piperita* leaf extract have used for the synthesis of silver nanoparticles, in the size range of 10-25 nm were obtained within 15 min of the reaction (Parashar *et al*, 2009). The latex of *Jatropha curcas*, has been used for the synthesis of silver nanoparticles. The latex of *Jatropha curcas* is shown to have curcain, curacycline A and curacycline B. (Bar *et al.*, 2009).The silver nanoparticles obtained using this source had two broad distributions- one having particles in the range of 20- 40 nm and the other having larger and uneven particles. Molecular modeling studies of the peptides in the latex revealed that the silver ions were first entrapped in the core structure of the cyclic structure of the protein and were then reduced and stabilized *in situ* by the amide group of the peptide. This resulted in particles with radius similar to the peptides. It was also found that the larger particles with uneven shapes were stabilized by the enzyme curcain (Bar *et al.*, 2009). Synthesis of plant mediated silver nanoparticles using papaya fruit extract and it's evaluation for the antimicrobial activity is also reported, where nanoparticle of size 15 nm was synthesized (Jain *et al*, 2009). Biosynthesis of silver nanoparticles was reported from the leaf extract of *Coriandrum sativum* where nanoparticles of average size 26 nm was synthesized (Satyavathi *et al.*, 2010). Silver nanoparticles were successfully reported from the leaf extract of *Argemone mexicana*. These nanoparticles were cubic and hexagonal in shape with an average size of 10-50 nm (Jain *et al.*, 2010). Biological synthesis of silver nanoparticles were also reported in onion –*Allium cepa*, which exhibit antimicrobial activities against *E. coli.* and *Salmonella tyhpimurium*. The silver nanoparticle synthesized were of 33.6 nm in size (Saxena *et al.*, 2010).Use of biological entities for the synthesis of various nanoparticles *Azadirachta indica* leaf extract has also been used for the synthesis of silver, gold and bimetallic (silver and gold) nanoparticles. Studies indicated that the reducing phytochemicals in the neem leaf consisted mainly of terpenoids. It was found that these reducing components also served as capping and stabilizing agents in addition to reduction as revealed from FTIR studies.

The major advantage of using the neem leaves is that it is a commonly available medicinal plant and the antibacterial activity of the biosynthesized silver nanoparticle might have been enhanced as it was capped with the neem leaf extract.The major chemical constituents in the extract were identified as nimbin and quercetin (Shankar *et al.*, 2004, Tripathy *et al.*, 2008, 2009; Thirumurugan *et al.*, 2010). Leaf extract of *Nicotiana tobaccum* was also reported to be potent source of silver nanoparticle synthesis where nanoparticles of 8 nm in size and showed antimicrobial behavior against *Pseudomonas aeruginosa* and *Escherichia coli* (Prasad *et al,* 2011). Leaf broth *Ocimum sanctum* was reported to be a potent source for the green synthesis of silver nanoparticles where nanoparticles of 3-20 nm in size was synthesized (Mallikarjuna *et al.,* 2011). Phytofabrication of silver nanoparticles using pomegranate fruit seed was also reported where ranging from 10-50 nm and spherical shape was synthesized and also exhibit antimicrobial behaviour against human pathogen *E. coli.* and fungus *Aspergillus flavus* (Chauhan and Upadhyay, 2011). Plant mediated synthesis of silver nanoparticles were also reported using leaf extract of *Citrullus colocynthesis* where nanoparticles of 30nm size was obtained (Satyavani *et al.,* 2011). *Euphorbia hirta* and *Neriun indicum* were also reported as a potent source for the green synthesis of silver nanoparticles giving silver nanoparticles of the particle size 29 nm (Priya et *al.,* 2011). Silver nanoparticles of an average particle size of 48-67 nm were also reported in *Catharanthus roseus.* These synthesized nanoparticles shows positive test against *Staphylococcus aurens, Escherichia coli, Klebsiella pneumonia, Bacillus cereus and Pseudonomas aeruginosa* (Mukunthan, 2011). Silver nanoparticles were also succesfully reported from the dried powder extract of various plants. The stem bark dried powder of *Boswellia ovalofoliolata* was reported to be potent source of silver nanoparticle synthesis resulting in the particle size of 30-40 nm of size.They were also tested for their antimicrobial assay using *E.coli, Pseudomonas aeruginosa, Bacillus subtilis, Proteus* and *Klebsiella* (Savithramma *et al.,* 2011). Green syhthesis of silver nanoparticle were also evident from the callus of *Citrullus colocynthis* resulting in the silver nanoparticles of the particle size 75 nm and exhibit amtimicrobila property against *E.coli, V. paraheamolyticus, P. aeruginosa, Proteus vulgaris, L.mirabilis, Salmonella enteritidis* and *Staphyl!ococcus aureus* (Satyavani *et al.*, 2011). Ecofriendly silver nanoparticles were also reported from the latex of plants which is also used as an important taxanomic tool for the phylogenetic inter-relationship (Mondal *et al.,* 2011).

Table 1: Instrumental methods used for nanoparticle characterization

S. No.	Name of Instruments	Characterization parameters
1	Particle Size Analyzer: Dynamic Light Scattering	Measuring the size distribution
2	Transmission Electron Microscopy	To measure the size and shape of entire particle
3	Scanning Electron Microscopy	To measure the space image of the particle
4	Atomic Force Microscopy	To measure the size, morphology, surface texture and roughness of the particles. It is capable of 3D visualization.
5	Fourier Transform Infrared Spectroscopy	To measure the chemical functional groups of the particle
6	X-ray Diffraction	To determine the arrangement of atomswithin a particle, their chemical bonds and their disorder

Continued...

S. No.	Name of Instruments	Characterization parameters
7	Lithography	To transfer previously determined patterns on a substrate
8	Inductively coupled Plasma Mass Spectroscopy	Capable of sub ppt (parts per trillion) detection limit, which help to measure the nanoparticle movement
9	Energy Dispersive X-ray Spectroscopy	To Identify the purity of the particle

Source: Tarafdar, 2012

Types and Effects of Nanoparticles on Plant Growth

Various types of nanoparticles can be used and readily up-taken by the plants. But the increased applications of these engineered nano-materials create great concern regarding their toxicity to humans and animals (Panessa *et al.*, 2006; Pacurari *et al.*, 2008). In this review the nano-materials used in crop improvement were classified as follows:

Carbon Nano-materials

Carbon at a nano-scale is used for various purposes such as to improve mechanical properties of tyres, photographic film development and nanometer particle development which are used as a catalyst. Nanotubes are the tube like structures which are formed due to the spherical arrangement of carbon atoms. These tubes are made up of single layer of graphite. Nanotubes are basically made up of carbon in a honeycomb manner and rolled up to form a cylindrical structure which has the diameter of one nanometer. They are successfully used for the early germination of seeds and enhancing the growth of seedlings in tomato. The cylindrical structure of the nanotubes helps in the rapid uptake of water (Khotakosky *et al.*, 2009). Clearly, the CNT's can be effectively used to deliver desired molecules that even control pests at the time of germination. On the contrary they also promote growth of a plant and therefore, do not have any toxic or inhibitory effect on the plant. In recent years CNT's have gained much importance in nanotechnology. Diverse types of nanotubes are available and there use amounts to $26 trillionth worth manufacture goods by the year 2014. CNT's are effectively used for the rapid delivery of therapeutically active molecules as they have the capability to effectively cross the cell barriers. Functional CNT's are found to be more effective as they can be easily taken up by the cells and can traffic intracellularly through different cellular barriers (Kastarelos *et al.*, 2007).Virus neutralizing antibodies can be easily conjugated with CNT's and therefore can be used to treat viral diseases (Pantarotto *et al.*, 2003). They are used in crop improvement on a broader scale but are of great interest due to some toxic effects. These carbon nano-materials were again classified as single walled carbon nanotubes, multi walled carbon nanotubes and lignin walled carbon nanotubes.

Single walled carbon nanotubes

The effects of functionalized SWCNTs (fCNTs; functionalized with poly-3-aminobenzenesulfonic for high dispersibility) and non-functionalized SWCNTs (CNTs) on root elongation of 6 different crop species [cabbage (*Brassica oleracea*), carrot (*Daucus carota*), cucumber (*Cucumis sativus*), lettuce (*Lactuca sativa*), onion (*Allium cepa*) and tomato (*Solanum lycopersicum*)] were studied to understand their toxicity in crops. Single walled CNT's are responsible for root elongation. Experiments have been performed to detect the effect of functional and non-functional nanotubes on root elongation in as many as six crop species, which includes cabbage, carrot, lettuce, cucumber, onion and tomato. Nanotubes are functionalized using poly 3-aminobenzene sulphonic acid. With the help of SEM root elongation was measured at 0, 24 and 48 hours of exposure. This helps to measure the uptake of CNT as well as the effect of interaction between CNT and roots. The data revealed that CNT's do not show any effect on carrot and cabbage in either of the form but functionalized CNT's inhibited root elongation in lettuce while enhanced root elongation in onion and cucumber (Canas *et al.*, 2008). The ability of carbon nanomaterials to easily penetrate the cell wall and cell membrane of intact plant cells were also recently reported using *Nicotiana tabacum* L. BY-2 cells as models for the uptake of oxidized SWCNTs (less than 500nm length) labelled with fluorescein isothiocyanate (FITC). During analysis it was observed that FITC alone was not taken up by the plant cells whereas SWCNT/FITC was translocated effectively into intact cells by fluid phase endocytosis (Samaj *et al.*, 2004). This result clearly pointed out the ability of SWCNTs to penetrate the cell wall of plant cells. Further studies also showed the ability of nanotubes to carry single-stranded DNA into intact plant cells by incubating the BY-2 cells in a medium with SWCNT/ss DNA-FITC (Liu *et al.*,_2009). A pronounced difference was observed in the distribution of SWCNT/FITC and SWCNT/ssDNA-FITC in the cell. The former nanotube was localized in the vacuolar region and latter in the cytoplasmic strands thus showing their ability to deliver different components to different cellular compartments. However the exact mechanism or mode of entry of nanotubes through the cell wall still remains a controversial point. But it is assumed that the orientation of nanotubes with respect to the plant cell wall might be important for their penetration.

Multiwalled carbon nano-tubes

Multi-walled carbon nanotubes are effectively used to determine the uptake and functioning of atrazine and simazine which are triazine herbicides. These multi-walled CNT's act as absorbents and show greater efficiency towards these two chemicals. Samples are collected from the river water, reservoirs water, tap water and waste water after the primary treatment (Zhou *et al.*, 2006). CNT's are also used for the biomolecular sensing which is based on the phenomenon of reverse quenching. In this method single walled CNT's are conjugated with dye ligand conjugates (dye is redox active dye molecule, which is covalently bound to the biological receptors ligand such as biotin). When the receptor ligand binds with the conjugate and target analyzed, the process of reverse quenching occurs. Use of CNT's have become an important area of biochemistry, thermodynamics, electronics, biomechanics (Daxiang, 2007). Multi-walled CNT's are effectively used as absorbents for the determination of pesticides. These CNT's are in the form of Matrix Solid phase dispersion extraction absorbents.(Fang *et al.* 2009).This gives rise to nano-biosensors which help to keep a check on contamination and food quality (Charych *et al.*, 1996). Recent reports focus on the effects of MWCNTs on the seed germination and growth of tomato plants. Their interaction with nanotubes showed positive responses (Lin *et al.*, 2009). The results showed an increased water uptake by seeds in the presence of MWCNTs which enhances the process of germination. Some other studies were also reported to have positive effects of suspensions of MWCNTs on seed germination and root growth with six different crop species [radish

(*Raphanus sativus*), rape (*Brassica napus*), rye grass (*Lolium perenne*), lettuce (*Lactuca sativa*), corn (*Zea mays*) and cucumber (*Cucumis sativus*)] (Lin and Xing, 2007). The studies were performed to observe the effect of both and MWCNTs on the germination of rice seeds (Fig. 3). In zucchini plants, positive results were observed on seed germination and root elongation using MWCNTs whereas a decrease in the biomass of plants was observed during further growth in the presence of SWCNTs (Stampoulis and Sinha, 2009). This clearly states that the response of plants to nanomaterials varies with the type of plant species, their growth stages and the nature of nanomaterials. Besides, CNT's also show the negative effect on the plant growth. An experiment was performed where CNT's were found to increase ROS concentration in plants. The reactive oxygen species are the main cause of oxidative stress causing cell death. This shows the overall effect as the stunting of roots and shoots. The ill effect appears to be due to the toxicity effect caused by the CNT's. This has opened a new area of research to investigate and evaluate the ill effects of CNT's.

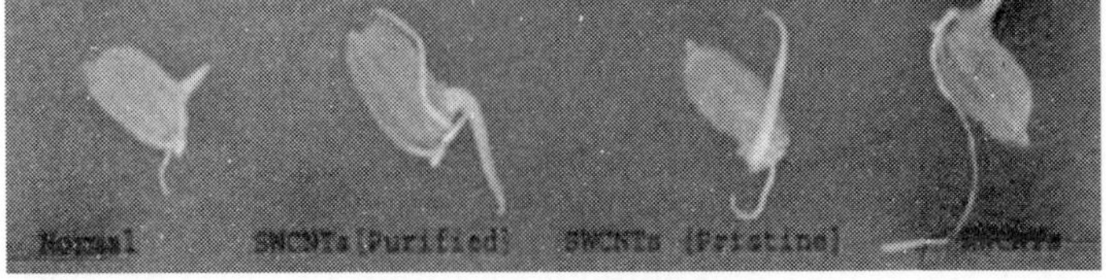

Fig. 3. Effects of different carbon nanotubes on the germination of rice seeds. Both SWCNTs and MWCNTs are added to plant growth media @ 30_g/ml of the media. Enhanced germination was observed for seeds germinated in presence of carbon nanotubes compared to the control [unpublished data]. (Nair *et al.*, 2010)

The cells having immediate interaction with nanotubes act as a barrier thus saving the rest of cells. The attachment of MWCNTs to the proteins and polysaccharides of cell wall act as signaling cascade which increases the production of certain compounds which are necessary for cell wall thickening. Such hypersensitive responses are very common in plants during biotic and abiotic stresses adversely affect plant growth. Reports were also present where the uptake, accumulation and transmission of natural organic matter (NOM) suspended carbon nanomaterials in rice plants was explained (Li *et al.*, 2009). For the experimentation, fullerene C_{70} or MWCNTs were suspended in NOM solution to produce C_{70}-NOM and MWCNTs-NOM to reduce the hydrophobicity of engineered nanomaterials. Rice seeds were then germinated in different concentrations of C_{70}-NOM or MWCNTs-NOM and then the seedlings were grown in greenhouse till maturity. During the experiment first generation seeds were used to raise the second generation plants without any addition of nanomaterials. Microscopic images were developed for various tissues of both first and second generation rice plants at different developmental stages. The results showed the occurrence of C_{70} nanomaterials whereas little MWCNTs were observed inside the cells (Fig. 4). These results revealed the transmission of C_{70} to the progeny through the seeds of first generation. C_{70} nanoparticles enter the plant roots either via osmotic pressure and capillary forces and enter through the cell wall pores for translocation through intercellular plasmadesmata. It is difficult to state that how different nanomaterials interact with plants and leads to the modification of the

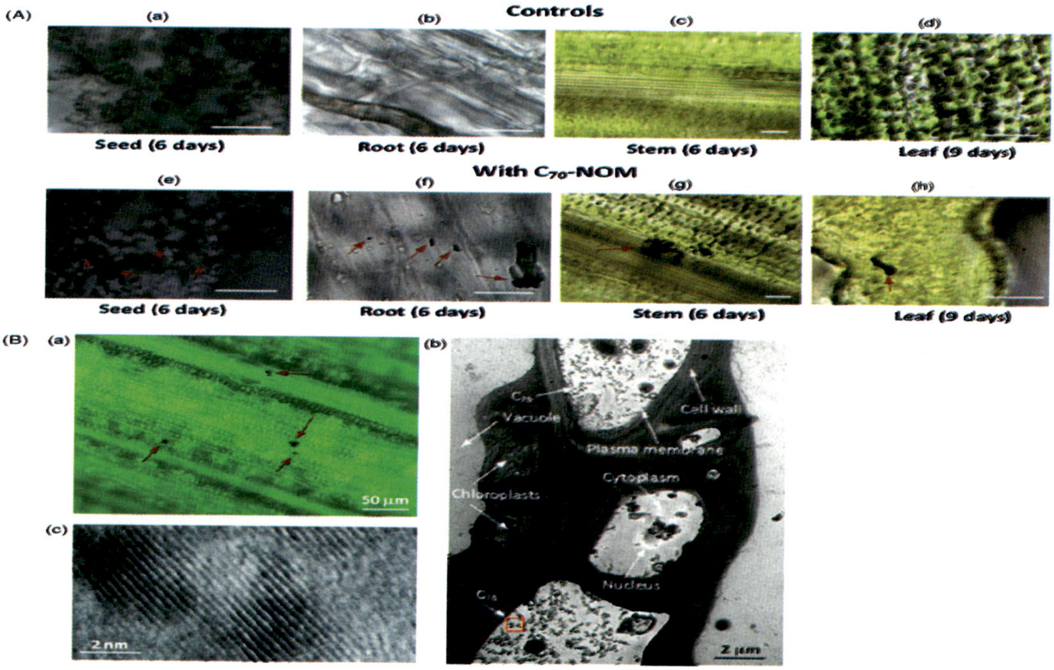

Fig. 7.4. Bright field images of rice plants showing C70 uptake. (A) Bright field images of root and leaf portions of 1-week-old rice seedlings. Control plants without any C70 (a–d) and treated plants showing C70 uptake (e and f). Arrows indicate the aggregation of nano-particles in corresponding C70 treated plant tissues (scale bar: 20_m). (B) (a) Bright field image of the leaf portion of a second generation rice plant. C70 aggregates were mostly found near the leaf vascular system. (b) TEM image of the leaf cells showing C70 particles (C70: 20mg/L). (c) TEM image of C70 particles with higher magnification (adopted from Ref. Li *et al.*, 2009)

genetic and molecular mechanisms of plants. It is therefore clear that the interaction varies with type and time of exposure to nanomaterials and hence these points need to be considered while assessing nano toxicity studies.

Lignin based carbon nano-tubes

Lignin nanotubes which are plant based and effectively used for biofuel production by creating biorefineries from lignin and lead to the production of bioethanol which is a renewable alternative of fossil fuel. Lignin nano-tubes are also used successfully for the treatment of cancer being smaller then viruses and hence easily travel through the body of the cancerous patient. They are advantageous over chemotherapy as these nanotubes have shell outside them and do not allow the anticancerous drugs to affect the normal body cells. They are highly target specific (Kushwaha and Malik, 2012).

Effects of magnetic nanoparticles

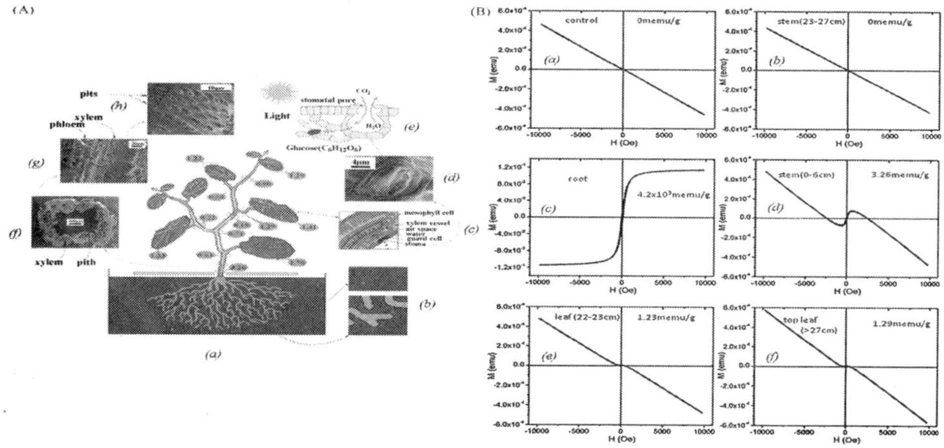

Fig. 5. Uptake, translocation and accumulation of manufactured Fe_3O_4 nanoparticles by pumpkin plants. (A) A schematic pumpkin plant and VSM measured magnetization at various sampling locations, marked as numbers in the unit of 10"3 emug"1. Strong magnetic signals (>1.0 memug"1) were detected in all leaf specimens regardless of their distances from the roots, while signals were much weaker from the stem tissues samples except those close to the roots. Although large numbers of the suspended particles accumulated on root surfaces, it is clear that Fe_3O_4 particles were taken up by the pumpkin plants and moved into the various tissues. The strongest magnetization (3.26 memug"1) was detected right above the roots which might be due to nanoparticle agglomeration. (B) Selected VSM (Vibrating Sample Magnetometer) measurements of pumpkin plant tissues for Fe3O4 nanoparticles. (a) Control plant (root), (b) treatment plant stem at height of 23–27 cm, (c) treatment plant root, (d) treatment plant stem at height of 0–6 cm, (e) treatment plant leaf at height of 22–23 cm, and (f) treatment plant leaf above 27 cm. The background diamagnetic signal is linear with magnetic field. The magnetization at H= 10,000 Oe is determined after the background signal is removed. No magnetization signal was detected from the control plants whereas most of the tissue samples of the treatment plants showed magnetization of various strengths (Nair *et al.*, 2010)

Magnetic nanoparticles are more preferred as they are highly site specific in releasing their loads and thereby prevent any ill effect to the nearby cells. Studies have been reported where these nanoparticles were readily up-taken, translocated and specifically localized using pumpkin plant (Gonzalez-Melendi *et al.*, 2008; Zhuet *et al.*, 2008; Corredor *et al.*, 2009). There after magnetized signals of variable strengths were observed from different parts (ranging from roots to leaves) of the treated plants (Fig. 5). The results clearly indicate the successful translocation of nanoparticles in the entire plant system irrespective of the area of application with no toxic effects on plant growth.

In last decade, great attention has been paid to genetic effects of ferrofluids which are responsible for causing chromosomal aberrations in young plants (Racuciu and Creanga, 2009; 2007; Pavel and Creanga, 2005; Pavel *et al.*, 1999). Reports are there which stated the effect of magnetic nanoparticles coated with tetramethylammonium hydroxide (TMA-OH) (as a stabilizing agent) on the growth of maize plants in early growth stages (Racuciu and Creanga, 2006). It was found that 'Chlorophyll a' level was increased with low ferrofluid concentrations while at higher concentrations reverse result was obtained. In another experimentation effects were recorded using maize seed and water based magnetic fluid obtained by coating magnetic nanoparticles with perchloric acid (Racuciu and Creanga, 2009). As a result slight inhibitory effect was observed on the growth of the plantlets leading to brown spots on leaves at higher volume fractions of magnetic fluid. This is because excess iron treatment generates some oxidative stress in leaf cells which in turn affect the process of photosynthesis and lead to the decrease in rate of metabolic process.

Effects of metal oxide nanoparticles

Metal nanoparticles are found to be potent source to deliver the specified doses of chemicals at a specific site in highly controlled manner. These nanoparticles are characterized into various categories but overall the size of the nanoparticles was considered as one of the critical factor.

Nano-TiO_2

The effect of nano-TiO_2 was studied in spinach plant where it resulted in the increase in the Rubisco activase activity and hence effect the overall growth of the plant (Zheng *et al.*, 2005; Hong *et al.*, 2005; Gao *et al.*, 2006; Lei *et al.*, 2007; Gao *et al.*, 2008; Linglan *et al.*, 2008; Xuming *et al.*, 2008; Mingyu *et al.*, 2008). This also increases the fresh and dry weight of spinach plant by effecting the nitrogen metabolism (Yang *et al.*, 2006) and also showed the effects of nitrogen photoreduction on the improved growth of treated spinach plant (Yang *et al.*, 2007; Mingyu *et al.*, 2007). Studies were also reported where these nano particle also promoted the energy transfer and oxygen evolution in photosystem II (PS II) of spinach (Mingyu *et al.*, 2007). Using UV-B radiation it has also been found that nano-anatase TiO_2 promotes antioxidant stress by decreasing the accumulation of superoxide radicals, hydrogen peroxide, malonyldialdehyde content and enhancing the activities of superoxide dismutase, catalase, ascorbate peroxidase, guaiacol peroxidase which ultimately increases the oxygen evolution rate in spinach chloroplasts (Lei *et al.*, 2008). Solar energy trapping efficiency can be increased by using TiO_2 quantum dots.

Effect of aluminium based nanoparticles

Reports were there which states the use of pure alumina nanoparticles (13 nm) without any modifications reduces the root elongation in several plants [corn (*Zea mays*), cucumber (*Cucumis sativus*), soybean (*Glycine max*), carrot (*Daucus carota*) and cabbage (*Brassica oleracea*)] (Yang and Watts, 205). However when these were modified it was observed that when loaded with phenanthrene (a major constituent of polycyclic aromatic hydrocarbons), their toxicity significantly decreased with on adverse effects on roots of plants. Studies were made where the impact of aluminium oxide and aluminum oxide with carboxylate ligand coating particles (100nm in size) on plants (California red kidney bean (*Phaseolus vulgaris*) and rye grass (*Lolium perenne*) was reported showing no ill effect on the plant growth (Lin and Xing, 2007; Doshi *et al.*, 2008).

Effects of zinc based nanoparticles

Nanoparticles made up of ZnO helps to increase the amount of polysaccharide in microorganism which ultimately leads to the increase in the moisture retention capacity of the soil (Tarafdar *et al.*, 2012). Various studies were also reported where some shows no negative effect on plant growth whereas some shows negative effect on plant growth. Studies showing no negative effect was the seed germination and root growth study of Zucchini seeds in hydroponic solution containing ZnO nanoparticles (Stampoulis and Sinha, 2009) whereas those showing negative effect was the seed germination of rye grass and corn was inhibited by nano-scale zinc (nano-Zn (35 nm)) and zinc oxide (nano-ZnO (15–25 nm)) respectively (Lin and Xing, 2007). On the other hand the root growth of radish and rape incubated in nano-Zn suspension was significantly inhibited; however such an inhibition was not detected while soaked in nano-ZnO suspension due to the selective permeability of seed coat.

Effect of Silver Nanoparticles

As silver nanoparticles (Ag NPs) exhibit have several antimicrobial functions as controlling various phytopathogens they are effectively used for pathogen treatment (Jo and Kim, 2009; Park *et al.*, 2006; Min *et al.*, 2009; Kim *et al.*, 2009). Studies have been made on the seed germination and root growth using Zucchini plants in hydroponic solution supplemented with Ag NPs but no negative effects were observed whereas a decrease in plant biomass and transpiration was observed on prolonged exposure to Ag NPs (Stampoulis and Sinha, 2009). The cytotoxic and genotoxic impacts of Ag NPs were also studied using root tips of onion. The investigation revealed that Ag NPs causes impairment in the stages of cell division and caused cell disintegration (Kumari *et al.*, 2009). All such studies states that there is a need for a more cytotoxic and genotoxic evaluations by considering the properties of nanoparticles, their uptake, translocation and distribution in different plant tissues.

Effect of Copper Nanoparticles

The effect of copper nanoparticles was studied using mung beans and wheat considering the effect on seedling growth. The results state that the mung beans were more sensitive than wheat to copper nanoparticles with inhibition on seedling growth parameters. Studies were also made using Zucchini plants to observe the effects of Cu NPs on the growth. The results reveal reduction in the length of emerging roots (Stampoulis and Sinha, 2009). However, increase in shoot to root ratio was observed in lettuce seeds in the presence of Cu NPs as compared to control plants (Shah and Belozerova, 2009). It is thereby clear that different flora and fauna responds differently to nanomaterials and hence, it is necessary to evaluate the safe effective concentration of each group of nanoparticles before their application to reduce the risks of ecotoxicity to a great extend.

Effect of Phosphorous Nanoparticles

Phosphorous nanoparticles when applied to the field application reduces the Ca, Fe, Al and other organic matter. The activity of mega P particles were observed, they nullify the competition with Si in the soil and thus increases the efficiency of P from 15% to at least 50 %. Positive impact of P nanoparticles were observed when they were compiled with oleic acid and leads to the increase in crop yield

Table 2: Effect of nano P fertilizer on cluster beans and pearl millet yiels under arid field conditions (Tarafdar, 2012).

Treatment	Grain yiels kg ha-1	
	Cluster beans	*Pearl millet*
Control	625	910
Ordinary P (2 WAG)**	650 (4.0)*	950 (4.4)
Nano P (2 WAG)	900 (44.0)	11.5 (21.4)
LSD (p=0.05)	28.7	20.9

Manipulation of agricultural crops using nanotechnology

Manipulation in crops at genetic level can be easily obtained using nanotechnology by means of nanoparticles, nanofibres and nanocapsules. Properly functionalized nanomaterials can serve as a platform to transport large number of genes as well as chemicals to trigger gene expression in plants.

Nanofibres

Nanofibre is the arrays that can be used for delivering the genetic material to cells quickly and efficiently. These have the potential applications in drug delivery, crop engineering and environmental monitoring. Reports are there which reveals the integration of carbon nanofibres that are surface modified with plasmid DNA with viable cells for controlled biochemical manipulations in cells (McKnight *et al.*, 2003; 2004). The process is almost similar to microinjection method of gene delivery (Segura and Shea, 2001; Neuhaus and Spangerberg, 1990; Bolik and Koop,1991) and hence can be possibly used for the plant cells in which the treated cells could be used to regenerate the whole plant, that would actively express the introduced trait.

Nanoparticles

Fluorescent labelled starch-nanoparticles was used as plant transgenic vehicle in which the nanoparticle biomaterial was designed in such a way that they were capable of binding and transporting genes across the cell wall of plant cells. This could be achieved by inducing instantaneous pore channels in cell wall, cell membrane and nuclear membrane with the help of ultrasound (Jun *et al.*, 2008). It is now possible to integrate different genes on the nanoparticle and at the same time imaging can be done by fluorescent labelled nanoparticle with the help of fluorescence microscope which thereby helps in understanding the movement of exterior genes along with the expression of transferred genes. Surface functionalized mesoporous silica nanoparticles (MSNs) shows their potential to penetrate plant cell walls which ultimately opens up the new ways to precisely manipulate gene expression at single cell level in a controlled fashion (Torney *et al.*, 2007).

Molecular biology complementing nanotechnology (Nanobiotechnology)

The term "nanobiotechnology" was given by Lynn W. Jelinski, a biophysicist at Cornell University. It is a connecting link between nanotechnology and molecular biology. Nanotechnologists can achieve many goals by exploiting the extraordinary properties of biological molecules and cell processes that are difficult or impossible to achieve by other means. Nanotechnologists greatly rely on the self-assembling properties of biological molecules which ultimately lead to nanostructures, such as lipids that spontaneously form liquid crystals. DNA has been used not only to build nanostructures but also as an essential component of nanomachines that may even serve as the basis of the next generation of computers. As microprocessors and microcircuits, nanoprocessors and nanocircuits came into existence. DNA molecules can be mounted onto a silicon chips which may replace the microchips with electron flow-channels etched in silicon. Such biochips can work as DNA-based processors and uses DNA's extraordinary information storage capacity. Biochips on the other hand exploit the properties of DNA to solve computational problems. Other biological molecules are assisting in our continual quest to store and transmit more information in smaller places. Nanobiotechnology is therefore an emerging area of opportunity by relating all the applications at genomics level including mammalian, plant and microbial. It provides the basic tools and technology that helps in gathering sequence information and designing innovative devices to explore questions related to the biological importance at genomic level and the application of this knowledge in diverse fields, particularly medicine and agriculture. The impact of nanobiotechnology may be immediately felt in the following areas -

DNA Sequencing

The processes available for DNA sequencing are complex as well as time consuming. To make the approach of DNA sequencing more feasible and simple, nanofabrication technology has proven to be very beneficial. Nanofabrication not only improves the existing method but also proves to be a novel approach in sequence detection. For this nanofabrication-gel free system are used which are coupled with more powerful approaches such as genetic analysis, finding the sequence of the crop germplasm which allows the rapid DNA sequencing. This process also provides the gene pool of a cultivated crop and their wild relatives. This helps in providing useful information about the important molecular markers which are beneficial in agriculture and economy. This is bound to enhance the molecular marker assisted breeding programs and will lead to crop improvement.

Biochips

Biochips are considered to be the first product available in the market based on Nanobiotechnology. Other most advanced technology in this area is the micro-fluid biochip which are based on the manipulation of the minute biological-objects which are immersed in the fluids. This is based on the principle of on chip biochemical processing. Both these chips are used in a wide range of field including drug discovery, drug delivery, cell analysis and complex surgery.

Nano-fuels

Nanotechnology also aims to produce biofuel which are obtained from vegetable oil in the presence of nanocatalyst. The main intent behind the concept of nanofuel production is to produce fuel at lesser cost and the one that will provide more energy output utilizing less energy input. On the other hand these nano-fuels will not have any ill effect on the environment while utilization.

Quantum dots

These are the only few nm in diameter, spherical, fluorescent, crystalline particles of semiconducting nature. They are used for biological detection and also reveal the photochemical properties (Bruchz *et al.* 1998; Chan and Nie, 1998). They are robust and very stable light emitters. Using this property they are used in various areas like: cell labeling, cell tracking (Parak *et al.*, 2002), *in vivo* imagining (Dubertret *et al.*, 2002) and DNA detection.

Atomically modified seed

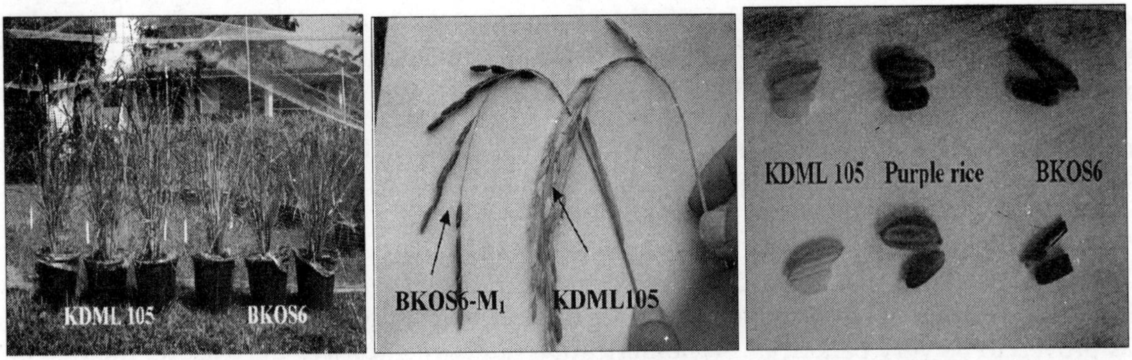

Fig. 6: The rice mutants designated BKOS6 was derived by bombardment with N++N2+ ions from KDML 105 rice embryos.

Experiment was performed 2004 by ETC Group at Chiang Mai University's nuclear physics laboratory, in Thailand on rice. They drilled a hole on the cell wall of the rice and introduce nitrogen into it that will stimulate the rearrangement of the DNA molecules. As a result alteration in the rice colour from purple to green was obtained. Their next trials were on Thai Jasmine Rice where they want to generate rice with reduced stem and improved grain colour. They introduced low energy beams through bombardment at energy levels in the range of 60–125 keV and ion fluences (dose) of 1×1016–5×1017 ions/cm^2 to introduce mutation in Thai jasmine rice.

The mutation was obtained from KDML 105 rice embryos bombarded with $N^{++}N_2^+$ ions at an energy level of 60 keV and ion fluence of 2×1016 ions/cm^2. The phenotypic variations occurred was BKOS6 which was characterized by short in stature, red/purple color in leaf sheath, collar, auricles, ligule, and dark brown stripes on leaf blade, dark brown seed coat and pericarp. For analyzing the genomic mutation in muatant HAT-RAPD (High Annealing Temperature- Random Amplified Polymorphic DNA) was applied using 10 primers out of which two primers detected two additional DNA bands at 450 bp and 400 bp. DNA sequencing revealed that the 450 bp and the 400 bp fragments were 60% and 61% identity to amino acid sequence of flavanoid 32 hydroxylase and cytochrome P450 of *O. sativa and japonica*, respectively.

Synthetic Tree

Wheeler and Stroock report tiny 'synthetic tree' through whose trunk water flows at pressures of around -10 atmospheres. For this they used hydrogel sysytem, which mimics the mesophyll by holding water in molecular-scale pores, smaller than those of other porous solids. For their respective 'root' and 'leaf', two networks of channels, 10 micrometres in diameter, in a sheet of poly(hydroxyethyl methacrylate)was formed and connected them by a single channel, the 'trunk'. As the 'root' was exposed to a source of water and the 'leaf' to a stream of damp air, water flows through the system powered solely by 'leaf' evaporation. By this method the pressures developed in the trunk was 15 times more negative than in any reported pumping system. The synthetic tree proves as a test device for theories of tree physiology and, scaled-up, the technology where evaporation makes the 'leaf' a heat sink. On the other hand large negative pressures developed in synthetic trees might be used to drag out water even from dry soils which simultaneously filter out the impurities by passage through the 'root' hydrogel. The process was termed "reverse reverse osmosis" and could form the basis of solar-powered mining of pure water in arid or contaminated environments.

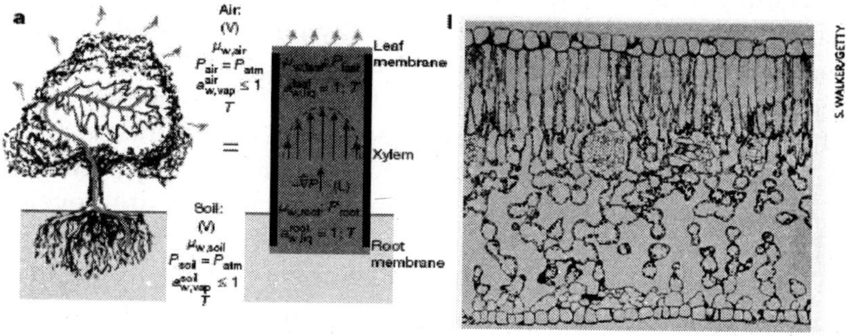

Fig. 7: The synthetic hydrogel mimics the transpiration pattern of a typical plant system.

Hormone and antibiotics delivery in plants

Protein and nucleic acid drugs exhibit poor stability in physiological conditions. It is therefore required to sustain the stability of the drug at the time of delivery at a specific site. Sustainable release of these drugs involves the combination of a biocompatible material or device with a drug that is to be delivered and its release at target sites in a designed manner. Drug-delivery system protects the potential drug candidates by increasing their solubility and stability by the application of coating of polymer drug conjugates, polymeric micelles, polymeric nanospheres and nanocapsules, and polyplexes. Polymer-drug conjugates (520 nm) are the smallest nanoparticulate delivery system. The polymers used for such purposes are usually highly water-soluble and which includes synthetic polymers (for example, polyethylene glycol) (PEG)) as well as natural polymers (such as dextran).

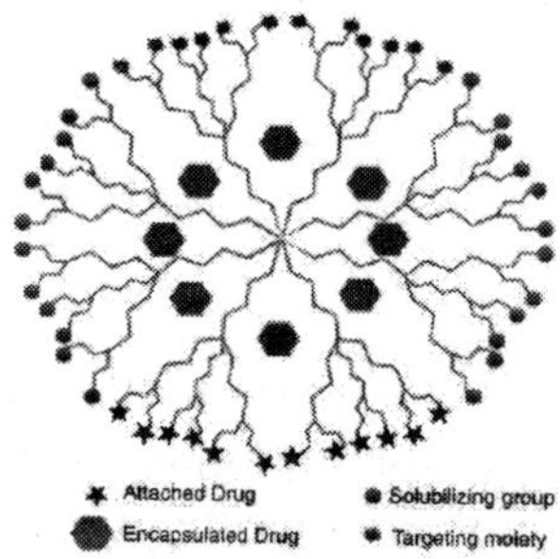

Conclusion

Crop improvement is a necessary continual process and also a major concern to increase food security and crop production by improving yields and product quality. Advances in life science have enabled plant scientists to develop a host of tools that have expanded our knowledge of the genetics and physiological processes involved in many crop traits of interest. Nanoscience is leading to the development of a range of inexpensive nanotech applications for enhanced plant growth. Nanomaterials like nanoparticles provide an efficient means for the distribution of pesticides and fertilizers in a controlled manner and highly site specific fashion and thus reducing the collateral damages. Nanoparticle technology enables the efficient transportation of substances, such as systemic chemicals, to specific sites and thereby solving the ill effects of plant–pathogen interaction. This could help to carry on several studies at physiological, biochemical and genetic levels. Plant mediated synthesis of metal nanoparticles provide a safe green synthesis route with a control over morphology of nanoparticles and proves to be beneficial in medical sciences due to their antimicrobial behaviour. Plant growth can easily be effected by the interaction of plant cell with the nanoparticles resulting in modification of plant gene expression and associated biological pathways and ultimately affecting plant growth and development. To date, nanoscientists are developing techniques for atom-by-atom construction of objects that have potential applications not just in agriculture but also in medicine, electronics, information technology, and environmental monitoring and remediation. Indian scientists have abundant talent in computational science, biology, physics, chemistry and engineering. It would be worthwhile to link all the mentioned disciplines and then launch effective Nanobiotechnology programs for the benefits to human kinds in future. Hence future studies should include the highlight like the needs to clarify the nanotoxicity to plants, possible uptake and translocation of nanoparticles by plants, and physical and chemical properties of nanoparticles in rhizosphere and on root surfaces.

REFERENCES

Arora S, Sharma P, Kumar S, Nayan R, Khanna PK and Zaidi MGH (2012) Gold-nanoparticle induced enhancement in growth and seed yield of *Brassica juncea. Plant Growth Regulation* 66(3): 303-310

Athanassiou CG, Kavallieratos NG and Meletsis CM (2007) Insecticidal effect of three diatomaceous earth formulations, applied alone or in combination, against three stored-product beetle species on wheat and maize. *J Stored Prod Res* 43: 330–334

Bar H, Bhui DK, Sahoo GP, Sarkar P, De SP and Misra A (2009) Green synthesis of silver nanoparticles using latex of *Jatropha curcas. Colloids and surfaces A: Physicochemical and Engineering aspects* 339: 134- 139

Barik TK, Sahu B and Swain V (2008) Nanosilica-from medicine to pest control. *Parasitol Res* 103(2): 253-258

Boehm AL, Martinon I, Zerrouk R, Rump E and Fessi H (2003) Nanoprecipitation technique for the encapsulation of agrochemical active ingredients *J. Microencapsul* 20 : 433–441

Bolik M and Koop HU (1991) Identification of embryogenic microspores of barley (Hordeum vulgare) by individual selection and culture and their potential for transformation by microinjection. *Protoplasma* 162: 61–68

Brecht M, Datnoff L, Nagata R and Kucharek T (2003) The role of silicon in suppressing tray leaf spot development in St.Augustine grass. *Publication in University of Florida* 1-4

Bruchez M, Moronne M, Gin P, Weiss S and Alivisatos AP (1998) Semiconductor nanocrystals as fluorescent biological labels. *Science* 281:2013-2016

Cañas JE, Long N, Vadan S, Dai R, Lou L, Ambikapathi M, Lee R and Olszyk D (2008) Effects of functionalized and nonfunctionalized single-walled carbon nanotubes on root elongation of select crop species. *Envir Toxicology and Chem.* 27(9): 1922-1931

Cao YWC, Jin R and Mirkin CA (2002) Nanoparticles with Raman spectroscopic fingerprints for DNA and RNA detection. *Science* 297: 1536-1540

Caruthers SD, Wickline SA and Lanza GM (2007) Nanotechnological applications in medicine. *Curr Opin Biotechnol* 18: 26–30

Chan WCW and Nie SM (1998) Quantum dot bioconjugates for ultrasensitive nonisotopic detection. *Science* 281: 2016 -2018

Charych D, Cheng Q, Reichert A, Uziemko G, Stroh N, Nagy J, Spevak W and Stevens R (1996) Principle of formation of clay monolayer containing nanocomposites *Chem Biol* 3: 113

Chauhan S, Upadhyay MK, Rishi N and Rishi S (2011) Phytofabrication of silver nanoparticles using pomegranate fruit seeds. *Intern Jour of Nanomaterials and Biostructures* 1(2): 17-21

Corredor E et al., (2009) Nanoparticle penetration and transport in living pumpkin plants: in situ subcellular identification. *BMC Plant Biol* 9: 45

Da Silva LC, Oliva MA, Azevedo AA and De Araujo JM (2006) Responses of resting plant species to pollution from an iron pelletization factory. *Water Air Soil Pollut* 175: 241–256

Daxiang C (2007) Advances and prospects on biomolecules functionalized carbon nanotubes. *Jour of Nanosci and Nanotechnology* 7(4-5):1298-1314

Donald HA, Wang H, Petersen EJ, Cleveland D, Holbrook RD, Jaruga J, Dizdaroglu M, Xing B and Nelson BC (2012) Copper oxide nanoparticle mediated DNA damage in terrestrial plant models. *Environmental Science & Technology* 46 (3): 1819

Doshi R, Braida W, Christodoulatos C, Wazne M and O'Connor G (2008) Nanoaluminium: transport through sand columns and environmental effects on plants and soil communities, *Environ Res* 106:296–303

Dubertret B, Skourides P, Norris DJ, Noireaux V, Brivanlou AH and Libchaber A (2002) *In vivo* imaging of quantum dots encapsulated in phospholipid micelles. *Science* 298:1759–1762

Eichert T, Kurtz A, Steiner U and Goldbach HE (2008) Size exclusion limits and lateral heterogeneity of the stomatal foliar uptake pathway for aqueous solutes and water-suspended nanoparticles. *Physiol Plant* 134: 151–160

Fang G, Min G, He J, Zhang C, Qian K and Wang S (2009) Multiwalled carbon nanotubes as matrix solid-phase dispersion extraction absorbents to determine 31 pesticides in agriculture samples by gas chromatography"mass spectrometry. *Jour Agric Food Chem 57* (8): 3040–3045

Feiner LF (2006) Nanoelectronics: crossing boundaries and borders. *Nat Nanotechnol* 1: 91–92

Fernandez V and Eichert T (2009) Uptake of hydrophilic solutes through plant leaves: current state of knowledge and perspectives of foliar fertilization. *Crit Rev Plant Sci* 28:36–68

Fleischer MA O'Neill and Ehwald R (1999) The pore size of non-graminaceous plant cell wall is rapidly decreased by borate ester cross-linking of the pectic polysaccharide rhamnogalacturon II. *Plant Physiol* 121: 829–838

Gao F et al., (2006) Mechanism of nano-anatase TiO_2 on promoting photosynthetic carbon reaction of spinach: inducing complex of rubisco–rubisco activase. *Biol Trace Elem Res* 111: 239–253

Gao F et al., (2008) Was improvement of spinach growth by nano-TiO_2 treatment related to the changes of rubisco activase? *Biometals* 21: 211–217

Gardea-Torresdey JL, Parsons LG, Gomez E, Peralta-Videa J, Troiani HE, Santiago P and Yacaman MJ (2002) Formation and Growth of Au Nanoparticles inside Live Alfalfa Plants *Nano Lett* 2 (4): 397-401

Gonzalez-Melendi P et al., (2008) Nanoparticles as smart treatment delivery systems in plants: assessment of different techniques of microscopy for their visualization in plant tissues. *Ann Bot* 101: 187–195

Green JM and Beestman GB (2007) Recently patented and commercialized formulation and adjuvant technology. *Crop Prot* 26: 320–327

Hong F et al., (2005) Effect of nano-TiO_2 on photochemical reaction of chloroplasts of spinach. *Biol. Trace Elem Res* 105: 269–280

Hu L and Chen G (2007) Analysis of optical absorption in silicon nanowire arrays for photovoltaic applications. *Nano Let* 7: 3249–3252

Hu J, Choi E, Tamanoi F JI Zink and (2008) Light activated nanoimpeller-controlled drug release in cancer cells. *Small* 4 421–426

Jain D, Daima K, Kachhwaha S and Kothari SL (2009) Synthesis of plant-mediated silver nanoparticles using Papaya fruit extract and evaluation of their antimicrobial activities. *Digest Jour. of Nanomaterials and Biostructures* 4(3):557-563

Jain PK, Huang X, El Sayed LH and El Sayed MA (2007) Review of some interesting Surface Plasmon Resonance- enhanced properties of noble metal Nanoparticles and their applications to biosystems. *Plasmonics* 2: 107- 118

Jain S, Jain R and Singh R (2010) Micropropagation of *Verbesina encelioides*-An invasive weed. *Indian Jour of Biotech* 9:333-335

Jo YK and Kim BH (2009) Antifungal activity of silver ions and nanoparticles on phytopathogenic fungi. *Plant Dis* 93: 1037–1043

Jun L, et al. (2008) Preparation of fluorescence starch-nanoparticle and its application as plant transgenic vehicle. *J Cent South Univ Technol* 15:768–773

Kausch AP and Bruce BD (1994) Isolation and immobilization of various plastid subtypes by magnetic immunoabsorption. *Plant J* 6(5):767-779

Khodakovskaya M, Dervishi E, Mahmood M, Xu Y, Li Z, Watanabe F and Biris AS (2009) Carbon nanotubes are able to penetrate plant seed coat and dramatically affect seed germination and plant growth. *ACS Nano* 3 (10):3221-3227

Kim SW, et al. (2009) An in vitro study of the antifungal effect of silver nanoparticles on oak wilt pathogen *Raffaelea* sp. *J Microbiol Biotechnol* 19 : 760–764

Kostarelos K, Lacerda L, Pastorin G, Wu W, Wieckowski S, Luangsivilay J, Godefroy SD, Pantarotto D, Briand JP, Muller S, Prato M and Bianco A (2007) Cellular uptake of functionalized carbon nanotubes is independent of functional group and cell type. *Nature Nanotechnology* 2:108 -113

Kumari M, Mukherjee A and Chandrasekaran N (2009) Genotoxicity of silver nanoparticles in *Allium cepa*. *Sci Total Environ* 407: 5243–5246

Kushwaha HB and Malik CP (2012) Contributions of Nanotechnology in Agriculture and Food Processing. *Phytomorphology* 62(1 & 2): 57-67

Kushwaha HB and Malik CP (2013) Nanotechnology: A Hope of New Revolution in Agriculture. In Crop Improvement: Biotechnological, Physiological and Nanotechnological Approaches EDT Malik CP, Aavishkar Publishers, Jaipur, pp 266-285

Kuzma J (2006) The Nanotechnology-Biology Interface: Exploring Models for Oversight. Workshop Report. September 15, 2005.

Lei Z, et al. (2007) Effects of nanoanatase TiO_2 on the photosynthesis of spinach chloroplasts under different light illumination. *Biol. Trace Elem Res* 11 : 68–76

Lei Z et al., (2008) Antioxidant stress is promoted by nano-anatase in spinach chloroplasts under UV-Beta radiation. *Biol. Trace Elem Res* 121 : 69–79

Lewis JD, Destito LG, Zijlstra A, Gonzalez MJ, Quigley JP, Manchester M and Stuhlmann H (2006) Viral nanoparticles as tools for intravital vascular imaging. *Nat. Med* 12:354-360

Li S, Shen Y, Xie A, Yu X, Qui L, Zhang L and Zhang Q (2007) Green synthesis of silver nanoparticles using *Capsicum annum* L. extract. *Green Chemistry* 9 : 852-858

Li Y, Hong Cu Y and Luo D (2005) Multiplexed detection of pathogen DNA with DNA-based fluorescence nanobarcodes. *Nat. Biotechnol* 23:885-889

Li ZZ, Chen JF, Liu F, Lu AQ, Wang Q, Sun HY and Wen LX (2007) Study of UVshielding properties of novel porous hollow silica nanoparticle carriers for avermectin. *Pest Manag Sci* **63:** 241–246

Lin D and Xing B (2007) Phytotoxicity of nanoparticles: inhibition of seed germination and root growth. *Environ. Pollut* 150: 243–250

Lin S, *et al.* (2009) Uptake, translocation, and transmission of carbon nanomaterials in rice plants. *Small* 5 : 1128–1132

Linglan M, *et al.* (2008) Rubisco activase m RNA expression in spinach: modulation by nano anatase treatment. *Biol. Trace Elem Res* 122 : 168–178

Liu J et al., (2009) Nano-silver pulse treatments inhibit stem-end bacteria on cut gerbera cv. Ruikou flowers. *Postharvest Biol. Technol* 54 : 59–62.

Liu Q, Chen B, Wang Q, Shi X, Xiao Z, Lin J and Fang X (2009) Carbon nanotubes as molecular transporters for walled plant cells. *Nano Lett* 9 : 1007–1010

Liu X, Feng Z, Zhang F, Zhang S and He X (2006) Preparation and testing of cementing and coating nano-subnanocomposites of slow/controlled-release fertilizer. *Agric. Sci. China* 5: 700–706

Liu Y, Laks P and Heiden P (2002) Controlled release of biocides in solid wood. Part 1. Efficacy against Gloeophyllum trabeum, a brown rot wood decay fungus. *J Appl Polym Sci* 86 : 596–607

Liu Y, Laks P and Heiden P (2002) Controlled release of biocides in solid wood. Part 2. *Efficacy against Trametes versicolor* and *Gloeophyllum trabeum wood decay fungi. J. Appl Polym Sci* 86 : 608–614

Liu Y, Laks P and Heiden P (2002) Controlled release of biocides in solid wood. Part 3. Preparation and characterization of surfactant-free nanoparticles. *J. Appl. Polym. Sci* 86 : 615–621

Liu Y, Laks P and Heiden P (2003) Nanoparticles for the controlled release of fungicides in *wood: soil Jar studies using Gloeophyllum trabeum* and *Trametes versicolor wood decay fungi. Holzforschung* 57: 135–139

Liu Y. Yan L, Heiden P and Laks P (2001) Use of nanoparticles for controlled release of biocides in solid wood. *J Appl Polym Sci* 79: 458–465

Lokko Y, Mba C, Spencer M, Till B and Lagoda P (2011) Nanotechnology and synthetic biology - Potential in crop improvement. *Journal of Food, Agriculture & Environment* 9: (3& 4) : 599 - 604

Mallikarjuna M, Narasimha G, Dillip R, Praveen B, Shreedhar B, Shree Lakshmi C, Reddy BVS and Deva Prasad Babu B (2011) Green synthesis of silver nanoparticles using Ocium leaf extract and their characterization.*Digest Jour. of Nanomaterial and Biostructures* 6:181-186

Mano Priya P, Karunai Selvi V and John Paul JA (2011) Green synthesis of silver nanoparticles from the leaf extract of *Euphorbia hirta* and *Nerium indicum. Digest Jour. of Nanomaterials and Biostructures* 6(2): 869-877

McKnight TE, Melechko AV, Griffin GD, Guillorn MA, Merkulov VI, Serna F, Hensley DK, Doktycz MJ, Lowndes DH and Simpson ML (2003) Intracellular integration of synthetic nanostructures with viable cells for controlled biochemical manipulation. *Nanotechnology* 14: 551–556

McKnight TE, Melechko AV, Hensley DK, Mann DGJ, Griffin GD and Simpson ML (2004) Tracking gene expression after DNAdelivery using spatially indexed nanofiber arrays *Nano Lett* 4 :1213–1219

Mewis I and Ulrichs CH (2001) Action of amorphous diatomaceous earth against different stages of the stored product pests Tribolium confusum, Tenebrio molitor, Sitophilus granaries and Plodia interpunctella. *J Stored Prod. Res* 37 : 153–164

Min JS, *et al.* (2009) Effects of colloidal silver nanoparticles on sclerotium-forming phytopathogenic fungi. *Plant Pathol J* 25:376–380

Mingyu, S *et al.* (2007) Promotion of energy transfer and oxygen evolution in spinach photosystem II by nano-anatase TiO2. *Biol. Trace Elem. Res* 119 : 183–192

Mingyu S, Liu J, Yin S, Linglan M and Hong F (2008) Effects of nano-anatase on the photosynthetic improvement of chloroplast damaged by linolenic acid. *Biol. Trace Elem. Res* 124: 173–183

Mondal AK, Mondal S, Samanta S and Mallick S (2011) Syntheiss of ecofriendly silver nanoparticles from the plant latex used as an important Taxonimic Tool for Phylogenetic Interrelationship. *Advances in Bioresearch* 2(1):122-133

Moore MN (2006) Do nanoparticles present ecotoxicological risks for the health of the aquatic environment *Environ Int* 32: 967–976

Mukunthan KS, Elumalai EK, Patel TN and Murty VR (2011) *Catharanthus roseus*: a natural source for the synthesis of silver nanoparticles. *Asian Pacific Jour of Tropical Biomedicine* 1691(11) : 60041-6005

Nair R, Varghese SH, Nair BG, Maekawa T, Yoshida Y and Kumar DS (2010) Nanoparticulate material delivery to plants. *Plant Sciences* 179: 154-163

Nam JM, Thaxton CS and Mirkin CA (2003) Nanoparticle-based bio-bar codes for the ultrasensitive detection of proteins. *Science* 301(5641):884-1886

Navarro E, *et al.* (2008) Environmental behaviour and ecotoxicity of engineered nanoparticles to algae, plants and fungi. *Ecotoxicology* 17 : 372–386

Neuhaus G and Spangerberg G (1990) Plant transformation by microinjection techniques, *Physiol Plant* 79: 213–217

Pacurari M, *et al.* (2008) Raw single-wall carbon nanotubes induce oxidative stress and activate MAPKs, AP-1, NF-KB and Akt in normal and malignant human mesothelial cells, Environ. *Health Perspect* 116: 1211–1217

Panacek A, Kolar M, Vecerova R, Prucek R, Soukupova J, Krystof V, Hamal P, Zboril R and Kvitek L (2009) Antifungal activity of silver nanoparticles against *Candida* spp. *Biometals* 30 : 6333–6340

Panessa-Warren BJ, Warren JB, Wong SS and Misewich JA (2006) Biological cellular response to carbon nanoparticle toxicity. *J Phy Condens Matter* 18 : S2185–S2201

Pantarotto D, Partidos CD, Hoebeke J, Brown F, Kramer E, Briand JP, Muller S, Prato M and **Bianco A** (2003) Immunization with peptide-functionalized carbon nano-tubes enhances virus-specific neutralizing antibody responses. *Chemistry & Biology* 10(10):961-966

Parashar V, Prashar R, Sharma B and Pandey AC (2009) Parthenium leaf extract mediated synthesis of silver nanoparticles: a novel approach towards weed utilization. *Digest Jour of Nanomaterials and Biostructures* 4(1): 45-50

Park HJ, Sung HK, Hwa JK and Seong HC (2006) A New Composition of Nanosized Silica-Silver for Control of Various Plant Diseases. *Plant Pathol Jour* 22(3) : 295-302

Park HJ, Kim SH, Kim HJ and Choi SH (2006) A new composition of nanosized silica–silver for control of various plant diseases *Plant Pathol. J* 22 : 295–302

Park HJ, Taton TA and Mirkin CA (2002) Array-based electrical detection of DNA with nanoparticle probes. *Science* 295: 1503

Patolsky F, Zheng G and Lieber CM (2006) Nanowire sensors for medicine and life sciences *Nanomedicine* 1: 51–65

Pavel A and Creanga V (2005) Chromosomal aberrations in plants under magnetic fluid influence. *J Magn Mater* 289:469–472

Pavel A, Trifan M, Bara II, Creanga D and Cotae C (1999) Accumulation dynamics and some cytogenetical tests at *C. majus* and *P. somniferum* callus under the magnetic liquid effect. *J Magn Magn Mater* 201: 443–445

Prasad SK, Pathak D, Patel A, Dalwadi P, Prasad R, Patel P and Selvaraj K (2011) Biogenic synthesis of silver nanoparticles using *Nicotiana tobaccum* leaf extract and the study of their antibacterial effect. *African Jour of Biotech* 10(41): 8122-8130

Racuciu M and Creanga D (2006) TMA-OH coated magnetic nanoparticles internalize in vegetal tissue. *Romanian J Phys* 52 : 395–402

Racuciu M and Creanga D (2007) Cytogenetic changes induced by aqueous ferrofluids in agricultural plants. *J Magn Magn Mater* 311 : 288–290

Racuciu M and Creanga D (2009) Biocompatible magnetic fluid nanoparticles internalized in vegetal tissues *Romanian J Phys* 54: 115–124

Racuciu M and Creanga D (2009) Cytogenetic changes induced by beta-cyclodextrin coated nanoparticles in plant seeds. *Romanian J Phys* 54: 125–131

Rahman A, Seth D, Mukhopadhyaya SK, Brahmachary RL, Ulrichs C and Goswami A (2009) Surface functionalized amorphous nanosilica and microsilica with nanopores as promising tools in biomedicine. *Naturwissenschaften* 96: 31–38

Ravindran S, Kim S, Martin R, Lord EM and Ozkan CS (2005) Quantum dots as bio-labels for the localization of a small plant adhesion protein. *Nanotechnology* 16:1-5

Roco MC (2003) Nanotechnology convergence with modern biology and medicine. *Curr Opin Biotechnol* 14: 337–346

Roy I, Ohulchanskyy TY, Bharali DJ, Pudavar HE, Mistretta RA, Kaur N and Prasad PN (2005) Optical tracking of organically modified silica nanoparticles as DNA carriers: A nonviral, nanomedicine approach for gene delivery. *Proc Nat Acad Sci USA* 102(2): 279-284

Salata OV (2004) Applications of nanoparticles in biology and medicine. *J Nanobiotechnology* 2: 3

Samaj J, Baluska F, Voigt B, Schlicht M, Volkmann D and Menzel D (2004) Endocytosis,actin cytoskeleton, and signalling. *Plant Physiol* 135 : 1150–1161

Sathyavani R Balamuali KM Rao SV Saritha R and Rao N (2010) Biosynthesis of silver nanoparticles using *Corriandrum sativum* leaf extract and their application in nonlinear optics. *Adv Sci Letters* 3:1

Satyavani K, Ramanathan T and Gurudeeban S (2011) Green synthesis of silver nanoparticles by using stem derived callus extract of bitter apple (*Citrullus colocynthis*). *Digest Jour of Nanomaterials and Biostructures* 6:1019-1024

Savithramma N, Rao L and Suvarnalatha DP (2011) Antibacterial efficacy of biologically synthesized silver nanoparticles using stem barks of *Boswellia ovalifoliolata. Jour of Biol Sci* 11:39-45

Scrinis G and Lyons K (2007) The emerging nano-corporate paradigm: nanotechnology and the transformation of nature, food and agri-food systems. *Int J Sociol Food Agric* 15 : 22–44

Segura T and Shea LD (2001) Materials for non viral gene delivery. *Annu Rev Mater Res* 31: 25–46

Shah V and Belozerova I (2009) Influence of metal nanoparticles on the soil microbial community and germination of lettuce seeds. *Water Air Soil Pollut* 197 : 143–148

Shankar SS, Rai A, Ahmad A and Sastry M (2004) Rapid synthesis of Au, Ag and bimetallic Au core- Ag shell nanoparticles using neem (*Azadirachta indica*) leaf broth. *Jour of Colloid and Interface Sci* 275: 496-502

Sharma NC, Sahi SV, Nath S, Parsons JG, Gardea-Torresdey JL and Pal T (2007) Synthesis of plant mediated gold nanoparticles and catalytic role of biomatrix-embedded nanomaterials *Environ Sci Technol* 41 : 5137–5142.

Singh M, Singh S, Prasad S and Gambhir IS (2008) Nanotechnology in medicine and antibacterial effect of silver nanoparticles. *Digest J Nanomater Biostruct* 3 : 115–122

Solgi M, Kafi M, Taghavi TS and Naderi R (2009) Essential oils and silver nanoparticles (SNP) as novel agents to extend vase-life of gerbera (*Gerbera jamesonii* cv. 'Dune') flowers. *Postharvest Biol Technol* 53 : 155 158

Stampoulis D, Sinha SK and White JC (2009) Assay-dependent phytotoxicity of nanoparticles to plants. *Environ Sci Technol* 43 : 9473–9479

Storhoff JJ, Marla SS, Bao P, Hagenow S, Mehta H, Lucas A, Garimella V, Patno T, Buckingham W, Cork W and Muller UR (2004) Gold nanoparticle-based detection of genomic DNA targets on microarrays using a novel optical detection system. *Biosens Bioelectron* 19(8):875-83

Thirumurugan A, Jiflin GJ, Rajagomathi G, Tomy NA, Ramachandran S and Jaiganesh R (2010) Biotechnological synthesis of gold nanoparticles of *Azardirachta indica* leaf extract. *Inter Jour of Biol Techn* 1(1):75-77

Thomas S and McCubin P (2003) A comparison of the antimicrobial effects of four silvercontaining dressings on three organisms. *Jour Wound Care* 12:101-107

Torney F, Trewyn BG, Lin SY and Wang K (2007) Mesoporous silica nanoparticles deliver DNA and chemicals into plants. *Nat Nanotechnol* 2: 295–300

Tripathy A, Raichur AM, Chandrasekaran N, Prathna TC and Mukherjee A (2009) Process variables in biomimetic synthesis of silver nanoparticles by aqueous extract of *Azadirachta indica* (Neem) leaves. *Jour of Nanoparticle Res* Article in press.

Tripathy A, Chandrasekaran N, Raichur AM and Mukherjee A (2008) Antibacterial applications of silver nanoparticles synthesized by aqueous extract of *Azadirachta indica* (Neem) leaves. *Jour of Biomedical Nanotechnology* 4(3): 1- 6

Tsuji K (2001) Microencapsulation of pesticides and their improved handling safety. *J. Microencapsul* 18 : 137–147

Uzu G, Sobanska S, Sarret G, Munoz M and Dumat C (2010) Foliar lead uptake by lettuce exposed to atmospheric pollution *Environ Sci Technol* 44 : 1036–1042

Vdayasoorian C, Kumar VRM and Jayabalahrishnan (2011) Extracellular synthesis of silver nanoparticles using leaf extract of *Cascia auriculate. Digest Jour of Nanomaterials and Biostructures* 6 (1):279-283

Wang L, Li X, Zhang G, Dong J and Eastoe J (2007) Oil-in-water nanoemulsions for pesticide formulations. *J Colloid Interface Sci* 314 : 230–235

Wang YA, Li JJ, Chen HY and Peng XG (2002) Stabilization of inorganic nanocrystals by organic dendrons. *J Am Chem Soc* 124: 2293-8

Wang YA, Li JJ, Chen HY and Peng XG (2002) Stabilization of inorganic nanocrystals by organic dendrons. *J Am Chem Soc* 124: 2293-2298

Werner S, Marillonnet S, Hause G, Klimyuk V and Gleba Y (2006) Immunoabsorbent nanoparticles based on a tobamovirus displaying protein A. *Proc Natl Acad Sci USA* 103(47):17678-17683

Willner I, Baron R and Willner B (2006) Growing metal nanoparticles by enzymes. *Adv Mater* 18:1109-1120

Xuming W, *et al.* (2008) Effects of nano-anatase on ribulose-1, 5-biphosphate carboxylase/oxygenase mRNA expression in spinach. *Biol Trace Elem Res* 126 : 280–289

Yang F, *et al.* (2006) Influences of nano-anatase TiO_2 on the nitrogen metabolism of growing spinach. *Biol. Trace Elem. Res* 110 : 179–190

Yang L and Watts DJ (2005) Particle surface characteristics may play an important role in phytotoxicity of alumina nanoparticles. *Toxicol Lett* 158 : 122–132

Zhang F, Wang R, Xiao Q, Wang Y and Zhang J (2006) Effects of slow/controlled-release fertilizer cemented and coated by nano-materials on biology. II. Effects of slow/controlled-release fertilizer cemented and coated by nano-materials on plants. *Nanoscience* 11 : 18–26

Zheng L, Hong F, Lu S and Liu C (2005) Effects of nano-TiO2 on strength of naturally aged seeds and growth of spinach. *Biol Trace Elem Res* 104 : 83–92

Zhou Q, Xiao J, Wang W, Liu G, Shi Q and Wang J (2006) Determination of atrazine and simazine in environmental water samples using multiwalled carbonnanotubes as the adsorbents for preconcentration prior to high performance liquid chromatography with diode array detector. *Epub* 68(4):1309-1315

Zhu H, Han J, Xiao JQ and Jin Y (2008) Uptake, translocation and accumulation of manufactured iron oxide nanoparticles by pumpkin plants. *J Environ Monit* 10 : 713–717

8

Assessing Morphological, Biochemical Utility and Molecular Diversity in Coriander (*Coriandrum sativum* L.)

Nisha Pareek and *C P Malik*

Introduction

Historically, India has been known as a land of spices. According to The International Standards Organization (ISO) about 112 plant species in the country are recognized as spices. The later form an important group of agricultural commodities, which since time immemorial have been considered indispensable in the culinary arts of flavoring foods. The seed spices were introduced to India from mediterranean and central asian region perhaps a long time ago. A total of 20 seed spices are grown in the country, which are divided into major and minor groups. Coriander is covered under the major seed spices. Rajasthan and Gujarat are the seed spices bowl and contribute more than 80% of the total seed spices produced in the country (Malhotra and Vashishtha, 2000). Spices and aromatic plants are mainly used for imparting flavor, aroma, pungency and for seasoning the food. Besides boosting flavor, they are also known for their preservative and medicinal value. Generally, spices are dietary constituents consumed daily to enhance the flavor or taste of the human food. A group of well known plants which are used as spices and condiments belong to the Umbelliferae family. The species *Coriandrum sativum* L. commonly known as Dhania is widely distributed throughout the world. It is an annual, tender hollow stemmed and herbaceous plant with fine and lacy foliage. Its occupies unique position and is believed to be one of the earliest spices used by mankind. Coriander is a minor aromatic annual condiment and spice crop. The unripe fruits have a smell that has been compared to that of bedbugs. The plant is named after *koris*, the Greek word for bug. However when ripe, the seeds have a distinctive sweet citrus/mint/musty aroma that has been valued over the centuries. One of the important attributes of coriander is aroma quality, and this quality as paramount to the culinary value of the fresh herb, usually decreasing before the visual quality decreases (Loaiza and Cantwell, 2001). Coriander leaves have become popular worldwide because of their pleasant and delicate aroma. Coriander is valued for its characteristic and flavour, as an ingredient in food preparations and also as a condiment. It is prized for its flavour and medicinal properties. Cilantro or Chinese parsley refers to the foliage while coriander usually refers to the seeds (Okut and Yidirim, 2005) but in terms of international trade, it is mainly grown for the round, aromatic seeds, which are used as a spice. It is an important cash crop of India (Farhani *et al.*, 2008). It has been reported as a cultivated crop in Russia, Central Europe, Asia and Morocco. It thrives in black soil and arid regions. India is the world's largest producer and exporter to countries like EU, USA, Middle east, south east Asia (Ravi *et al.*, 2006). Coriander occupy important place in the diet of common man. Its supply is reduced during summer season due to reduction in acreage and irrigation supply and due to its temperate nature the production during summer season having high temperature is problematic in open fields.

History

The use of coriander can be traced back to 5,000 BC, making it one of the world's oldest spices. Coriander is considered to have originated in the Mediterranean and Middle Eastern regions and transferred eastward to Asia. Coriander was cultivated in ancient Egypt and given mention in the Old Testament. Several authors have named coriander as a wild plant. It was reported as a weed in cereals by Linnaeus (1780). Alefeld (1866) mentioned it as a common weed spread from southeastern Europe to southern Russia. Stoletova (1930) also reported on wild *Coriandrum sativum* from Armenia. Coriander historically comes from Morocco and Romania. Greeks and Romans used it as a spice to preserve meats and flavor breads. The early physicians, including Hippocrates, used coriander for its medicinal properties, including as an aromatic stimulant. It was taken by Europeans to Asia and North and South America, where the Spanish name cilantro is used. The seeds have been used medicinally and as a food flavouring since ancient times, and were introduced into Britain by the Romans. It is believed that the plant was known to India since the Vedic Period, but had turned more popular for its fresh leaves. Indians did not make use of its seeds as spice till the Muslims arrived in the Indian scenario to introduce its seeds as a customary spice. That is perhaps the reason why the bulk of coriander seeds is consumed in Mughlai preparations, which are unquestionably of Muslim origin. Fresh coriander is now grown throughout the world.

Economic importance

It is considered both as an herb and a spice since both of its leaves and seeds are used as a seasoning condiment. Powdered coriander fruits are used as spice in some special foods with cumin. The seeds, in whole form, or the powder are extensively used as condiment as well as a flavoring agent in different foods namely pastries, cookies, buns, cakes and breads (Bhuiyan *et al.*, 2009). The seeds are used in dishes as varied as sauces, pickles, chutneys, cakes and some confectionery and candy. In spiced sausages and meat products, coriander seeds add flavor and pungency, making the food more spicy. Coriander is used as an appetizer, a common food seasoning in Mediterranean dishes. Its aromatic leaves are used as a herb in many oriental dishes and soups. The green leaves of coriander are consumed as fresh herb in preparing *chutneys* (Ilyas, 1980), sauces, in flavoring curries and soups. The fresh leaves of cilantro are highly regarded in the cuisine of China, Mexico, South America, India, and Southeast Asia. In Indian Subcontinent, both leaves and seeds are utilized commonly utilized, but roots are not utilized, while in China and Thailand leaves and roots are main parts of utilization. Several conservation techniques, e. g. drying, are applied to ensure a constant supply of the herb (Prakash, 1990; Ivanova and Stoletova, 1990). Coriander is always used fresh, as its leaves lose the flavor and aroma on drying. Coriander is never cooked since its flavor diminishes. It is used in the end for flavoring and garnishing the dish. Coriander is used also to flavor several alcoholic beverages, e. g. gin (Jansen, 1981). Its bold, lingering flavor is often used in Asian, Indian, Hispanic and Middle Eastern cuisine.

Table 1 : Chemical composition of *Coriandrum sativum* L. (Diederichsen, 1996)

Components	Content %
Water	11.37
Crude protein	11.49
Fat	19.15
Crude fibre	28.43
Starch	10.53
Pentosans	10.29
Sugars	1.9
Mineral constitutes	4.98
Essential oils	0.84

Essential Oil and Fatty acids

Coriander essential oil is used as a flavor ingredient. Besides its traditional use as a spice, coriander fruit has now raised a deep interest due to the appreciable characteristics of the oil distilled from it. It is obtained by steam distillation of the dried fully ripe fruits (seeds) of *Coriandrum sativum* L. The oil is a colorless or pale yellow liquid with a characteristic odor and mild, sweet, warm and aromatic flavor; linalool is the major constituent (70%). The main compounds of the oil are hydrocarbons or terpenes oxygenated derivatives of hydrocarbons (e.g., linalool, α-pinene and γ-terpinene citronellol, geraniol), benzene compounds (alcohols, acids, phenols, esters and lactones) and nitrogen or sulfur containing compounds (indole, hydrogen sulfide, methyl propyl disulfide and sinapine hydrogen sulfate). Essential oils lose their aroma with age. The seed essential oil is used to flavor alcoholic beverages, liqueur industry, chocolate making and pharmaceutics (Jansen, 1981) and in processing perfumes and soaps (Prakash, 1990), candies, meat, sauces and tobacco (Maroufi *et al.*, 2010). The oil is also fungicidal and bactericidal (Agrifacts, 2008). Like the fruits, it is also employed in medicine as a carminative or as a flavoring agent. It has an advantage of being more stable and of retaining its agreeable odour longer than any other oil of its class. After extraction of the essential oil, the residues are used as ruminant feed, since their composition is nearly the same as that of the whole fruits, and therefore still contains digestible fat and protein (Stoletova, 1931). Because of the high crude fiber content, coriander oil cake can only be fed to ruminants, but the nutritional value of this feed is limited (Rohr *et al.*, 1990).

The fatty oil is obtained either by pressing or by extraction of the fruits. The main component of the fatty seed oil, petroselinic acid [C18:1(6c)] which is an isomer of the oleic acid [C18:1(9c)]. The high content of petroselinic acid gives the oil peculiar physicochemical properties that make it potentially suitable for use in surfactants, polymers or as oleochemical raw material (Barclay and Earle, 1965). Petroselinic acid is a characteristic component of the fatty oil of umbelliferous plants.

Variation in oil content and compositions and other characteristics of Coriander accessions from different parts of the world has been studied in extensive analysis (Diederichsen, 1996; Diederichsen and Hammer, 2003; López *et al.*, 2008).

Table 2 : The composition of Essential oil in ripe fruits of *C. sativum* is given (Gildemeister and Hoffmann, 1931)

Main Components	**% of total essential oil**	**Minor components** (all with less than 2%)
Linalool	67. 7	β-pinene
α-pinene	10.5	Camphene
γ-terpinene	9.0	Myrcene
Geranylacetate	4.0	Limonene
Camphor	3.0	p-cymol
Geraniol	1.9	Dipentene
		α-terpinene
		γ-decylaldehyde
		Borenol
		Acetic acid esters

Table 3 : Composition of fatty acids in seeds of coriander (Diederichsen, 1996a)

Main components	**% of all fatty acids**	**Minor components**
Petroselinic acid	68.8	Stearic acid
Linoleic acid	16.6	Vaccenic acid
Oleic acid	7.5	Myristic acid
Palmitic acid	3.8	

Fig. 8.1: (a) Coriander leaves

(b) Coriander fruits

Medicinal importance

Coriander is an important medicinal herb. In addition to its use as spice and condiment, it has also been credited with long list of medicinal uses. It is used to treat liver complaints, flatulence, jaundice, rheumatism, piles and anemia in Indian systems of medicines (Aired, Sridhar and Unani). Great Ayurvedic physicians, Hippocrates and other Greek physicians employed it in digestive formularies. Coriander seeds have a health-supporting reputation that is high on the list of the healing spices. It has traditionally been referred to as an anti-diabetic, anti-inflammatory and recently been studied for its cholesterol-lowering effects. Coriander seeds are used in traditional Indian medicine as a diuretic by boiling equal amounts of coriander seeds and cumin seeds, then cooling and consuming the resulting liquid. In addition, its leaves have memory-enhancing effect, therefore appears to be a promising candidate for improving memory. Coriander has been used as a folk medicine for the relief of anxiety and insomnia in Iran. In holistic and traditional medicine, it is used as a carminative and as a digestive.

Common Indian names

Hindi	Dhaniya, Dhanya
Bengali	Dhane, Dhania
Gujrati	Kothmiri, Konphir, Libdhane
Sanskrit	Dhanyaka
Punjabi	Dhania
Kashmiri	Daaniwal, Kothambalari
Malayalam	Kothumpkalari bija, Kothumpalari
Marathi	Dhana, Kothimber
Oriya	Dhania
Tamil	Kothamali
Telgu	Dhaniyalu
Nepali	Dhania

Coriander in international languages

Spanish	Culantro
French	Corriandre
Arabic	Kuzhbare
Chinese	Hu-sui
Swedish	Koriander
Portuguese	Coentro
Russian	Koriandr
Japanese	Koendoro
Dutch	Koriander
German	Koriander

Portuguese	Coentro
Russian	Koriandr
Latin	*Coriandrum sativum* L.

Botany, Taxonomy, Systematics

Kingdom	Plantae
Sub Kingdom	Tracheobionta
Super Division	Spermatophyta
Division	Magnoliophyta
Class	Magnoliopsida
Sub class	Rosidae
Order	Apiales
Family	Apiaceae
Genus	*Coriandrum* L.
Species	*Coriandrum sativum* L.

Morphology

It is a weak-stemmed glabrous plant. The plant is a thin stemmed, small, bushy herb, 25 to 50 cm in height with many branches and umbels. The plant forms a tap root. Leaves are alternate, compound. The leaves are variable in shape, broadly lobed at the base of the plant and slender and feathery. The leaves are of two types, lower with leaflets and upper divided into narrow linear segments. The higher the leaves are inserted, the more pinnate they are. Thus, the upper leaves are deeply incised with narrow lanceolate or even filiform-shaped blades. The lower leaves are stalked, while the petiole of the upper leaves is reduced to a small, nearly amplexicaul leaf sheath (Diederichsen, 1996). Some varieties form a rosette of leaves at the base. Inflorescence is a compound umbel comprises 5 smaller umbels. Inner flowers are smaller and sterile, outer ones with longer petals are fertile. Hermaphrodite and staminate flowers may occur in each umbel (Rangahau, 2001). The pentamerous flowers are borne in small loose umbels, pink or white with a bicarpellary ovary. The fruit is a globular dry schizocarp which commonly called seed (Omidbaigi, 1997). Fruit, when pressed break into two locules each having one seed. It has delicate fragrance; seeds are pale white to light brown in colour. Fruit size is an indication of volatile oil content and suitability for particular end uses. Depending upon the fruit size and oil yield there are two varieties of coriander released from the Crop Development Centre (CDC) at the University of Saskatchewan. CDC Major is medium-large seeded, with a 1,000-fruit weight of around 9 grams. The essential oil content is 0.91 per cent, with linalool making up 65.6 per cent of the essential oils. CDC Minor is small seeded, with a 1,000-fruit weight around 7 grams. The essential oil content is higher than CDC Major, at 1.22 per cent, with 68.4 per cent linalool. Large fruited types are grown mainly by tropical and subtropical countries, e.g. Morocco, India and Australia. They are used extensively for grinding and blending purposes in the spice trade. Types with smaller fruit are produced in temperate regionsand are highly valued as a raw material for the preparation of essential oil. The whole plant has a pleasant aroma.

Medicinal benefits

The raw seed is chewed to stimulate the flow of gastric juices, to cure foul breath. Some caution is advised, however, because if used too freely, the seeds can have a narcotic effect. Seeds are also used as a purgative and roasted seeds are useful in dyspepsia. Coriander seed decoction is used in sore throat, common cold and bilious complaints. Decoction may also be used in eye wash and in chronic conjunctivitis. Coriander seeds are rich source of vitamin A and C and are considered as appetizer, digestive, carminative, tonic, antipyretic and diuretic. The paste of seeds is applied to relieve pain in cephalgia. The seeds have been used medicinally since ancient times. One pharmaceutical use of coriander seed is to mask or disguise the tastes of other medicinal compounds (active purgatives) or to calm the irritating effects on the stomach that some medicines cause, such as their tendency to cause gastric or intestinal pain. An infusion of the seeds in combination with cardamom and caraway seeds is used against flatulence, indigestion, vomiting and intestinal disorders. The decoction with milk and sugar is beneficial during bleeding piles. The juice of fresh plant is applied to erythema. The leaves are carminative, antibilious, diuretics, tonic, stomachic and aphrodisiac. Coriander is a commonly used domestic remedy, valued especially for its effect on the digestive system, treating flatulence, diarrhoea and colic. It settles spasms in the gut and counters the effects of nervous tension. Used externally, the seeds have been applied as a lotion or have been bruised and used as a poultice to treat rheumatic pains. The oil is also fungicidal and bactericidal. Coriander is also used to reduce intoxicating effects of spirituous liquors, prepare drink with dry seeds of coriander ground with water, add little water to it and keep for one hour then add sugar crystals and honey. The drink should be taken in short intervals in polydipsia and burning condition. Equal parts of powder of coriander, cumin and krishna jeera, boiled with jaggery should be taken to cure gout. Many of healing properties of coriander can be attributed to its exceptional phytonutrients and hence, it is often referred to as store house for bioactive compounds (Rajeshwari and Andallu, 2011).

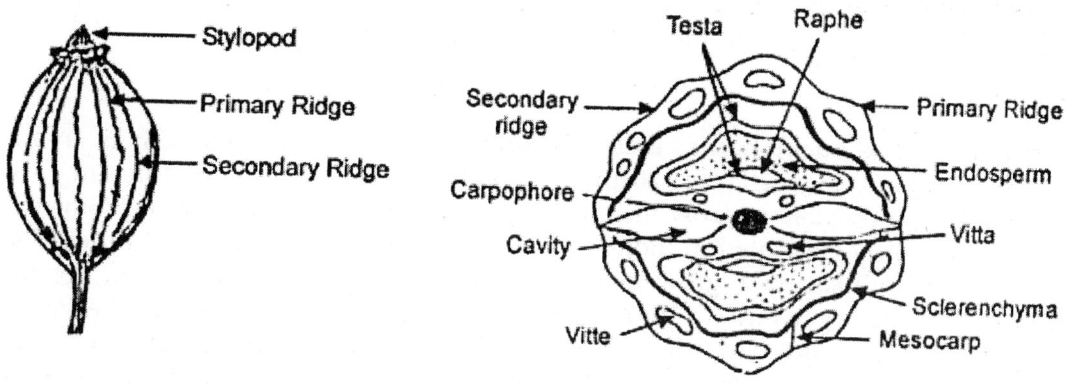

Fig. 2 (a) Coriander fruit **(b)** T.S. of coriander fruit (cremocarp) (Kokate *et al.*, 2008)

Indian production scenario

Export of coriander from India

India is the largest-producer and exporter of coriander in the global market. The exports have increased significantly in the past few years due to strong demand from the overseas markets. The major importers of coriander from India are Europe, the US, Singapore and the Gulf countries. Coriander is mainly grown in the northwest and central parts of the country. The major producers are Rajasthan and Madhya Pradesh. The strengths of Rajasthan in coriander are:

- High area, production and productivity in the country
- Internationally acceptable quality
- Climatic conditions suitable for quality production

The other coriander producing states are Punjab, Madhya Pradesh, Gujarat, Uttar Pradesh, Uttaranchal, Assam, Andhra Pradesh, Tamil Nadu and Orissa.

Global coriander scenario

India is the largest producer of coriander with a production share of more than 70% of the total world output. Other major producers are Bulgaria, Romania, Russia, Iran, Morocco, Canada and Australia. The global production of coriander seed is estimated to be around 6 lakh tonnes. India is world's largest producer and a potent exporter to countries like USA, Europe, Middle East and South East Asia (Ravi *et al.*, 2007). The other producers are Morocco, Canada, Romania, Russia and Ukraine, Iran, Turkey, Israel, Egypt, China, US, Argentina and Mexico. The global trade in coriander is estimated to be around 0.85-1 lakh tonnes per year.

Arrivals & Major Trading Centres

Coriander arrivals begin in February and extends until July-August. However, peak arrivals are seen only during March-May in the spot markets. The major markets for coriander are located in Rajasthan and Madhya Pradesh. The three largest mandis, namely, Kota, Ramganj mandi and Baran are located in the south-eastern parts of Rajasthan.

Cultivation

It displays broad adaptation as a crop around the world, cultivated both as an annual and a perennial herb (Maroufi *et al.*, 2010). Its rapid life allows it to fit into different growing seasons, making it possible to grow the crop under a wide range of conditions (Singh *et al.*, 2012) It is extensively cultivated in India, Russia, Central Europe, Asia and Morocco (Ravi *et al.*, 2007) and is well suited to grow under different types of soils and weather conditions (Guenther, 1952, Purseglove *et al.*, 1981). In India, it is cultivated in nearly all the states but largely cultivated in Rajasthan, Madhya Pradesh, Punjab, Andhra Pradesh, Tamil Nadu, Orissa, Uttar Pradesh, and Uttaranchal states. Rajasthan is major coriander growing state with its share of about 60% in the total area and production of the country. It produces various types of coriander like Badami, Eagle, Scooter, Double pet, Big type, Single, Green,

Special green, small size of Kumbhraj quality etc. India is the largest producer of coriander in the world with a market share of 70%. In Rajasthan, south east districts are traditionally known for the spice and are harvested in March-April. Coriander quality is evaluated for its color, size, aroma, number of broken seeds, and after taste. Although coriander is a tropical crop, it is successfully grown in a wide range of conditions. In India it is cultivated as a Rabi crop but can be grown throughout the year (except very hot season i.e. March-May).

Sowing

The time of sowing however varies in different locations in India. In West Bengal, Karnataka and Uttar Pradesh, the seeds are sown during the rabi season, while sowing is done during the kharif in Maharashtra and Tamil Nadu. The propagation of coriander is through fruits. Before sowing, coriander fruits are rubbed, to separate the mericarps. Seeds can be sown directly in the garden or in a container or pot. Seeds are planted 1/2" to 1" deep, 2-3 feet apart in rows and rows should be 15" apart. Sowing is done during October - November and new crop arrivals are seen in February-March. Coriander traditionally germinates very slowly and can take as long as 21 days to emerge. The plants will grow to about 30-90 cm in height. It requires a cool climate in the early stages of growth and warm dry weather at maturity. For leaf purpose, it can be grown in small beds around urban areas throughout the year. As an irrigated crop, it is grown during June-July and September-October at the onset of northeastern monsoon and harvested on maturity during January-February (Nimish *et al.*, 2011). It can be successfully grown on a wide range of climatic conditions but it thrives well in sunny locations. A temperature range between 15°C to 25°C is suitable for its seed germination and between 10°C to 25°C is favourable for its vegetative growth. A dry and cold weather favours higher seed production. Cloudy conditions at flowering and fruiting stage increases pest and disease incidence. Heavy rains are harmful for the crop and continuous cloudy weather increases the risk of diseases and aphids. It does not grow well in spring summer season for greens, because it switches over within short time from vegetative growth to reproductive phase as soon as temperature raises beyond 20°C. The crop duration is about 110-140 days.

Harvesting

The crop is and harvested in March-April. Fruits need a long season to ripe. Harvesting of coriander for seed is done when 75 to 80 per cent of seed capsules in the umbel turn yellow. Delay in harvesting should be avoided, since it will cause shattering of seeds. Seeds are extracted, dried in the shade to retain the green color and aroma to minimize moisture content. Harvesting of seeds can be done by cutting the clusters as soon as the seeds start to ripe. They are hanged upside-down in the room to dry, then shaken on a sheet of paper. Seeds are collected and stored in airtight container away from sunlight and heat.

Soil

Coriander can be on a wide range of soils from sandy loam to black clayey soils, but well-drained loamy soil suits the crop well. The pH of the soil should be near 7.0. Saline, alkaline and sandy soils are not suitable for its cultivation. Coriander is cultivated both as an irrigated and rain-fed crop. Under irrigation when sufficient organic matter is available in the soil and under rainfed conditions, black cotton soils, having high water holding capacity are the best soils suitable for its growth.

Pests

Young coriander plants can be attacked by rabbits or sparrows, they can be protected by using nets. Slugs can be combated by hand picking, parasitic nematodes, encouraging biological controls such as ground beetles, using baits such as citrus halves or beer, or ferric phosphate based slug pellets. At the seedling stage coriander is often attacked by the leaf eating caterpillars and semi-loopers.

Aphids (*Hyadaphis coriandra*) attack the plants at the flowering stage. They damage the crop by sucking the cell sap from inflorescence (umbel). They can be controlled by growing relatively resistant varieties like UD-20, Pusa Selection 360 or by spraying the crop before flowering with methyl demeton (0.05%), Endosulfan or Phosphamidon or Monocrotophos @ 0.03 %.

Brown wheat mite (*Petrobia latens*) appears during the first week of March. It can be controlled by spraying before flowering with Endosulfan or Phosphamidon or Monocrotophos 0.03%.

Other insects, causing minor damage to coriander crops are (a) pentatomid bug (*Agnoscalis nubila*), (b) surface grass hopper (*Chrotogonus trachypterus*), (c) Lucerne caterpillar (*Spodoptera exigua*), (d) white fly (*Bemisia tabica*), (e) green peach aphid (*Myzus persicae*) and (f) thrips (*Thrips tobaci*). Spraying with Monocrotophos or Nuvacron I0.05 %, Kelthane (Dicofol) or spraying 0.05 % Dimecron or Rogor can be effective in controlling the insects.

Diseases

Under cool damp conditions, coriander may suffer from the seedborne disease, bacterial leafspot (*Pseudomonas syringae* pv. coriandricola). It can be minimised by using certified, clean seed from reputable seed suppliers, cleaning up plant debris, and watering at the base of plants where possible rather than overhead watering.

Powdery mildew - Powdery mildew (*Erysiphe polygoni*) (Singh and Banwasi, 2011) is a serious disease, which ruin the crop if allowed unchecked in the initial stage itself. Spraying wettable sulphur 0.25% or 0.2% solution of Karathane twice at 10 to 15 days interval or dusting the crop with 20-25 kg sulphur per hectare is recommended.

Wilt disease - There are no direct control measures available for wilt disease. However, selection of disease free seeds, seed treatment with fungicides, use of crop rotation, etc. are some of the preventive measures suggested to control the wilt disease.

Stem-gall - Stem gall is also difficult to control. Some preventive measures suggested to control this disease are (i) sowing seed of resistant varieties like UD-41 or Karan, (2) treating seed with Agrosan GN @ 2.5 g/ kg seed before sowing, and (3) use of 3-4 years crop rotation.

Coriander can be affected by fungal diseases common to many broadleaf crops including *Alternaria* spp. and *Septoria* spp. These diseases are particularly prevalent in warm and humid weather. Grain mould is caused by *Helminthosporium* sp., *Alternaria* sp., *Carvularia* sp. and *Fusarium* sp. It can be controlled by spraying carbendazim 0.1% 20 days after grain set.

Mixed Cropping

The plants of coriander do not create any problem in space and competition for sunlight, hence can very well be grown as mixed crop, or sow a few lines in crops like wheat, linseed, rapeseed, mustard, sugarcane, potatoes, winter vegetables etc.

Crop Rotation

Coriander crop can be sown after the harvest of kharif crops. Therefore, it can be grown in various combinations.

1. Paddy-coriander-wheat
2. Maize-coriander-sorghum
3. Cotton-coriander-maize
4. Maize-coriander-green manuring -potato –sugarcane (ratoon)
5. Bajra-coriander-wheat-maize

Since it is a short duration crop, it makes adoption of cropping with most of the crops.

1. Maize-coriander-moong
2. Soyabean-coriander-iobia
3. Maize-coriander-been.

Varieties

Some important and improved varieties of coriander are described below:-

CO 1

This variety was released by TNAU, Coimbatore as a pure line selection. It is tall, produces many umbels per plant. It matures in 110 days.

CO 2

It was released by TANU Coimbatore as a reselection from culture P_2 of Gujarat. High yielding, dual purpose variety, tolerant to drought, oil content is 0.3%. Matures in 90-110 days and yield is 600-700 kg per hectare.

CO 3

It was obtained from reselection from Acc. No. 695, developed at Coimbatore. High yielding, dual purpose variety with medium grain size. It is less susceptible to wilt and grain mould. The seed oil content is 0.38-0.41%. Duration of maturity is 103 days and yield is 640 kg per hectare.

CS-287

This variety was developed at Coimbatore by recurrent selection from CS-6 variety of Andhra Pradesh. It is a short duration variety having small bold grains. It is tolerant to wilt and grain mould. Suitable for rainfed tracts, it matures in 78 days and yield is 510 kg per hectare.

Lam Sel. CS-2

This variety was developed at Guntur by mass selection. It is a medium tall, bushy type variety with more number of branches. It is a grain purpose variety, tolerant to pests and diseases with good quality grain having 0.4% oil. It matures in 110 days yield is 1300 kg per hectare.

UD-20

This variety was developed at Jobner (RAU). It is a medium tall plant with bold grains. It matures in 100-110 days yielding 1175 kg per hectare.

UD-21

It is a medium tall variety developed at Jobner (RAU). Matures in 90 days, produces about 21 umbels with 75 umbel lets per plant. It yields about 700 kg per hectare.

Gujarat Coriander 1

It was developed in GAU, Jagudan by the selection of a local release. It has large number of branches and seeds are bolder and greenish in colour. High yield, more number of branches, seeds bolder and greenish in colour. Maturity duration is 112 days. It has a high yield of 1100 kg per hectare.

Gujarat Coriander 2

It has large number of branches, dense foliage, large sized umbels and bold seeds. It is used for grain purpose. It is a high yielding variety with a yield of 1500 kg per hectare. It was developed by a selection from CO_2 released by GAU, Jagudan. Matures in 110-115 days.

Pusa Selection-360

It is recommended for Rajasthan and Maharashtra states. Plant height is about 60 cm. It produces large number of umbellets per plant and yields about 40 g of grains per plant. It yields about 1250 kg per hectare.

Rajendra Swati

It originated by the mass selection from germplasm type, released by RAU, Dholi. It is suitable for intercropping. It is a fine seeded, high yielding variety, rich in essential oil, resistant to stem gall disease. Duration of maturity is 110 days. Yield is 1200-1400 kg per hectare.

RCr-41

It was obtained by the recurrent selection from UD 41, released by RAU, Jobner. It is tall erect, high yielding variety, suitable for irrigated areas and resistant to stem gall. Maturity duration is 130-140 days. Yield is 1200kg per hectare.

Swati

Released by APAU, Guntur, produced from mass selection. A semi erect high yielding variety, suitable for delayed sowing. It matures in 80-90 days. Yield is 885 kg per hectare.

Sadhana

Released by APAU, Guntur obtained through mass selection. A semi erect high yielding variety, suitable for rainfed areas, resistant to aphid and mites. Maturity duration is 95-105 days. Yield is 1000 kg per hectare.

Gwalior No-5365

It is a medium-tall short duration variety, maturing in 85 days with bushy type branches. It is free from powdery mildew disease due to its early maturing habit. Hence it is recommended for areas where dew fall is high.

Cimpo 8-33

This is a new promising variety introduced from Bulgaria. It is late maturing with taller growth habit. It has fine seeds, rich in essential oil-content (1.3 %). It matures in 180-190 days yielding 18-20 quintals of grains per hectare.

Morrocan

Medium sized seeds with high oil content. Yields about 10-12 quintals grains per hectare, maturing in 150 days.

Russian varieties

Strains like A-26, A-247, A- 704 and Roose have been developed in Russia. These are high yielding varieties, with high essential oil contents.

Therapeutic and Pharmacological potential of *C. sativum*

Antioxidant Activity

Coriander extracts have phenolic compounds and flavonoids, these compounds contribute to the antioxidative activity (Helle Wangensteen, 2004). Phenolic substances such as flavonoids, cumarins, cinnamic acid and caffeic acids are believed to have antioxidant properties, which may play an important role in protecting cells and any organ from oxidative degeneration (Wiseman *et al.*, 2000). The antioxidative property of coriander seeds is related to the large amount of tocopherols, carotenoids and phospholipids which act through different mechanisms (Kansal *et al.*, 2012). Radical Scavenging Activity (RSA) among the seed oils of of Black Cumin (*Nigella sativa* L.), Coriander (*Coriandrum sativum* L.) and Niger (*Guizotia abyssinica* Cass.) was compared by Ramadan *et al.*, 2003. Coriander seed oil and its fractions exhibited the strongest RSA compared to black cumin and niger seed oils. Extracts of different polarity from leaves and seeds of coriander and coriander oil are reported to have antioxidant activity (Wangensteen *et al.*, 2004). Melo and coworkers (2004) have characterized antioxidant compounds in aqueous extract of coriander.

Cholesterol lowering

Coriander has been documented as a traditional treatment for cholesterol and diabetes patients. Dhanapakiam and coworkers (2006) studied the effect of the administration of coriander seeds on the metabolism of lipids in rats, fed with high fat diet and added cholesterol. The seeds were shown to have a significant hypolipidemic action.

Antidiabetic activity

The seeds of coriander showed significant hypoglyemic activity by enhanced glycogenesis, glycolysis and decreased glycogenolysis and gluconeogenesis and may be due to increased utilization of glucose in liver glycogen synthesis and decreased degradation of glycogen to ive blood sugar. Waheed *et al*., 2003 investigated clinically the hypoglycemic effect of Coriander in Type-2 diabetes mellitus and found that it to have a significant hypoglycemic activity in high dose and which can be successfully combined with oral hypoglycemic agents in type-2 diabetic patients.

Aflatoxin control

Basilico and Basilico (1999) have studied inhibitory effects of essential oil of coriander on the mycelial growth and ochratoxin A production by *A. ochraceus* NRRL 3174. Efficacy of coriander essential oil on seed mycoflora and seedling quality of some crop species have also been reported.

Insecticidal effect

Pascual Villalobos (2003) found the potential of plant essential oils against stored-product beetle pests. Coriander oil (10 μl) showed insecticidal activity against the bruchid *Callosobruchus maculates*, the cereal storage pest.

Anti anxiety

Hydroalcoholic extract of *Coriandrum sativum* (Linn.) are reported to possess anti-anxiety activity (Mahendra and Bisht, 2011).

Antibacterial activity

Essential oils extracted from the fruits of *Coriandrum sativum* were found to possess antibacterial activity against *Escherichia coli* and *Bacillus megaterium* (Cantore *et al*., 2004) Aliphatic (2*E*)-alkenals and alkanals characterized from the fresh leaves of the coriander are shown to possess bactericidal activity against *Salmonella choleraesuis* ssp. *choleraesuis* ATCC 35640 (Kubo *et al*., 2004).

Hepatoprotective activity

C. sativum extract protects liver from oxidative stress induced by carbon-tetrachloride (CCl_4) and thus helps in evaluation of traditional claim on this plant. Pretreatment of rats with different doses of plant extract (100 and 200 mg/kg) significantly lowered serum glutamate oxaloacetate transaminase (SGOT), serum glutamate pyruvate transaminase (SGPT), and TBARS levels against CCl_4 treated rats. Hepatic enzymes like superoxide dismutase (SOD), catalase (CAT), and glutathione peroxidase (GPx) were significantly increased by treatment with plant extract, against CCl_4 treated rats. Oral administration of the leaf extract at a dose of 200 mg/kg significantly reduced the toxic effects of CCl_4. The activity of leaf extract at this dose was comparable to the standard drug, silymarin (Sreelatha *et al*., 2009).

Anthelmentic activity

In vitro anthelmentic activities of crude aqueous and hydro-alcoholic extracts of the seeds of coriander were investigated on the egg and adult nematode parasite *Haemonchus contortus* Momin *et al.*, (2012). The aqueous extract of coriander was also investigated for *in vivo* anthelmintic activity in sheep infected with *Haemonchus contortus.* Both extract types of *Coriandrum sativum* inhibited hatching of eggs completely at a concentration less than 0.5 mg/ml. ED(50) of aqueous extract of *Coriandrum sativum* was 0.12 mg/ml while that of hydro-alcoholic extract was 0.18 mg/ml. The hydro-alcoholic extract showed better in vitro activity against adult parasites than the aqueous one.

Cardioprotective effects

The preventive effect of *Coriandrum sativum L.* (CS) on cardiac damage was evaluated by Isoproterenol (IP) induced cardiotoxicity model in male *Wistar* rats by Patel *et al* (2012).

Biosynthesis of Nanoparticles

There are reports on phytosynthesis of gold nanoparticles by employing coriander leaves (Narayanan and Sakthivel, 2008). Extracellular biological synthesis of gold nanoparticles was achieved by a simple biological procedure using coriander extract as the reducing agent. The aqueous gold ions when exposed to coriander leaf extract are reduced and resulted in the biosynthesis of gold nanoparticles in the size range from 20.65-7.09 nm. FTIR spectroscopic analysis showed the presence of Amide I and II bands of protein that capped and stabilized the nanoparticles, and diffraction studies like XRD and SAED confirmed the FCC structure of crystalline gold nanoparticles.

Cytological studies in C. sativum

The somatic chromosome number of coriandrum is reported to be 2n = 22. Das and Mallick (1989) carried out karyotypic analysis of 8 varieties of *C. sativum* collected from India and abroad and indicated the importance of structural alterations of chromosomes in the origin of new varieties.

Genetic resources conservation strategies

Conservation of genetic resources is extremely important in context of rapid gene erosion taking place due to a variety of biotic, abiotic, socio political and economic factors. The loss of land races and traditional varieties is rapid in certain crops due to devastating diseases, spread of improved cultivars, deforestation etc. Coriander (*Coriandrum sativum* L.) was domesticated at some point of early history and now has no wild ancestors, so collections of accessions in genebanks are the only source of new genes for breeders (Bashtanova and Flowers, 2011). To meet the demand of an ever-increasing population there is a need to explore conservation strategies to tap the potential of minor and underutilized crop species like *C. sativum.* For many plant species, *ex situ* and *in situ* conservation strategies have been developed to safeguard the extent of genetic diversity.

In situ conservation

It refers to conservation of germplasm under natural conditions which allow the maintenance of organisms in their natural habitat. For cultivated crops, conservation of genetic resources *in situ* refers to the continued cultivation and management of crop populations by farmers in the open. Coriander is widely grown in home gardens on a small scale. In the Near East and Indian subcontinent, the production of coriander has recently become widespread, and collecting has also been carried out in these regions (Hammer, 1993b).

Ex situ conservation

Ex situ conservation of plant genetic resources is preservation of germplasm in gene banks. This is the most practical method of germplasm conservation. The All India Co-ordinated Research Project on Spices (AICRPS) is the largest spices research network in the country. Research on 12 mandate crops is carried out under the AICRPS in 20 coordinating centres, located in 15 states with the respective agriculture universities, representing the major agro climatic regions. A good number of germplasm collections are also being maintained at the AICRPS Centers. One thousand nine hundred and ninety (1990) germplasm of coriander have been conserved at various AICRPS centers for further evaluation and characterization. The National Research Centre on Seed Spices Tabiji, Ajmer is actively implementing conservation of crop species. So far the Institute holds about 288 accessions of coriander comprising 270 indigenous and 18 from exotic sources. Other coriander germplasm collection centers under ICAR (Indian Council of Agricultural Research) are Jobner (Rajasthan), Coimbatore (Tamil Nadu), Jagudan (Gujarat), Dholi (Bihar) and Guntur (Andhra Pradesh).

International Institutions with coriander germplasm collections

- Genetic Resources Centre, Bangladesh Agricultural Research Institute — Bangladesh
- Institute of Crop Germplasm Resources (CAAS) Beijing — China
- Institute for Plant Genetics and Crop Plant Research (IPK), Genebank, Gatersleben — Germany
- Plant Genetic Resources Centre — Ethiopia
- North Central Regional Plant Introduction Station — USA

(Diederichsen, 1996a)

Genetic Diversity

Genetic diversity is the sum total of genetic variations found in a species or population. Quantification of genetic diversity existing within and between groups of germplasm is important and particularly useful in proper choice of parents for realizing higher heterosis and obtaining useful recombinants and its fruitful utilization for its systematic collection, evaluation, description and grouping based on economic descriptors. Evaluation of genetic diversity is of key importance in order to identify and select suitable genotypes for their proper utilization in crop improvement program. It is a source of continuing advances in yield, pest resistance and quality improvement. It is widely accepted that great varietal and species diversity would enable agricultural system to maintain productivity over a wide range of conditions. It is a suitable precursor for improvement of the crop because it generates baseline data to guide selection of parental lines and design of a breeding scheme. Knowledge of genetic diversity of genetic resources is crucial for breeders to understand evolution and gene relation among accessions, to select germplasm in a more systematic and effective fashion, and to develop strategies to incorporate useful diversity in their breeding program.

Sources

Genetic diversity between and within plant populations can be due to physical remoteness, population size, type of mating system (selfing or out crossing), mode of dispersal of seed and pollen, and speed of gene flow (Loveless and Hamrick, 1984). Species that are largely out crossing show lower inter-population and higher intra-population difference in genetic variation compared to the species where self-fertilization predominates. Species that are self pollinated are less genetically diverse as compared to out crossing species.

Significance

- It allows species to adapt to ever changing environmental conditions by providing a range of possible phenotypes thus protecting it against extinction.
- Genetic diversity is crucial for survival of a species by producing certain anatomical or physiological changes in it.
- It plays important role in evolution of a species.
- It is essential for the development of improved cultivars with wide adaptability and broad genetic base.

A thorough assessment of genetic diversity, extent of variation, genetic makeup of the plant and hereditary of characters is desirable for the purpose of better management and conservation of the genetic resources and for planning the breeding strategies.

Characterization of Genetic Diversity

Markers

A marker can be a visually assessable trait, protein molecule or a DNA segment that lie near each other on a chromosome and tend to be inherited together. Markers are used in tracking the inheritance pattern of a gene that has not yet been identified, but whose approximate location is known. They are important tools in characterization of genetic diversity as they help to obtain more precise and detailed information regarding polymorphism.

Morphological

Any observable traits that can be measured as a difference between any two individuals e.g. plant height, seed shape, flower color, etc serves as a morphological marker. Conventionally, genetic diversity has been assessed on the basis of dissimilarity in morphological and agronomic characters. Several different phenotypic markers have been identified in several important crops like wheat (Khodadadi *et al.*, 2011), soybean (Khan *et al.*, 2011) and tea (Rajanna *et al.*, 2011). Morphological traits are easy to mark, offer straight forward detection and measurement, they are significant in identification of varieties and in plant breeding programs and permit plant species classification. They provide a partial knowledge of the average of the functional variants of genes carried by a given species or population. However, they are sensitive to environment; different individuals of the same species may present a variation in their morphology either naturally or in connection with local adaptation. Moreover, evaluation of traits requires growing the plants to full maturity prior to identification and large tracts of land to carry out the experiments, making it more expensive. These markers are few in number, insufficiently polymorphic, show epistatic and pleiotropic effects and are mostly dominant and less genome coverage hinders their usage. They have complex genetic control thus making it difficult to relate patterns of phenotypic variation to their genetic bases, and environmental modification (Lombard *et al.*, 2001; Nuel *et al.*, 2001). The number of suitable phenotypic markers linked to genes of interest is limited, which makes it difficult to construct a dense genetic map only with phenotypic markers. All these drawbacks limit the wide application of phenotypic markers in plant breeding.

There are investigations of genetic divergence among coriander populations based on phenotypic traits like plant height, leaf shape, and seed weight (Jindla *et al.*, 1985; Sastri *et al.*, 1989; Angelini *et al.*, 1997; Patel *et al.*, 2000).

Diederichsen (1996) conducted an extensive analysis of coriander populations collected from six ecogeographic types: Near Eastern, Indian, Central Asian, Syrian, Caucasian, and Ethiopian using morphological variation, complemented by some major constituents from essential oil. Similar studies based on phenological and morphological traits have been reported by Diederichsen and Hammer (2003).

Table 4 : Morpho-agronomic traits correlation studies in *C. sativum*

Characters studied	Material used	Correlation with grain yield	References
Umbellets/plant, Umbels/plant, Number of effective branches, Straw yield/plant, Number of primary branches, Plant height, Number of grains/umbellet, Harvest Index	200 genotypes of (29 entries from U.S.A and 171 from different states of India)	Highly significant and Positive	Bhandari and Gupta (1991)
Plant height, Umbels/plant, leaves/plant, seeds/umbel,	70 accessions of *C. sativum* collected country-wide	Positive	Singh *et al.*, (2005)
Plant height, Umbels/ plant, leaves/plant, seeds/umbel,	49 Ethiopian *C.sativum* accessions	Positive	Mengesha and Alemaw, (2010)

Molecular characterization

The productivity of spice crops is considerably low due to various factors such as inadequate availability of high yielding varieties, absence of genotypes resistant to pests and diseases and absence of variability in many of the introduced crops. Biotechnology with its apparently unlimited potential offers new and exciting opportunities to address the above crop specific problems. In recent times there is increased emphasis in molecular markers for characterization of the genotypes, genetic fingerprinting, in identification and cloning of important genes, marker assisted selection and in understanding of inter relationships at molecular level. Molecular markers like RAPD, AFLP and ISSR polymorphism have been used for assessment of genetic variability in coriander to develop finger prints and to study the inter relationships.

AFLP

A study was conducted to assess phenotypic, biochemical, and molecular diversity in coriander accessions from the North Central Regional Plant Introduction Station, in Ames, IA. Lopez *et al.* (2008) assessed the phenotypic, biochemical and molecular diversity in *C. sativum* germplasm. They reported that weak correlations between phenotypic and biochemical matrices and between biochemical and AFLP matrices and reported no correlation between phenotypic and AFLP matrices. They concluded that this incongruence is due to phenotypuc plasticity and wide spread trade of coriander seeds as spice, which might have resulted in poorly differentiated molecular variation, even when phenotypic and biochemical differentiation is easily documented.

RAPD

Genetic variability is highly significant for the improvement of many crop species including Coriandrum. The genetic divergence among 10 varieties of Coriander (*Coriandrum sativum*) grown in Rajasthan UD 29, 182, 262, 358, RCr 20, 41, 435, 436, 480 and RCr 684 was assessed employing RAPD and ISSR markers (Pareek *et al.*, 2012 and 2013). For RAPD- PCR, 30 random primers of GCC series were tested, out of which 8 gave good amplification and polymorphism. RAPD generated a total of 74 clear bands, out of which 43.2% were polymorphic. The total number of markers varied from 6 to 13 with a mean of 9.25 markers per primer. The number of polymorphic markers for each primer varied from 1 to 11 with a mean of 4 polymorphic markers per primer. PIC value ranged from 0.05 to 0.22 with a mean of 0.12. The Jaccard's similarity coefficient values ranged 0.70 to 0.95 with a mean of 0.81. A dendrogram constructed based on the UPGMA clustering method, revealed two major clusters.

ISSR

A set of 25 primers were screened, out of which 10 showed polymorphism. A total of 71 amplicons were obtained among which 18 were polymorphic. The number of bands obtained for each primer ranged from 5 (UBC 841, 900) to 10 (UBC 827). The percentage of polymorphic bands was 25.35%. Highest number of polymorphic loci was 3 obtained with primer UBC 810 and 827 (30%). The discrimination power of each locus was estimated by the PIC

(Polymorphism Information Content) value. It ranged from 0.04 to 0.13. The similarity coefficient ranged from 0.79-1 with an average of 0.91. The two most divergent varieties based on similarity coefficient were 3 and 1 (0.79) while the varieties 8 and 5 were 100% similar.

Coupling of RAPD and ISSR markers

RAPD and ISSR allelic profiles of coriander germplasms were coupled to infer genetic relationships among them. Based on the combined data, maximum similarity index was recorded for genotypes 8 and 6 (0.95) while the minimum was observed for 3 and 1 (0.76). On an average polymorphism of 34.3% was obtained.

The dendrogram constructed based on RAPD and ISSR marker data revealed two major clusters at 82% similarity.

Further, the dendrogram obtained on coupling the RAPD and ISSR data exhibited some similarity with RAPD dendrogram. Varieties UD 358 and RCr 20 were clustered together both in RAPD as well as in RAPD + ISSR analyses. This shows high similarity between the two genotypes.

Genetic diversity using both the dominant markers (RAPD + ISSR) has been described in many species, especially in *Apium* (Castellini *et al.*, 2006), *Chamomile* (Solouki *et al.*, 2008), *Argemone* (Karnawat and Malik, 2011, Karnawat *et al.*, 2013) and *Jatropha* (Basha *et al.*, 2009). These data permit establishment of core, non-redundant germplasm collections and help to guide breeders who wish to use germplasm assets for crop improvement. Comparative studies in plant species involving combinations of molecular markers have been successfully used by some researches (Ravi *et al.*, 2003, Gillaspie *et al.*, 2005, Dikshit *et al.*, 2008). This the first study involving combination of RAPD and ISSR analyses to estimate genetic variability of coriander genotypes grown in Rajasthan. Variation in DNA sequences leads to polymorphism which is indicative of genetic diversity. The present results indicated the occurrence of meager genetic variability among the studied genotypes of coriander. The variability in the number of fragments produced by RAPD and ISSR primers could be attributed to the differences in the binding sites throughout genome of the genotypes included. Primers exhibiting higher polymorphic bands are more efficient for studying genetic diversity and discrimination of the genotypes. Among the RAPD primers used by us, primer GCC 81 was found to be highly polymorphic (84.6%) and among ISSR primers, primer UBC 827 was found to be highly polymorphic (30%). PIC values reveal allelic diversity for a specific locus. These values can be used to estimate relative differences in the degree of heterozygosity for each of the population. The higher the PIC value for a locus, the elevated is the probability that polymorphism will exist between two accessions at that given locus. The ISSR markers exhibited a relatively low PIC average value (0.08) compared with RAPD markers (0.12). It may be stated that PIC values range from 0, when a given marker is present in all the populations under study, to a maximum value of 0.5 when a given marker is present in exactly half of the studied populations. Low PIC values suggest that some coriander varieties may be homogenous, either because of their breeding history or because of small sample size during acquiring or regeneration. Similar studies reporting lower PIC values are expected for inbred lines or for self pollinated populations than for cross-pollinated populations. It may be mentioned that PIC values are vital estimate of the intra-population diversity. Consequently, they can be used as a decisive factor when selecting base germplasm for selection in breeding programs in coriander.

However, the results obtained from cluster analysis based on RAPD and ISSR data sets were different. Cluster analysis based on RAPD and ISSR data showed distinct separation of genotypes at 76% and 87% of variation respectively, while grouping based on RAPD + ISSR data showed separation of genotypes at 82% of variation. Clustering of coriander genotypes based on the combined RAPD and ISSR data showed some similarities with the RAPD clustering. UD 358 and RCr 20 were clustered together both in RAPD as well as in RAPD + ISSR analysis. This indicates high similarity between the two genotypes.

On the basis of percent of polymorphism (RAPD = 43.24%; ISSR = 25.35%) and similarity matrix, the RAPD markers were marginally more informative than ISSR. Present results are in congruence with those reported in *Cladesia grandis*, castor and canola (Abdelmigid *et al.*, 2012).

The efficiency of a molecular marker technique depends on the amount of polymorphism that could be detected among the set of accessions investigated. In the present study, RAPD fingerprinting was more efficient than the ISSR assay. While RAPD scans the entire genome in coding and noncoding regions including repeated or single-copy sequences, ISSR markers disclose polymorphism in the repeat regions. Therefore, RAPD markers still provide a rapid and useful technique to investigate genetic diversities at the whole genome level (Ahmad *et al.*, 2010). Markers differ in their ability to differentiate among individuals, in the mechanism of detecting polymorphism, in genome coverage, and in the ease of application. The low correlation of grouping patterns obtained from RAPD and ISSR data could be attributed to the different genomic constitution of two markers (Ravi *et al.*, 2003). Polymorphism obtained in the study indicates the existence of genetic variation among the coriander genotypes. This variation is attributed to the differences in the number of alleles per locus/or loci and their distribution within the population. Our investigation provides new information at the molecular level for the genetic variability of *C. sativum* and for a general phylogenetic relationship among the 10 varieties of this species. Current study highlights meager genetic diversity among various varieties and attributed to recent human selection pertinent to few genes.

ISSR

Melo and coworkers (2011) evaluated the genetic similarity between ten coriander genotypes (nine cultivars and one line) using ISSR markers. The cultivars used were: Americano, Asteca, Palmeira, Português, Santo, Supéria, Tabocas, Tapacurá, Verdão and the experimental line HTV-9299. The genetic similarity between the cultivars was estimated using 227 banded regions of ISSR molecular markers. The UBC 897 primer generated the highest number of fragments (16), resulting in a higher polymorphism. The results indicated that the 29 primers chosen were satisfactory for detecting polymorphism. Based on the cluster analysis determined from the similarity data, the genotypes were clustered into two groups and two sub groups. The calculated similarity for the genotypes varied from 52 to 75%. The lowest similarity was observed between Português and Verdão, at 52%. The highest similarity was found between Português and Palmeira, at 75%. In conclusion ISSR proved to be an efficient marker for identifying DNA polymorphism in coriander.

Internal Transcribed Spacer Regions of rRNA (ITS)

Singh *et al.* (2012) estimated genetic divergence among 22 Indian varieties of coriander viz. Hisar sugandh, Rajendra swathi, Hisar surbhi, Azad Dhania, NRCSS-Acr 1, CO 1, CO 2, CO 3, CO 4, GCr 1, GCr 2, RCr 20, RCr 41, RCr 435, RCr 436, RCr 684, Sudha, H. Anand, Sindhu, JDI, P. Haritma, Swathi, using phenotypic and genetic markers. Initially 20 random primers were screened out of which 9 RAPD primers generated 68 bands out of which 45 were polymorphic. An average polymorphism of 66.18% was obtained. On the basis of delineation, one representative variety from each RAPD sub-group and all distinct varieties were selected. Genomic DNA of these 12 varieties viz. Hisar sugandh, Rajendra swathi, Hisar surbhi, Azad Dhania, NRCSS-Acr 1, CO 3, CO 4, GCr 1, GCr 2, RCr 435, RCr 436 and RCr 20 was used for amplification and sequencing of 5.8S gene region. Total internal transcribed spacer (ITS) length variations and single nucleotide polymorphisms, insertions/ deletions (INDELS) were detected at seven sites in ITS-1 region. Multiple sequence alignment of 12 sequenced varieties revealed 100% identities of 5.8S gene region that validates its conserved nature. Multiple sequence alignment of ITS-1 region may be of phylogenetic significance in distinguishing and cataloguing of coriander germplasm. It was found that the measures of relative genetic distances among the varieties of coriander did not completely correlate the geographical places of their development.

REFERENCES

Abdelmigid HM (2012) Efficiency of random amplified polymorphic DNA (RAPD) and inter-simple sequence repeats (ISSR) markers for genotype fingerprinting and genetic diversity studies in canola (*Brassica napus*). *African Journal of Biotechnology* 11(24): 6409-6419

Agri-facts Coriander (2008) Practical information for Alberta's agriculture industry. Accessed online at: http://agricgovabca/agdex/100/147_20-2html#.

Ahmad G, Mudasir Kudesia R, Shikha Srivastava MK (2010) Evaluation of genetic diversity in pea (*Pisum sativum*) using RAPD analysis *Genetic Engineering and Biotechnology Journal* -16

Akbar W, Miana GA, Ahmad SI, Khan MA (2006) Clinical investigation of hypoglycemic effect of *Coriandrum sativum* in type-2 (niddm) diabetic patients. *Pakistan Journal of Pharmacology* 23(1): 7-11

Alefeld F (1866) Landwirthschaftliche Flora. Wiegandt and Hempel, Berline, pp 165

Angelini LG, Moscheni E, Colonna G, Belloni P, Bonari E (1997) Variation in agronomic characteristics and seed oil composition of new oilseed crops in Central Italy. *Industrial Crops & Products* 6: 313-323

Asgarpanah J and Kazemivash N (2012) Phytochemistry, pharmacology and medicinal properties of *Coriandrum sativum* L. *African Journal of Pharmacy and Pharmacology* 6(31): 2340-2345.

Barclay AS and Earle FR (1965) The search for new industrial crops. V. The South African Calenduleae (Compositae) as source of new oil seeds. *Econ Bot* 25: 33-43

Basha SD, Francis G, Makkar HPS, Becker K, Sujatha M (2009) A comparative study of biochemical traits and molecular markers for assessment of genetic relationships between *Jatropha curcas* L. germplasm from different countries. *Plant Science* 176: 812-823

Basilico MZ and Basilico JC (1999) Inhibitory effects of some spice essential oils on Aspergillus ochraceus NRRL 3174 growth and ochratoxin A production. *Letters in applied Microbiology* 29(4): 238-241

Bhandari MM and Gupta A (1991) Variation and association analysis in coriander *Euphytica* 58: 1-4

Bhuiyan MNI, Begum J, Sultana M (2009) Chemical composition of leaf and seed essential oil of *Coriandrum sativum* L. from Bangladesh. *Bangladesh J Pharmacol* 4: 150-153

Cantore PL, Iacobellis NS, Marco AD, Capasso F, Senatore F (2004) Antibacterial activity of *Coriandrum sativum* L. and *Foeniculum vulgare* Miller Var. vulgare (Miller) Essential Oils. *J Agric Food Chem* 52(26): 7862–7866

Castellini G, Torricelli R, Albertini E, Falcinelli M (2006) Morphological and molecular characterization of celery landrace from central Italy, *Apium graveolens* L. Var Dulce (Miller) Pers. Proceedings of the 50th Indian Society of Agricultural Genetics Annual Congress Ischia, Italy - 10/14 September. Poster Abstract - A 33.

Das A and R Mallick (1989) Variation in 4C DNA content and chromosome characteristics in different varieties of *Coriandrum sativum* L. *Cytologia* 54: 609-616

Dhanapakiam P, Joseph JP, Ramaswamy VK, Moorthi M, Kumar AS (2008) The cholesterol lowering property of coriander seeds (*Coriandrum sativum*): Mechanism of action. *Journal of Environmental Biology* 29(1): 53-56

Diederichsen A (1996) Coriander (*Coriandrum sativum* L.) Promoting the conservation and use of underutilized and neglected crops 3 Institute of Plant Genetics and Crop Plant Research, Gatersleben/International Plant Genetic Resources Institute, Rome.

Diederichsen A and Hammer K (2003) The infraspecific taxa of coriander (*Coriandrum sativum* L.). *Genetic Resources and Crop Evolution* 50: 33-63

Dikshit HK, Jhang T, Singh NK, Koundal KR, Bansal KC, Chandra N, Tickoo JL, Sharma TR (2008) Genetic differentiation of *Vigna* species by RAPD, URP and SSR markers. *Biologia Plantarum* 51(3): 451-457

Gildemeister E and Hoffmann F (1931) Corianderöl Die ätherischen Öle. Aufl. (E. Gildemeister ed.). *Verlag der Schimmel & Co Aktiengesellschaft, Miltitz bei Leipzig*. 3: 455-461

Gillaspie AG, Hopkins MS, Dean RE (2005) Determining genetic diversity between lines of *Vigna unguiculata* subspecies by AFLP and SSR markers. *Genetic Resources and Crop Evolution* 52(3): 245-247

Guenther E (1952) The essential oils *D. van Nostrand Co.* pp: 602-615.

Gupta SK, Hamal IA, Koul AK (1986) Reproductive biology of *Coriandrum sativum* L. *Journal of Plant Science Research* 2: 81-95

Hammer K (1993) The 50th anniversary of the Gatersleben Genebank FAO/IBPGR. *Plant Genet Resour Newsl* 91/92:1-8

Ilyas M (1980) Spices in India 3. *Econ Bot* 34: 236-259

Ivanova KV and Stoletova EA (1990) History of cultivation and intraspecific classification of coriander (*Coriandrum sativum* L.). *Sbornik Nauchnykh Trudov Prikladnoi Botanike Genetike i Selectsii* 133: 26-40

Jansen PCM (1981) Spices, condiments and medicinal plants in Ethiopia, their taxonomy and agricultural significance. *Centre for Agricultural Publishing and Documentation. Wageningen, the Netherlands* pp: 56-67

Jindla LN, Singh TH, Rang A, Bansal ML (1985) Stability for seed yield and its components in coriander (*Coriandrum sativum* L.). *Indian J Genet* 45: 358-361

Kansal L, Sharma A, Lodi S (2011) Remedial effect of *Coriandrum sativum* (coriander) extracts on lead induced oxidative damage in soft tissues of swiss albino mice. *International Journal of Pharmacy and Pharmaceutical Sciences* 4(3): 10-20

Karnawat M and Malik CP (2011) Determination of nature of polyploidy in *Argemone ochroleuca ssp. ochroleuca Sweet. Nucleus* 54(3): 153-158

Khan S, Latif A, Ahmad SQ, Ahmad F, Fida M (2011) Genetic variability analysis in some advanced lines of soybean (*Glycine max* L.). *Asian Journal of Agricultural Sciences* 3(2): 138-141

Khodadadi M, Fotokian MH, Miransari M (2011) Genetic diversity of wheat (*Triticum aestivum* L.) genotypes based on cluster and principal component analyses for breeding strategies. *Australian Journal of Crop Science* 5(1): 17-24

Koul P (1985) Reproductive biology of some Umbellifers. PhD thesis, Jammu University, Jammu

Kubo I, Fujita K, Kubo A, Nihei K, Ogura T (2004) Antibacterial activity of coriander volatile compounds against *Salmonella choleraesuis. J Agric Food Chem* 52(11): 3329–3332

Linnaeus C (1780) Ded Ritters Carl von Linnaeus, koniglich schwedischen Leibarztes u. u. vollstandiges pflanzensystem nach der dreyzehnten lateinischen Auusgabeund nach Anleitung des hollanddischen Houttuynischen Werks ubersetzt und miteiner ausfuhrlichen Erklarung ausgefertiget. *Von den Krautern Gabriel Nicolaus Raspe Nurnberg* 6: 152-154

Loaiza J and Cantwell M (1997) Postharvest physiology and quality of cilantro. *Hort Science* 32: 104-107

Lombard V, Dubreuil P, Dillman C, Baril CP (2001) Genetic distance estimators based on molecular data for plant registration and protection: A review. *Acta Horticulturae* 546: 55-63

López PA, Widrlechner MP, Simon PW, Rai S, Boylston TD, Isbell TA, Bailey TB, Gardner CA, Wilson LA (2008) Assessing phenotypic, biochemical and molecular diversity in coriander (*Coriandrum sativum* L.) germplasm. *Genetic Resour Crop Evol* 55: 247-275

Loveless MD and Hamrick JL (1984) Ecological determinants of genetic structure of plant population. *Annual review of ecology and systematics* 15: 65-95

Mahendra P and Bisht S (2011) Anti-anxiety activity of *Coriandrum sativum* assessed using different experimental anxiety models. *Indian J Pharmacol* 43(5): 574–577

Maroufi K, Farahani HA, Darvishi HH (2010) Importance of coriander (*coriandrum sativum* l) between the medicinal and aromatic plants. *Advances in Environmental Biology* 4(3): 433-436

Melo RA, Resende LV, Menezes D, Beck APA, Costa JC, Coutinho AE, Nascimento AVS (2011) Genetic similarity between coriander genotypes using ISSR markers. *Horticultura Brasileira* 29: 526-530

Mengesha B and Alemaw G (2010) Variability in Ethiopian coriander accessions for agronomic and quality traits. *African Crop Science Journal* 18(2): 43-49

Momin AH, Acharya SS, Gajjar AV (2012) *Coriandrum sativum*- Review of advances in phytopharmacology. *IJPSR* 3(5): 1233-1239

Narayanan BK and Sakthivel N (2008) Coriander leaf mediated biosynthesis of gold nanoparticles. *Materials Letters* 62(30): 4588–4590

Nuel G, Baril C, Robin S (2001) Varietal distinctness assisted by molecular markers: A methodological approach *Acta Horticulturae* 546: 65-71

Pareek N, Jakhar ML, Malik CP (2011) Analysis of genetic diversity in coriander (*Coriandrum sativum* L.) varieties using random amplified polymorphic DNA (RAPD) markers. *Journal of Microbiology and Biotechnology Research* 1(4): 206-215

Pareek N, Jakhr ML, Malik CP (2013) A comparative analysis of RAPD and ISSR markers for studying genetic diversity among Coriander (*Coriandrum sativum*) varieties *Phytomorphology* In press.

Patel DK, Desai SN, Gandhi HP, Devkar RV, Ramachandran AV (2012) Cardio protective effect of *Coriandrum sativum* L. on isoproterenol induced myocardial necrosis in rats. *Food and Chemical Toxicology* 50(9): 3120–3125

Patel KC, Tiwari AS, Kushwaha HS (2000) Genetic divergence in coriander. *Agric Sci Digest* 20(1): 13-16

Pathak NL, Kasture SB, Bhatt NM, Rathod JD (2011) Phytopharmacological Properties of Coriander Sativum as a Potential Medicinal Tree: An Overview. *Journal of Applied Pharmaceutical Science* 1(4): 20-25

Prakash V (1990) Leafy Spices *CRC Press Inc. Boca Raton* pp: 31-32

Purseglove JW, Brown EG, Green CL, Robbins SRJ (1981) Spices. Vol 2 *Longman, New York* pp: 736-788

Rajanna L, Ramakrishnan M, Simon L (2011) Evaluation of morphological diversity in south Indian tea clones using statistical methods. *Maejo International Journal of Science and Technology* 5(1): 1-12

Rajeshwari U and Andallu B (2011) Medicinal benefits of coriander (*Coriandrum sativum* L.). *Spatula DD* 1(1): 51-58

Ramadan MF, Kroh LW, Mörsel JT (2003) Radical Scavenging Activity of Black Cumin (*Nigella sativa* L.), Coriander (*Coriandrum sativum* L.), and Niger (*Guizotia abyssinica* Cass.) Crude Seed Oils and Oil. Fractions *J Agric Food Chem* 51(24): 6961–6969

Ravi M, Geethanjali S, Sammeyafarheen F, Maheswaran M (2003) Molecular marker based genetic diversity analysis in rice (*Oryza sativa* L.) using RAPD and SSR markers. *Euphytica* 133(2): 243-253

Ravi R, Prakash M, Bhatt KK (2007) Aroma characterization of coriander oil samples. *Eur Food Res Tech* 225: 367-374

Rohr K, Engling FP, Lebzien P, Oslage HJ (1990) Analyse und Bewertung von Korianderkuchen für Futterzwecke bei Wiederkäuern. *Landbauforsch Völkenrode* 40: 133-137

Sastri EVD, Singh D, Sharma KC, Sharma RK (1989) Stability analysis in coriander (*Coriandrum sativum* L.) *Indian J Genet* 49: 151-153

Singh AK, Banwasi R (2012) Effect of fungicides and bio-agents on powdery mildew of coriander (*Coriandrum sativum* L.) *Journal of Spices and Aromatic Crops* 21(1): 76–77

Singh AK, Singh M, Singh AK, Singh R, Kumar S, Kalloo G (2005) Genetic diversity within the genus *Solanum* (Solanaceae) as revealed by RAPD markers. *Curr Sci* 90: 711-714

Singh SK, Kakani RK, Meena RS, Pancholy A, Pathak R, Raturi A (2012) Study on genetic divergence among Indian varieties of a spice herb, *Coriandrum sativum J. Environ Biol* 33: 781-789

Solouki M, Mehdikhani H, Zeinali H, Emamjomeh AA (2008) Study of genetic diversity in Chamomile (*Matricaria chamomilla)* based on morphological traits and molecular markers *Sci Hort* 117: 281-287

Stoletova EA (1930) Polevye I ogorodnye kul'tury Armenni (in Russ.) Tr Poprikl bot, gen I sel 23: 290-291

Stoletova EA (1931) Koriandr. Vsesojuznaja akademija sel'skogo-chosjajstvennych naukimeni Lenina izdanie instituta rastenievodstva, Leningrad.

Verma A, Pandeya SN, Yadav SK, Singh S, Soni P (2011) A Review on *Coriandrum sativum* (Linn.): An Ayurvedic Medicinal Herb of Happiness. *JAPHR* (2011) 1(3): 28-48

Villalobos MJ (2003) Volatile activity of plant essential oils against stored product beetle pests. Advances in stored product protection. Proceedings of the 8th International Working Conference on Stored Product Protection, CAB International,Wallingford: UK; 648–650

Wangensteen H, Samuelsen AB, Malterud KE (2004) Antioxidant activity in extracts from coriander. *Food Chem* 88: 293-297

Wiseman H, Okeilly JD, Aldlercreutz H, Mallet AJ, Bowery EA, Sanders AB (2000) Isoflavones phytoestrogen consumed in soya decrease F2-isoprostane concentrations and increase resistance of low density lipoprotein to oxidation in humans. *Am Clin Nutr* 72: 397-400

9

Present Status and Strategies for Improvement of Seed Spices

M.L. Jakhar and S.S. Rajput

India is the largest producer, consumer and exporter of seed spices in the world. Seed spices are those annuals whose dried fruits and seeds are used as condiments. These are extensively used as flavouring agents in various food products and in pharmaceutical industry, especially in the preparation of Ayurvedic medicines. The enhancement in the production and productivity of seed spices is a challenging task under changing climate scenario. Availability of improved varieties fitting in different agro-climatic situation ensuring acceptable performance is the basic need for successful cultivation of seed spices. According to current estimate, the country would need to produce 17-20 lakh tons of seed spices by 2030 AD to meet the domestic as well as export demand in the world market. In this, situation crop improvement and breeding approaches are only the best solution to meet foresaid quantity of seed spices. However, for effective application of breeding strategies for crop improvement is possible only with sound knowledge of botanical characteristics of plants, their genetic constitution and association of different seed yield contributing characters with each other. All the seed spices crops are affected by diseases which reduce the yield levels of these crops which is already low. Some of the diseases like wilt in cumin, root rot in fenugreek and gummosis in fennel have so far evaded the effective control measures. Breeding programmes for resistance to these diseases need to be immediately initiated. Identification of pathogen and/ or its races causing the diseases, developing technique to create uniform and artificial disease epiphytotics, use of tissue culture as an aid to accelerate the resistance screening programme are some of the aspects on which research work needs to be initiated to make the resistance breeding effective. Therefore, concerned multi-disciplinary efforts are needed to make the seed spices more competitive with other crops like cereals and oilseeds in the country. Quality of the produce needs special emphasis in seed spices as these are exported to foreign markets where they have to compete with the produce of other countries. Volatile oil content, shape, size and luster of seeds and their cleanliness constitutes the important quality factors. Obviously, besides breeding efforts post-harvest technology to have clean seeds and preserve the luster needs to be developed.

Introduction

India is known as the land of spices and is the largest producer, consumer and exporter of seed spices and their products. Out of 109 spices listed by the International Organization for Standardization (ISO), India produces as many as 63 owing to its varied agro climatic regions out of which 20 are classified as seed spices with 36 per cent share in the area and 17 per cent share in production of total spices in India. The country produces a wide variety of seed spices like fennel, fenugreek, coriander, cumin, dill, nigella, ajwain, anise, celery, caraway, poppy seeds, rai (*Brassica nigra*), mustard (white and yellow), sesame, parsley and *Bunium persicum.*

The seed spices are those annuals whose dried fruits or seeds are used as condiments. Seed spices are grown in low rainfall areas and fewer inputs are given as compared to other crops. Rajasthan and Gujrat form the seed spices bowl on India. These crops are extensively grown in semi-arid and arid regions of these two states of India during rabi season. About 10% of the total production of seed spices in India is exported, particularly to USA, both raw seeds and their value added products; earning foreign exchange worth almost Rs. 775 crores annually (Gupta *et al.*, 2013). Export of seed spices from India accounts for over 50% of their world trade. Out of the total 20 seed spices grown in the country, nine (cumin, fennel, coriander, fenugreek, ajwain, dill, celery, aniseed and nigella) are prominent as they are cultivated on sizeable area and contribute to the economy. These important seed spices are further, grouped into major and minor groups. The diversity of the groups is given in Table 1.

Seed spices are extensively used as flavoring agents in various food products and in pharmaceutical industry, especially in the preparation of Ayurvedic medicines. These stimulate the appetite and augment the flow of the salivary and gastric juice (Murty and Sridhar, 2001). Seed spices are also used very frequently in homemade medicines for different ailments. The new group of value added products such as the volatile oils and oleoresins obtained from spices are also in large demand in the international market. The volatile oil is also used in flavoring liquors and obscuring unpleasant smell of medicines.

Brief description and uses of different seed spices

Coriander (*Coriandrum sativum*), also known as cilantro, Chinese parsley or dhania. It is a soft, hairless plant growing to 50 cm tall. The leaves are variable in shape, broadly lobed at the base of the plant, and slender and feathery higher on the flowering stems. The flowers are borne in small umbles, white or very pale pink, asymmetrical, with the petals pointing away from the centre of the umbel longer (5–6 mm) than those pointing towards it (only 1–3 mm long). The fruit is a globular, dry schizocarp 3–5 mm (0.12–0.20 in) in diameter. All parts of the plant are edible, but the fresh leaves and the dried seeds are the parts most traditionally used in cooking. Coriander leaves have much more potent volatile leaf oil and a stronger smell (Ramcharan, 1999). The leaves have a different taste from the seeds, with citrus overtones. However, many people experience an unpleasant "soapy" taste or a rank smell and avoid the leaves. Coriander seeds are used in brewing certain styles of beer, particularly some Belgian wheat beers. The coriander seeds are used with orange peel to add a citrus character.

Table 1 : Major and Minor Seed Spices Grown in India

Name	Botanical Name	Parts Used	Centre of origin	Family
Major Spices				
Coriander	*Coriandrum sativum* L.	Seed and leaf	Mediterranean region	Apiaceae
Cumin	*Cuminum cyminum* L.	Fruit	Egypt, Syria	Apiaceae
Fennel	*Foeniculum vulgare* Mill.	Fruit	South Europe and Mediterranean region	Apiaceae
Fenugreek	*Trigonella foenum graecum* L.	Seed	South East Europe and West Asia	Fabaceae
Minor Spices				
Ajwain	*Trachyspermum ammi* Sprague	Fruit	Egypt & India	Apiaceae
Dill(Sowa)	*Anethum graveolens* L. *Anethum sowa* Kurz	Fruit and seed	Europe, Africa & Asia	Apiaceae
Celery	*Apium graveolens* L.	Fruit	Mediterranean region	Apiaceae
Aniseed	*Pimpenella anisum* L.	Fruit	Eastern Mediterranean region	Apiaceae
Nigella	*Nigella sativa* L.	Seed	Eastern Mediterranean region	Ranuncul aceae
Caraway	*Carum carvi* L.	Fruit	Mediterranean region	Apiaceae
Black caraway	*Bunicum persicum*	Seed	Mediterranean region	Apiaceae
Black Cumin	*Caum bulbocastanum* L.	Seed	North India, Iran and Egypt	Apiaceae
Black mustard	*Brassica nigra* L.	Seed	Mediterranean region	Brassicaceae
Indian Dill	*Anethum graveolens* Kurz	Seed	Europe, Africa and Asia (India)	Apiaceae
Indian Muatard	*Brassica juncea* L.	Seed	Central Asia	Brassicaceae
Parsley	*Petroselinum cripum*	Leaf	Central Mediterranean region	Apiaceae
Poppy Seed	*Papaver somniferum*	Fruit	Western Mediterranean region of Europe	
Sesame	*Sesamum indicum*	Seed	Ethopia	Pedaliaceae
Sweet Fennel	*Foeniculum vulgare* Miller ssp. *Capilaceum* var. dulce	Seed	Egypt	Apiaceae
White Mustard	*Sinapsis alba* L.	Seed	Mediterranean region	Brassicaceae

Cumin: It is the dried seed of the herb *Cuminum cyminum*. The cumin plant grows to 30–50 cm tall and is harvested by hand. It is an annual herbaceous plant, with a slender, branched stem 20–30 cm tall. The leaves are 5–10 cm long, pinnate or bipinnate, with thread-like leaflets. The flowers are small, white or pink, and borne in umbles. The fruit is a lateral fusiform or ovoid achene 4–5 mm long, containing a single seed. Cumin seeds resemble caraway seeds, being oblong in shape, longitudinally ridged, and yellow-brown in color, like other members of the Apiaceae family such as caraway, parsley and dill. Cumin seeds are used as a spice for their distinctive flavour and aroma. It is globally popular and an essential flavouring in many preparations. Cumin seed can be used ground or as whole seeds. It helps to add an earthy and warming feeling to food, making it a staple in certain stews and soups, as well as spiced gravies such as chili.

Fennel: It is a hardy, perennial herb, with yellow flowers and feathery leaves. It is erect, glaucous green, and grows to heights of up to 2.5 m, with hollow stems. The leaves grow up to 40 cm long; they are finely dissected, with the ultimate segments filiform (threadlike), about 0.5 mm wide. Flowers are produced in terminal compound umblels 5–15 cm wide, each umbel section having 20–50 tiny yellow flowers on short pedicels. The fruit is a dry seed from 4–10 mm long, half as wide or less, and grooved. It is a highly aromatic and flavorful herb with culinary and medicinal uses and, along with the similar- test like anise. Fennel contains anethole, which can explain some of its medical effects: It, or its polymers, act as phytoestogens (Albert-Puleo, 1980). The essence of fennel can be used as a safe and effective herbal drug for primary dymenorrhea.

Fenugreek: It is a member of a genus of leguminous herbs very similar in habit and in most of their characters to the species of the genus *Medicago*. The leaves are formed of three obovate leaflets, the middle one of which is stalked; the flowers are solitary, or in clusters of two or three, and have a campanulate. 5-cleft calyx; and the pods are many-seeded, cylindrical or flattened, and straight or only slightly curved. It bears a sickle-shaped pod, containing from 10 to 20 seeds, from which 6% of a fetid, fatty and bitter oil can be extracted by ether. In India the fresh plant is employed as an esculent. The seed is an ingredient in curry powders, and is used for flavouring cattle foods. Fenugreek seeds are thought to be a galactagogue that is often used to increase milk supply in lactating women (Chantry *et al.*, 2004). It was formerly much esteemed as a medicine, and is still in repute in veterinary practice.

Dill: It grows to 40–60 cm, with slender stems and alternate, finely divided, softly delicate leaves 10–20 cm long. The ultimate leaf divisions are 1–2 mm broad, slightly broader than the similar leaves of fennel, which are threadlike, less than 1 mm broad, but harder in texture. The flowers are white to yellow, in small umbles 2–9 cm diameter. The seeds are 4–5 mm long and 1 mm thick, and straight to slightly curved with a longitudinally ridged surface. It is generally used in many traditional medicines, including medicines against jaundice, headache, boils, lack of appetite, stomach problems, nausea, liver problems, and much more (Sharma *et al.*, 2007).

Ajwain: It is the small seed-like fruit similar to that of the Bishop's Weed plant, egg-shaped and grayish in colour. The plant has a similarity to parsley. Because of their seed-like appearance, the fruit pods are sometimes called ajwain seeds or mistakenly as bishop's weed. It is also traditionally known as a digestive aid, relieves abdominal discomfort due to indigestion and antiseptic (Hill, 2004).

Nigella: It is a genus of about 14 species of annual plants. Common names applied to members of this genus are nigella, devil-in-a-bush or love in a mist. The species grow to 20–90 cm tall, with finely divided leaves; the leaf segments are narrowly linear to threadlike. The flowers are white, yellow, pink, pale blue or pale purple, with five to 10 petals. The fruit is a capsule composed of several united follicles, each containing numerous seeds. In India, the seeds are used as a carminative and stimulant to ease bowel and indigestion problems, and are given to treat intestinal worms, nerve defects, to reduce flatulence, and induce sweating. Dried pods are sniffed to restore a lost sense of smell. It is also used to repel some insects, much like mothballs (Ali and Blunden, 2003).

Aniseed: It is a herbaceous annual plant growing to 1 m or more tall. The leaves at the base of the plant are simple, 1-5 cm long and shallowly lobed, while leaves higher on the stems are feathery pinnate, divided into numerous leaves. The flowers are white, approximately 3 mm in diameter, produced in dense umbels. The fruit is an oblong dry schizocarp, 3–6 mm long, usually called "aniseed". Anise is sweet and very aromatic, distinguished by its characteristic flavor. The seeds, whole or powdered, are used in a wide variety of regional and ethnic confectioneries, including the black jelly bean.

Black caraway: It is also known as meridian fennel or Persian cumin. The plant is similar in appearance to other members of the carrot family, with finely divided, feathery leaves with thread-like divisions, growing on 20–30 cm stems. The main flower stem is 40–60 cm tall, with small white or pink flowers in umbles. Caraway fruits (erroneously called seed) are crescent-shaped acheses, around 2 mm long, with five pale ridges. Caraway fruit oil is also used as a fragrance component in soaps, lotions and perfumes. Caraway also has a long tradition of medical uses, primarily for stomach complaints.

Black cumin: Original black cumin is rarely available, so *N. sativa* is widely used instead; in India, *Carum carvi* is the substitute. Black cumin (not *N. sativa*) seeds come as paired or separate carpels, and are 3–4 mm long. They have a striped pattern of nine ridges and oil canals, and are fragrant, blackish in colour, boat-shaped, and tapering at each extremity, with tiny stalks attached. It has been used for medicinal purposes for centuries, both as a herb and pressed into oil.

Black mustard: It is an annual weedy plant cultivated for its seeds, which are commonly used as a spice. The spice is generally made from ground seeds of the plant. The seeds are small, hard and vary in color from dark brown to black. They are flavorful, although they have almost no aroma. The seeds are commonly used in curry, where it is known as *rai*. The seeds are usually thrown into hot oil or ghee, after which they pop, releasing a characteristic nutty flavor. The seeds have a significant amount of fatty oil. This oil is used often as cooking oil. Ground seeds of the plant mixed with honey are widely used as cough suppressant and to treat respiratory infections was popular before the advent of modern medicine (Zemede, 1995). It consisted in mixing ground mustard seeds with flour and water, and creating a cataplasm with the paste. This cataplasm was put on the chest or the back and left until the person felt a stinging sensation. The plant itself can grow from two to eight feet tall, with racemes of small yellow flowers. These flowers are usually up to 1/3" across, with four petals each. The leaves are covered in small hairs; they can wilt on hot days, but recover at night.

Celery: It is commonly known as celery or celeriac, depending on whether the petioles or roots are eaten: celery refers to the former and celeriac to the latter. *Apium graveolens* grows to 1 m tall. The leaves are pinnate to bipinnate leaves with rhombic leaflets 3–6 cm long and 2–4 cm broad. The flowers are creamy-white, 2–3 mm diameter, produced in dense compound umbles. The seeds are broad ovoid to globose, 1.5–2 mm long and wide. Celery is used in weight-loss diets, where it provides low-calorie dietary fiber bulk. Celery is often purported to be a "negative calorie food" based on the idea that the body will burn more calories during the digestion of the food than the body can extract from the food itself. The fact that the body uses very small amounts of energy in digestion compared to what can be extracted even from a low calorie food like celery disproves this theory.

White mustard: It is sometimes also referred to as *Brassica alba* or *B. hirta* grown for its seeds and fodder crop or as a green manure. The yellow flowers of the plant produce hairy seed pods, with each pod containing roughly a half dozen seeds. These seeds are harvested just prior to the pods becoming ripe and bursting. White mustard seeds are hard round seeds, usually around 1 to 1.5 millimetres in diameter (Balke, 2000) with a color ranging from beige or yellow to light brown. They can be used whole for pickling or toasted for use in dishes. When ground and mixed with other ingredients, a paste or more standard condiment can be produced. The seeds contain sinalbin, which is a thioglycoside responsible for their pungent taste. White mustard has fewer volatile oils and the flavor is considered to be milder than that produced by black mustard seeds.

Parsley: It is a bright green, hairless, biennial, herbaceous plant in temperate climates, or an annual herb in subtropical and tropical areas. Where it grows as a biennial, in the first year, it forms rosette of tripinnate leaves 10–25 cm long with numerous 1–3 cm leaflets, and a taproot used as a food store over the winter. In the second year, it grows a flowering stem to 75 cm tall with sparser leaves and flat-topped 3–10 cm diameter umbles with numerous 2 mm diameter yellow to yellowish-green flowers. The seeds are ovoid, 2–3mm long, with prominent style remnants at the apex. One of the compounds of the essential oil is apiol. The plant normally dies after seed maturation (Blamey and Grey-Wilson, 1989). Curly leaf parsley is often used as a granish. Green parsley is often used as a garnish on potato dishes (boiled or mashed potatoes), on rice dishes, on fish, fried chicken, lamb or goose, meat or vegetable stews (Meyer, 1998).

Poppy seed: It is an oilseed obtained from the *Papaver somniferum*. The tiny kidney-shaped seeds have been harvested from dried seed pods by various civilizations for thousands of years. The seeds are used, whole or ground, as an ingredient in many foods, and they are pressed to yield poppyseed oil. Poppy seeds are less than a millimeter in length, and minute: it takes 3,300 poppy seeds to make up a gram, and a pound contains between 1 and 2 million seeds (Harold McGee, 2004). To some extent harvesting for poppy seeds is in conflict with harvesting for opium. Poppy seeds of superior quality are harvested when they are ripe, after the seed pod has dried. Traditionally, opium is harvested while the seed pods are green and their latex is abundant, but the seeds have just begun to grow. Poppy seeds are highly nutritious and less allergenic than many other seeds and nuts. But has been reported to cause anaphylaxis (Keskin and Sekerel, 2006, Panasoff, 2008). Poppy seeds are a potential source of anti-cancer drugs (Aruna and Shivaramakrishanan, 1992).

Area and production in Indian perspective

Indian spices are famous world over and are integral part of Indian agriculture. To our credit, we are the leaders in spices production, consumption and export. India is also an important importer of seed spices. The flavor of Indian spices is spreading day by day across the globe. India is popularly known as the "Land of Spices" we have to continuously improve the productivity and quality of our spices to maintain the legacy. The seed spices are grown in different parts of the world covering mainly Mediterranean region, South Europe and Asia. Almost all of the seed spices are cultivated in India and most of the states in India grow one or more the seed spices, and has got the privilege to be called as the largest seed spices producing country in the world. The major growing belt spreads from arid to semi-arid regions covering large area in Rajasthan and Gujarat. Important seed spices growing states are listed in Table 2. During 2011-12, India produced 5.92 million tons of spices from an area of 3.16 million ha. and exported 575,270 tons of spices valued at Rs 6840.78 crores (Tamil Selvan and Cherian, 2013). The area and production of seed spices from the country are given in Table 3.

Table 2 : Major seed spices producing states of India

Crop	States where cultivated
Coriander	Rajasthan, M.P., A.P., Tamil Nadu, Karnataka, U.P., Orissa
Cumin	Rajasthan, Gujarat, U.P.
Fennel	Gujarat, Rajasthan, M.P., Haryana, U.P.
Fenugreek	Rajasthan, M.P., Maharastra, Haryana, U.P.
Ajwain	Rajasthan, Gujarat, U.P., Punjab, Tamil Nadu, A.P.
Dill	Rajasthan, Gujarat, J & K, U.P., Orissa, M.P., Punjab
Celery	Punjab, Haryana, U.P.
Aniseed	Rajasthan, Punjab, U.P., Orissa and M.P.
Nigella	U.P., Rajasthan, M.P., Tamilnadu & W. Bengal

Table 3 : Area and production of seed spices in India (2008-2011)

Crop	2008-09		2009-10		2010-11	
	Area (ha)	*Prodn. (tones)*	*Area (ha)*	*Prodn. (tones)*	*Area (ha)*	*Prodn. (tones)*
Coriander	537327	471515	530789	501485	530860	482230
Cumin	527132	283000	517183	303943	507850	314220
Fennel	74149	114277	53497	83576	61680	105320
Fenugreek	74512	97533	71985	88979	81220	118360
Ajwain	26148	18301	20628	8950	25850	22180
Dill seed	13139	13363	8537	10447	8537	10447
Celery	4117	5329	4321	5248	4312	5248

The major seed spices, namely coriander, cumin, fennel, fenugreek are predominantly grown in the arid and semi arid regions of Rajasthan and Gujarat states. A perusal of Table 3 and 4 indicates that among the Indian states, Rajasthan contributes sizeable area and production of all spices taken together. Rajasthan alone contributes 40-50% of total area as well as production of coriander and cumin. In case of fenugreek, the state contributes about 80% of total production in the country. The figures of area and production of these major seed spices in the state of Rajasthan are presented in Table 4.

Table 4 : Area, production and productivity of major seed spices in Rajasthan (2006-2011)

Crop	2006-07		2007-08		2008-09		2009-10		2010-11	
	Area (ha)	*Prod. (t)*	*Area (ha)*	*Prod. (t)*	*Area (ha)*	*Prod. (t)*	*Area (ha)*	*Prod. (t)*	*Area (ha)*	*Prod. (t)*
Coriander	131125	155088	212841	166033	245091	273573	232139	281076	197891	218899
Cumin	149816	23666	215474	66359	169142	42728	203855	80531	330637	114925
Fenugreek	40490	47222	49797	48914	62894	77319	58917	70328	80378	94199
Fennel	9095	7629	9154	4365	7500	6249	8755	5601	26973	26157

The area and production of the major seed spices in different districts of Rajasthan are presented in Table 5. It may be seen from the table that, while almost all the districts grow the seed spices on a limited scale, the cultivation is concentrated in specific districts. Coriander cultivation is concentrated in the three districts of Kota, Baran and Jhalawar constituting about 95% of the total area. Cumin is concentrated in the districts of Jodhpur, Jalore and Barmer together contributing more than to 70% of the total area of the state. Fennel is concentrated in the districts of Jodhpur, Nagaur, Tonk and Sirohi, together accounting for 70% of the area. Fenugreek is mainly cultivated in the districts of Sikar, Jhalawar, Nagaur and Chittor, together accounting about 60% of the area. As can be inferred from these figures, the fenugreek cultivation is more wide spread in the state in comparison to other crops, this is followed by cumin.

Table 5 : Area and production of different spices crops in the state of Rajasthan during the year 2010-11

(Source: Directorate of Horticulture, Govt. of Rajasthan, Jaipur)

Districts	Coriander		Cumin		Fenugreek		Fennel	
	Area (ha)	*Prodn. (t)*	*Area (ha)*	*Prodn. (t)*	*Area (ha)*	*Prodn. (t)*	*Area (ha)*	*Prodn. (t)*
Ajmer	88	97	7586	2235	167	196	128	100
Jaipur	49	54	348	121	3494	3622	124	74
Dausa	3	3	0	0	34	40	5247	5223

Tonk	335	371	2801	974	17	20	2323	2135
Sikar	18	20	25	9	12900	13674	0	0
Jhunjhnu	4	4	0	0	6916	9184	1	1
Nagour	6	7	33281	15153	4967	4916	10507	11116
Alwar	14	16	0	0	800	938	1	1
Bharatpur	13	14	0	0	11	13	42	13
Dholpur	0	0	0	0	0	0	5	5
Sawaimadhopur	268	297	1	0	1090	1277	1799	1799
Karoli	50	55	2	1	7	8	53	38
Bikaner	21	23	3369	1171	12314	15081	6	6
Churu	17	19	1276	444	9105	7302	1	0
Jaisalmer	9	10	16129	8254	122	143	2	0
Ganganagar	6	7	0	0	50	59	0	0
Hanumangarh	0	0	0	0	143	168	0	0
Jodhpur	214	237	58157	25379	5299	6210	1690	845
Barmer	0	0	104828	28410	282	330	0	0
Jalore	7	8	838851	27030	453	531	683	410
Pali	54	60	8627	3896	1222	1432	382	463
Sirohi	3	3	3591	1248	23	27	3665	3619
Kota	45452	57307	1	0	6388	10826	0	0
Baran	59979	86179	4	1	232	272	0	0
Bundi	2223	2253	6	2	3425	4014	5	3
Jhalawar	85795	69683	1	0	2323	2226	0	0
Banswara	0	0	0	0	9	10	0	0
Dungarpur	2	2	2	1	11	13	221	221
Udaipur	18	20	130	45	106	124	0	0
Pratapgarh	59	65	24	8	5096	7009	0	0
Bhilwara	8	9	1165	405	177	207	88	85
Chittor	3156	2054	394	137	3005	4104	0	0
Rajsamand	20	22	4	1	190	223	0	0
State	**197891**	**218899**	**330637**	**114925**	**80378**	**94199**	**26973**	**26157**

Perusal of the production and productivity of each of the crop indicates that there have been fluctuations in the area, production and productivity of these crops. Yet, when compared to the pan India statistics, there has been an increasing trend in production as well as productivity except for cumin. The productivity of fennel and fenugreek has been nearly constant over the years, while the productivity of the coriander has shown steady increase. The productivity of the cumin has shown decreasing trend. The changes are mainly attributable to environmental factors, particularly temperature, availability of irrigation and disease situation during the cropping season eg. the cumin productivity is highly linked to cloudiness particularly at the harvesting time. Cloudiness is conducive for the blight occurrence which damages the crop severely. Some of the other constraints which have been limiting the production and productivity of the spices are listed below.

1. Cultivation of these crops traditionally on marginal lands with low fertility.
2. Lack of availability of seed of improved varieties to suit different agro-climatic situations.
3. Lack of proper adoption of improved varieties, package of practices and control measures for diseases and pests.
4. Inherent weakness of the crops e.g. poor and slow germination in case of umbelliferous spices resulting in poor stand.
5. Slow initial growth of the crops resulting into severe weed problem.
6. High incidence of diseases e.g. wilt, blight, powdery mildew in cumin; wilt and stem gall in coriander; blight and gummosis in fennel and powdery mildew, downy mildew and mycoplasma like organism (MLO) in fenugreek.
7. Lack of proper adoption of harvesting and post harvest technology adversely affecting the quality.
8. Inadequate marketing facilities are other constraints to production and export of spices in the state.

Crop improvement approaches for major seed spices

Availability of improved varieties fitting in different agro-climatic situation ensuring acceptable performance is the basic need for successful cultivation of seed spices. According to current estimate, the country would need to produce 17-20 lakh tons of seed spices by 2030 AD to meet the domestic as well as export demand in the world market. In this, situation crop improvement and breeding approaches are only the best solution to meet foresaid quantity of seed spices. However, for effective application of breeding strategies for crop improvement is possible only with sound knowledge of botanical characteristics of plants, their genetic constitution and association of different seed yield contributing characters with each other. Thus for basic improvement in the seed spices firstly we have to spell out breeding objectives priority-wise (Table 6).

Table 6 : Major breeding objectives of seed spices

S. No.	Seed spices crop	Major breeding objectives
1	Coriander	1. Breeding for high seed yield 2. Early maturity for rainfed conditions 3. High volatile oil content 4. Powdery mildew resistance 5. Stem gall resistance

		6. Seed size, shape and lusture 7. Development of frost tolerant cultivars 8. Breeding for high leaf yield
2	Cumin	1. Breeding for high yield 2. Volatile oil content 3. Resistance to powdery mildew 4. Resistance to blight 5. Resistance to wilt 6. Seed size, colour and luster 7. Development of cultivars resistant to yellowing
3	Fennel	1. Breeding for short duration for mixed cropping 2. Breeding for high yield 3. Resistance to blight and gummosis 4. Volatile oil content
4	Fenugreek	1. Breeding for high seed yield 2. Early maturity 3. Resistance to root rot, powdery and downey mildew 4. Diosgenin content

Breeding Strategies

Important breeding strategies for development of sustainable cultivars with high yield, resistant to abiotic, biotic stresses and desired quality are as follows:

1. Germplasm collection, evaluation and conservation.
2. Population improvement and selection.
3. Exploitation of Heterosis.
4. Breeding for abiotic and biotic stresses.
5. Improvement of quality.
6. Mutation breeding.
7. Breeding varieties for mixed cropping.
8. Biotechnological approaches.

Germplasm collection, evaluation and conservation

Available literature indicates that variability in the germplasm particularly for yield, quality (volatile oil content) and reaction to different biotic and abiotic stresses is low in all the seed spices (Sharma, 1994). In the present germplasm collections variability for plant height maturity duration is high in coriander and fennel, medium in fenugreek. But in cumin it is low even for these characters obviously there is an urgent need to enhance the germplasm base to successfully support the crop improvement programme. Genetic enhancement of seed spices is possible only through accumulation of variability in the form of germplasm. India is endowed with germplasm of different kind of seed spices crops and a number of indigenous and exotic origin is available at different centre's (Table 7). Germplasm collection activities were initiated from Rajasthan and Gujarat, the major areas for the cultivation of cumin, coriander, fennel, fenugreek, ajwain, and dill. However, the main emphasis is given only to major seed spices and most of the minor seed spices remained untouched therefore, in order to accumulate the variability in these crops, diversified areas need to be revisited. There is tremendous scope for collection of valuable land races of seed spices. The development of improved cultivars has improved the production and productivity of seed spices to some extent, which has endangered the accumulated diversity among the traditional cultivars of these crops species. Therefore, the conservation of germplasm of the seed spices is of profound importance for sustaining their production. If not collected and conserved these valuable genetic resources may be lost forever, hence prime importance is required to be given in collection and conservation of the biological diversity of seed spices from all over the country and abroad.

Table 7 : Germplasm collection of seed spices in India

Crop	**Germplasm lines maintained**		**Total**
	Indigenous	***Exotic***	
Jobner, Rajasthan			
Cumin	275	1	276
Coriander	608	46	654
Fennel	248	4	252
Fenugreek	442	-	442
Jagudan, Gujrat			
Cumin	240	7	247
Coriander	74	21	95
Fennel	135	4	139
Fenugreek	64	-	64
Dill seed	68	-	68
Ajwain	59	-	59
NRCSS, Ajmer, Rajasthan			
Cumin	77	7	83
Coriander	117	3	120
Fennel	31	3	134
Fenugreek	134	59	193
Dill seed	102	5	107
Ajwain	86	1	87

Nigella	22	3	25
Celery	36	-	36
Anise	18	-	18
Caraway	2	2	4
Hissar, Haryana			
Coriander	251	-	251
Fennel	122	-	122
Fenugreek	240	-	240
Guntur,			
Coriander	248	-	248
Fenugreek	125	-	125
Dholi, Bihar			
Coriander	97	-	97
Fennel	31	-	31
Fenugreek	106	-	106
Kumarganj, Uttarpradesh			
Coriander	75	-	75
Fennel	44	-	44
Fenugreek	77	-	77
Coimbatore, Tamil Nadu			
Coriander	274	-	274

Evaluation and characterization of seed spices germplasm is required for their documentation and cataloguing crop wise for the further use by plant breeders and biotechnologists for improving in respect of yield, quality and resistant against biotic as well as abiotic stresses. Proper maintenance of the germplasm collections is essential to maintain the genetic structure of the original sampled population. This can be achieved by prevention of out crossing with other species and reduction the effect of natural selection in an environment other than the original one. Seed spices have both self and cross mode of reproduction. In the self pollinated seed spices gene and genotypic composition can be preserved in the absence of selection, whereas in cross pollinated it can be maintained through controlled pollination.

Population improvement and selection

Population improvement is the process of pyramiding of the positive genes for desirable characters in a variable population through selection or recurrent selection. The appropriate strategy in umbelliferous (cross-pollinated) spices would be to have short term and long term improvement programme. The short term programme would aim to improve the elite populations through recurrent selection based on the performance of individual plant progenies or even mass-selection. This approach has produced a number of good varieties in coriander (RCr-20, RCr-41, RCr- 435, RCr- 446, RCr-486 and RCr- 728), cumin (RZ- 19 and RZ-209), fennel (RF- 101, RF-143 RF-178, RF-205, RF-145 RF-281and RF-125) ajawain and sowa. In self pollinated spice like fenugreek pure line selection in the elite variable material has been quite successful in developing good varieties. Some of these are: RMt-1 and RMt-143 (Rajasthan); Lam-sel-1(Andhra Pradesh); Rajendra Kranti (Bihar); Pusa early branching (IARI-New Delhi), and Co-1 (TG-2336) (Tamil Nadu).

It is clear that recurrent selection in cross pollinated seed spices and pure line selection in self-pollinated crops like fenugreek, may be continued in the elite material as an effective method to develop varieties in a shorter time. Long term programme must be simultaneously started. The strategy here is to exploit both the variability existing within as well as between elite lines in cross pollinated spices. The steps involved include creation of a wide genetic base pool through intermating of selected elite lines/ varieties in a poly-cross nursery, followed by recurrent selection based on progeny performance which may be continued till a desirable level of performance and uniformity is achieved. This method is capable of exploiting both additive as well as non-additive effects as high heterozygosity is maintained in the varieties. In self-pollinated crop like fenugreek inter-varietal differences may be exploited only through hybridization followed by selection.

Exploitation of heterosis

Though recurrent selection, can be successfully employed in both intra as well as inter-population improvement, the best approach to exploit both the additive and non-additive gene effects would be the heterosis breeding. Manifestation of heterosis has been reported in coriander (Singh and Ramanujam, 1972), fennel (Ramanujam and Tiwari, 1970) and further studies may reveal its manifestation in other crops. Search for cytoplasmic - genetic types of system of controlling male sterility should be taken up to make the heterosis breeding a reality.

Breeding for biotic and abiotic stress resistance

The seed spices are affected by various diseases which reduce the yield level which is already low. Some of the diseases like cumin wilt and blight, root rot in fenugreek and gummosis in fennel have so far evaded the effective control measures. Breeding programmes for resistance to these diseases need to be immediately initiated. Identification of pathogen and/ or its races causing the diseases, developing techniques to create uniform and artificial disease epiphytotics, use of tissue culture as an aid to accelerate the resistance screening programmes are some of the aspects on which research work need to be initiated to make the resistance breeding effective. Resistance breeding is also directly related to quality of the produce as it obviates the necessity of application of pesticides which often leaves undesirable residues. In addition to biotic stresses these crops are also affected by large number of abiotic stresses, therefore screening programmes on these stresses need to be initiated eg. drought, salt and frost tolerant cultivars.

Improvement of quality

Quality of the produce needs special emphasis in seed spices as these are exported of foreign markets where the quality standards are very stringent. Volatile oil content, shape, size and luster of grains and their cleanliness constitute the important quality factors. Appropriate weightage has to be given to these quality attributes in the breeding programmes (Table 8).

Table 8 : Quality parameters in seed spices

S.No.	Seed Spice	Quality parameters
1.	Cumin	High volatile oil and high cuninaldehyde content.
2.	Coriander	High volatile oil, high linalool content and dual purpose use both as seed and leaves.
3	Fennel	High volatile oil and high anethol content.
4.	Fenugreek	High/low trigonellone content, high diosgenin content and dual purpose use both as seed and leaves.
5.	Ajwain	High volatile oil and high thymol content.
6.	Dill	High volatile oil, high carvone content, dillapiol toxin less, and splitting/non-splitting type.
7.	Nigella	High volatile oil, high nigellone content.
8.	Anise seed	High volatile oil and high anethol content.
9.	Caraway	High volatile oil, high carvone and limoene content.
10.	Celery	High volatile oil and high limeoene content.

(*Source:* Malhotra, 2013)

Mutation breeding

Mutations, induces with gamma rays, are useful in creating desirable variability in fenugreek and coriander (Sharma *et al.,* 1996). Noteworthy achievement is a RMt-305 spontaneous mutant of RMt-1. RMt-305 mutant line, which is resistant to powdery mildew and determinant in growth habit with high yield potential besides 4 others (UM 301 to 304) resistant to powdery mildew and possessing higher yield potential. The fenugreek varieties namely RMt-303, RMt-351 RMt-361 and RMt-365 also produced through physical mutagen. In coriander, mutation lines namely RCr-684 had been derived from RCr-20 through 20 Kr dose of gamma radiation. However, the results with fennel have not been so encouraging and in other crops the approaches are yet to be tried.

Breeding varieties for mixed cropping

Yield fluctuations over years as well as locations are very wide. Preliminary information obtained from evaluation of varieties/ or germplasm suggested that genetic differences among varieties for stability do exist. Factors responsible for differences need to be specially identified and utilized in breeding programme. In general these crops, particularly the cumin, respond very less to agronomic inputs *e.g.* fertilizers as well as irrigation. Attention therefore, needs to be given on identification of responsive types. The high genotype x environment interaction, noted for these crops, also suggests that breeding programmes may be targeted for specific environment and develops suitable varieties for each condition.

Biotechnological approaches

Different biotechnology techniques be used as the aids for improvement of seed spices such as somaclonal variations and production of transgenic lines and molecular characterization, at least of the elite germplasm needs to be immediately taken up, to facilitate their use in crop improvement programme in following ways.

A. Use of molecular markers in germplasm collection, evaluation and conservation.

B. Identification of resistant genes using isolated pathogenic strains and pure genotypes.

C. Understanding of genetic structure of population using molecular markers.

D. Tagging of genes of economic importance including genes for resistance to biotic and abiotic resistance and quality.

E. Enhancement of germplasm through soma-clonal variants, *in vitro* selection schemes for biotic/abiotic resistance.

F. Development of various molecular markers for marker assisted selection.

It may be concluded that in seed spices, where genetic variability assembled so far has been inadequate, well planned efforts be taken up to enhance the variability. Besides the systematic collection and maintenance of indigenous germplasm, exotic material particularly from centers of origin and secondary centers of variability must be introduced. Biotechnological approaches and mutation breeding should also be employed for enhancing, variability. Since the resistance to disease and improvement in quality has special significance in seed spices, a multi-disciplinary collaborative approach be initiated to achieve the objective of developing high yielding, disease resistant varieties with better quality. Breeding approaches must be refined to make the best use of existing variability. Search for cytoplasmic-genetic type of male sterility or temperature sensitive male sterility and development of inbreds be started as an effort targeted towards exploitation of heterosis. Different varieties so far produced in major seed spices are presented in Table no. 9 to 12.

Table 9 Coriander varieties released for cultivation in India

S.No.	Name of variety	Salient characteristics
1.	RCr-20	It is developed through recurrent half sib selection in a collection from Jaipur. The variety was for limited moisture conditions and heavy soils of Southern Rajasthan including Tonk, Bundi, Kota, Baran and Jhalawar districts. The variety matures in 100-110 days and produces an average yield of 10.0q/ha.

2.	RCr-41	It is developed through recurrent half sib selection in a local collection from Kota. The variety was released for irrigated conditions in the State of Rajasthan. It is tall, erect with small seeds (9.3g/ 1000 seeds). It is highly resistant to stem gall and wilt but only moderately resistant to powdery mildew. It matures in 130-140 days and produces an average yield of 9.2 q/ha.
3.	RCr- 435	The variety is developed through half sib-selection in a local collection from Jaipur district. Its plants are bushy with quick early growth and medium size seeds. It matures in 110-130 days and produces an average yield of 11.0q/ha.
4	RCr- 436	The variety is developed through half-sib recurrent selection in a collection from Kota (DC-86). The variety was released for the limited moisture conditions and heavy soils of Southern Rajasthan including Tonk, Bundi, Kota, Baran and Jhalawar districts. Plants are bushy with quick early growth and bold seeds. It matures in 90-100 days and produces an average yield of 11.0 q/ha under limited moisture condition.
5.	RCr- 446	It is developed through half-sib selection in a local collection from Jaipur district. The plants of this variety are leafy and erect with higher number of seeds per umbel. Seeds of the variety are medium in size and volatile oil contents 0.33%. It matures in 110-130 days and produces an average yield of 12 q/ha.
6.	RCr- 684	It is developed through mutation breeding by treating coriander variety RCr-20 to 5 Kr dose of gamma radiation. The plants are leafy, tall and erect with higher number of seeds per umbel. Seeds of the variety are bold in size and have volatile oil contents is 0.32%. It matures in 100-120 days and produces an average yield of 10 q/ha.
7.	RCr- 480	The variety has been developed through recurrent selection based on individual plant progeny (half sib) performance. The plants are bushy and erect with higher number of seeds per umbel and also have high volatile oil contents is 0.44%. It matures in 110-130 days and produces an average yield of 13.25q/ha.
8.	RCr- 728	The variety has been developed through recurrent selection based on individual plant progeny (half sib) performance in line UD-728 which was a local collection from Garoth, Mandsaur (MP). The plants are bushy and erect with higher number of seeds per umbel and volatile oil contents is 0.38%. It matures in 130-140 days and produces an average yield of 13.70 q/ha.
9.	Hisar Surabhi	Tolerant to frost and less susceptible to aphid with yield potential of 19.5 q/ha.

10.	Hisar Bhoomit	A high leaf yielding variety, suitable for leaf production with small seeded type and high oil content.
11.	DH-220	An early maturing type, seeds medium with high oil conten tand less susceptible to powdery mildew.
12.	DH-206	A variety with high yield potential, medium size seeds with high oil content, less susceptible to powdery mildew and tolerant to frost.
13.	Sudha	A variety with bold oblong medium sized grains less susceptible to pest and diseases.
15.	Suguna	A medium duration variety with high yield potential and high essential oil content.
16.	APHU Dhania-1	Grains with high volatile oil less susceptible to disease and pest. Suitable for rainfed conditios.
17.	Guj. Cori.-1	High yield (10.82q/ha) and stable.
18.	Guj. Cori.-2	High yield (14.43q/ha) and bold grain stable.
19.	**Other varieties :**	CO-1, CO-2, CO-3, CO-4, CS-287, Sadhana, Swathi, Sindhu, Hisar Anand and Hisar Sugandh

Table 10 : Cumin varieties released for cultivation in India

S.No.	Name of variety	Salient characteristics
1.	RZ-19	It is developed through recurrent single plant progeny selection from collection of Kekri (Ajmer).The plants are erect with pink flowers and bold grey pubescent seeds. As compared to local check the variety is more tolerant to wilt as well as blight. It matures in 140 to 150 days and produces an average yield of 5.6 q/ha (Figure 19).
2.	RZ-209	It is developed through recurrent single plant progeny selection in a local collection from Ahore (Jalore).The variety has shown higher resistance to wilt and blight diseases. It matures in 120 to 130 days and produces an average yield of 6.5 q/ha.
3.	RZ-223	It is developed through mutation breeding in UC-216. The variety has shown higher resistance to wilt and blight diseases. It matures in 120 to 130 days and produces an average yield of 6.0 q/ha.
4	RZ-341	The variety has been developed through polycross between high volatile oil content vs. low volatile oil content. The plants are bushy and semi-erect with long and bold seeds, lesser infestation of wilt, blight and powdery mildew and also have high volatile oil contents

		is 3.87%. It matures in 120-130 days and produces an average yield of 4.50 q/ha.
5.	RZ-345	The variety has been developed through recurrent selection based on individual plant progeny (half sib) performance in accession 345. The plants are bushy and semi-erect with long and bold seeds, lesser infestation of wilt, blight and powdery mildew and also have high volatile oil contents is 3.83%. It matures in 120-130 days and produces an average yield of 6.07 q/ha.
6.	Guj. Cumin-1	Yield potential (5.41q/ha).
7.	Guj. Cumin-2	Yield potential (6.22q/ha), better grain quality.
8.	Guj. Cumin-3	Yield potential (6.20q/ha), high oil content and resistant to wilt.
9.	Guj. Cumin-4	Yield potential (12.50q/ha). *Fusarium* wilt resistant variety with bold non splitting seeds and have high oil content of 4.2%.

Table 11 : Fennel varieties released for cultivation in India

S.No.	Name of variety	Salient characteristics
1.	RF-101	It is developed through recurrent half-sib selection from a local collection of Sohela (Tonk). It matures in 150-160 days and produces as average yield of15.5 q/ha.
2.	RF-125	It is developed through recurrent half-sib selection in an exotic collection EC-243380 from Italy. Its plants are early short statured with compact umbels and long bold seeds. When green, presents denser view of plants and the volatile oil of this variety is high so recommended for export purpose. It matures in 110-130 days and produces an average yield of17.3 q/ha.
3.	RF-143	It is developed through recurrent half-sib selection in local collection of Sarwad (Ajmer). Its plants are erect with compact umbels and long bold seeds. It matures in 140-150 days and produces an average yield of 18 q/ha.
4	RF-178	The variety has been developed through recurrent selection based on individual plant progeny (half-sib) from F_2 generation of a cross between UF-125 x UF-133. The plants are erect and medium tall with bold and attractive seeds and also have high volatile oil. It matures in 130-140 days and produces an average yield of 16.0 q/ha.
5.	RF-205	The variety has been developed through recurrent selection based on individual plant progeny (half-sib) from F_2 generation of a cross between JF-25 x RF-125. The plants are erect and medium tall with bold and attractive seeds and also have 2.48% volatile oil. It matures

		in 140-150 days and produces an average yield of 16.0 q/ha.
6.	RF-145	Erect and medium tall plants, Long and attractive seeds, Medium maturity grou. Average yield under normal conditions 2100 kg\ha. It possesses higher volatile oil content (2.89%).
7.	Guj. Fennel-1	Early maturity with yield potential of 17.24q/ha.
8.	Guj. Fennel-2	High oil content with yield potential of 19.4q/ha.
9.	Guj. Fennel-11	Heat tolerant, suitable for rabi cultivation and yield potential of 23.89q/ha.
10.	Guj. Fennel-12	Suitable for drill sowing in Rabi and transplant conditions in the state of Gujrat.
11.	**Other varieties** : PF-35, CO-1, Hisar Swarup and RF-281	

Table 12 : Fenugreek varieties released for cultivation in India

S.No.	Name of variety	Salient characteristics
1.	RMt-1	It is developed through pure-line selection in a local collection from Nagaur district. The plants are tall, moderately branched with bold typically yellow coloured seeds. It matures in 140-150 days and produces an average yield of14.7 q/ha.
2.	RMt-143	It is developed through pure line selection in a local collection from Jodhpur area. It is moderately resistant to powdery mildew. The seeds of this variety are bold with typical yellow colour. It takes 140-150 days to mature and produces an average yield of 16 q/ha. It is especially suitable for heavier soils of Chittor, Bhilwara, Jhalawar and Jodhpur area.
3.	RMt-303	It is developed through mutation breeding programme in a released variety RMt-1. It is moderately resistant to powdery mildew. The seeds of this variety are bold with typical yellow colour. It takes 140-150 days to mature and produces an average yield of 19 q/ha.
4	RMt-305	It is developed through mutation breeding from a released variety RMt-1This is the first determinate and multipoddded variety of fenugreek with high seed weight and harvest index. The variety has shown higher resistance to powdery mildew. It matures in 120 to 130 days and produces an average yield of 18.0 q/ha.
5.	RMt-351	The variety has been developed through irradiation of RMt-1 (a released variety of Rajasthan state) with gamma rays at 20 Kr. The plants are medium dwarf with bold and attractive seeds and also have more number of branches and seeds per pod. It matures in 140-150 days and produces an average yield of 18.41 q/ha.

6.	RMt-361	The variety has been developed through irradiation of RMt-1 with gamma rays. The plants are erect and bushy type with bold and attractive seeds and also have more number of branches and seeds per pod. It matures in 130-140 days and produces an average yield of17.70 q/ha.
7.	Guj. Methi-1	Seeds medium bold with yield of19.27q/ha.
8.	Guj. Methi-2	High yield (20.79q/ha.), bold lustrous grain and moderately resistant to powdery mildew.
9.	HM-219	Resistant to powdery mildew.
10.	APHU Methi-1	Early maturing variety.
11.	**Other varieties :**	CO-1, C0-2, Hisar Sonali, Hisar Suvarna, Hisar Madhavi, Hisar Mukta, Lam Sel.- 1, RMt-365 and Ajmer Fenugreek-3

Future thrust areas

Concerned multi-disciplinary efforts are needed on following areas to make the spices crops competitive with other crops e.g. cereals and oilseeds in the State and to improve the quality of the produce so that it may successfully compete in the International market (Sharma *et al.*, 1994).

1. **Breeding for high yield potential especially for export quality**: At present the average yield of cumin and coriander is around 5 q/ha and that of fenugreek and fennel is around 10 q/ha. To make these crops successfully compete with the cereals and oil seeds of Rabi season, the yield levels of these crops should be doubled. Efforts to breed for high yield potential, therefore, must be accelerated.
2. Efforts to update the existing germplasm through the addition of indigenous and exotic collection have to be intensified. This work may be supplemented with mutation breeding programme to enhance the variability for desirable characters.
3. **Breeding for resistance to diseases and pests:** All the seed spices crops are affected by diseases which reduce the yield levels of these crops which is already low. Some of the diseases like wilt in cumin, root rot in fenugreek and gummosis in fennel have so far evaded the effective control measures. Breeding programmes for resistance to these diseases need to be immediately initiated. Identification of pathogen and/ or its races causing the diseases, developing technique to create uniform and artificial disease epiphytotics, use of tissue culture as an aid to accelerate the resistance screening programme are some of the aspects on which research work needs to be initiated to make the resistance breeding effective.
4. **Improvement of quality:** Quality of the produce needs special emphasis in seed spices as these are exported to foreign markets where they have to compete with the produce of other countries. Volatile oil content, shape, size and luster of seeds and their cleanliness constitutes the important quality factors. Obviously, besides

breeding efforts post-harvest technology to have clean seeds and preserve the luster needs to be developed.

5. **Breeding for stability of performance:** Yield fluctuations over years as well as locations are very wide. Preliminary information obtained from evaluation of varieties/ or germplasm suggested that besides disease, genetic differences among varieties for stability do exist. Factors responsible for such differences need to be specifically identified and utilized in breeding programmes.
6. Determination of proper combination of fertilizer, irrigation, spacing and weed management with a view to maximize yield and quality of the produce needs emphasis.
7. **Efficient management of Pestilences:** Investigation towards better management of the disease pests and nematodes where resistant mechanism is not available, use of soil amendments, crop sequence and pesticides has to be accelerated.
8. Study on nematode problems particularly in fenugreek need to be stressed.
9. Extending research activities in Ajowain and Suwa in the state, as these crops are being grown in a sizeable area.
10. Research work on improving N_2 fixing ability of fenugreek through rhizobial symbiosis is inescapably needed to be initiated.

REFERENCES

Albert- Puleo M (1980) Fennel and anise, as estrogenic agent. *J Ethnopharmacology* 2(4): 337-344

Ali BH, Blunden G (2003) Pharmacological and toxicological properties of *Nigella sativa. Phytother Research* 17(4): 299-305

Aruna K, Shivaramakrishanan VM (1992) Anticarcinogenic effects of some Indian plant products. *Food and Chemical Toxocology* 30(11): 953-956

Balke D (2000) Rapid aqueous extraction of mucilage from whole white mustard seed. *Food Research International* 33(5): 347-356

Blamey M, Grey-Wilson C (1989) Illustrated flora of Britain and North Europe. *ISBNO*-340-40170-2

Chantry CJ, Howard CR, Mantogomary A, Wight N (2004) Use of galactogogues in initiating or augmenting maternal milk supply. ABM Protocols

Gupta OP, Singh R, Chaudhary GR (2013) Weed management status and strategies in seed spices in India. Presented in National Seminar on Production, Productivity and Quality of Spices held at Jaipur from 2-3 Feb., 2013, pp 129-142

Harold McGee (2004) On food and cocking. The science and lure of the kitchen. Simon and Schuster, p.513 ISBN 978-0-684-80001-1

Hill T (2004) Ajwain: Seasoning for global kitchen. Wiley, pp 21-23

Keskin O, Sekerel BE (2006) Poppy seed allergy: A case report and review of the literature.

Allergy and Asthma Proceeding 27(4): 396-398

Malhotra SK (2013) Plant genetic resources for crop improvement in seed spices crops. Presented in National Seminar on Production, Productivity and Quality of Spices held at Jaipur from 2-3 Feb., 2013, pp 5-17

Meyer J (1998). Authentic Hungarian Recipes Cookbook. Ed. 2 . Meyer and Assoc. ISBNO-9665062-0-0

Murty Rama A, Sridhar V (2001) Seed spices and Ayurveda. Seed Spices Production, Quality and Export. Edited by Agrawal,S.; Sastry, E.V.D. and Sharma, R.K. Pointer Publishers, Jaipur

Panasoff J (2008) Poppy seed anaphylaxis. *J Investigational Allergology and Clinical Immunology* 18(3): 224-225

Ramcharan C (1999) Culantro: A much utilized, little understood herb. J. Janic Prspectives on New Crop and New Uses. ASHS Press pp 504-509

Ramanujam S, Tiwari (1970) Heterosis in Fennel. *Indian J Genet* 30 (3): 732-737

Sharma RK (1994) Genetic resources of seed spices. In Advances in Horticulture Vol.1 (Ed. By Chadha, K.L. and Rethinam, P. Malhotra Pub. House New Delhi. Pp 193-207

Sharma RK, Dashora SL, Choudhary GR, Jain MP, Agrawal S, Singh D (1996) Seed Spices Research in Rajasthan. Directorate of Research. Rajasthan Agricultural University, Bikaner, Pub. No. DOR/1996/2

Singh VP, Ramanujam S (1972) Expression of andromonoecy in coriander, *Coriandrum sativum* L. *Euphytica* 22: 181-188

Singh RK, Meena SS, Vashishtha BB (2007) Medicinal Properties of seed spices. National Research Centre on seed spices, Tabiji, Ajmer

Tamil Selvan M, Cherian H (2013) Area, production and productivity and export scenario of spices in India. Presented in National Seminar on Production, Productivity and Quality of Spices held at Jaipur from 2-3 Feb., 2013, pp 1-4

Zemede A (1995) Conservation and uses of traditional vegetables in Ethopia. Proceedings of the IGPRI International Workshop on Genetic Resources of Traditional Vegetables in Africa held at Nairobi from 29-31 August, 1995

an important first step to study traits that are under the control of several to many genes (i.e., polygenic traits) and responses to external conditions [illegible] the [illegible] is measured simultaneously to provide a snapshot of [illegible] [illegible]

[illegible]

[illegible] gene expression (SAGE), massively parallel signature sequencing (MPSS) and next generation sequencing platforms (NGSPs) (Horak and Snyder, 2002; Rostoks *et al.*, 2005). Among these microarrays are widely used, although the sequencing of the transcriptome is rapidly increasing in popularity. Microarrays are solid-based platforms (e.g., glass slides), containing millions of copies for thousands of reporter probes, that comprise part of the sequences of the genes in the genome. [illegible] to the probes [illegible] these positions to the respective reporter probes, one can infer the presence and abundance of each transcript in the labelled RNA pool.

Proteomics

The term proteome refers to all proteins expressed by a genome in the targeted tissues at a defined time point. It encompasses a broad range of tools and techniques in determining the identity and quantity of the expressed proteins in cells/tissues, their 3D structure, and other interacting partners that help to disclose gene function. Thus, Proteomics deals with study of the structure and function of entire set of proteins, present in an organism and hence is essential for studying the whole metabolic pathway. This involves separation, identification, and determination of function and functional network of proteins allowing the integral study of many proteins at the same time (Agrawal and Rakwal, 2006). Proteomics not only enables the study of protein-protein interaction but also helps in identification of multisubunit complexes (Espagne *et al.*, 2007). Furthermore, proteomics can act as a powerful approach to organize and identify the proteome through development of 2DE-gel protein reference maps of subproteomes in different plant species.

Metabolomics

Metabolome refers to a set of metabolites that are formed within a biological system and their types and levels can be regarded as the ultimate response of biological systems to

10

Participatory Plant Breeding: Targeting the Needs of Resource-Poor Farmers in Marginal Areas

Gulzar S Sanghera, S H Dar, S C Kashyap and *G A Parray*

Participatory plant breeding (PPB) has developed over the past decade as an alternative and complementary breeding approach to conventional plant breeding. PPB involves scientists, farmers, and others, such as consumers, and rural cooperatives in plant breeding research. Diagnostic methods used in PPB help to create a more effective dialogue between researchers and farmers. This enables scientists to better understand the local farming conditions, the farmers' traditional diversity management as well as their specific needs and preferences. Therefore, participatory breeding programs work with a wider range of diverse breeding materials. Furthermore, participatory plant breeding implies decentralised selection, i.e. farmers grow crops in their own fields and make local selection decisions themselves. There are four main thrusts where farmers can contribute. The farmers can contribute their knowledge and experiences, they can contribute genetic materials, conduct trials, and can Select and evaluate germplasm. A wide range of impacts were achieved through use of participatory approaches, despite the short time that most of these projects have been in operation. Through the close collaboration of farmers and formal researchers, breeding progress was made in regions where variety improvement efforts of centralized programs with large uniform target regions have been less than successful. Farmer participation opened new opportunities for reducing the lag-time between variety testing, release, and adoption and, through making new germplasm available to farmers in experimental quantities. The improved understanding of local seed production and dissemination systems allowed researchers and project managers to overcome specific seed system constraints and to make targeted interventions to overcome seed supply problems quickly. This helped farmers gain access to the desired seeds rapidly, from known and reliable seed sources. Farmer participatory breeding programs also led to new alliances between research and development organizations. A number of PPB varieties have been already release for many crops e.g. rice, maize, sorghum, barley etc. Although the impact of PPB to date has been impressive PPB is having more widespread goals, focusing on a broader range of crops, working with explicitly more diverse user groups, and experimenting with organizational options may identify wider and more varied benefits that PPB can achieve.

Introduction

Participatory plant breeding (PPB) is the development of a plant breeding program in collaboration between breeders and farmers, marketers, processors and consumers and policy makers. But in the context of plant breeding PPB is breeding that involves close farmer-researcher collaboration to bring about plant genetic improvement within a species. Participatory Plant Breeding involves scientists, farmers, and others, such as consumers, extensionists, vendors, industry, and rural cooperatives in plant breeding research. It is termed 'participatory' because users can have a research role in all major stages of the breeding and selection process. Such 'users' become co-researchers as they can: help set overall goals, determine specific breeding priorities, make crosses, screen germplasm entries in the pre-adaptive phases of research, take charge of adaptive testing and lead the subsequent seed multiplication and diffusion process (Sperling and Ashby, 2001). The fundamental rationale for PPB programs is that joint efforts can deliver more than when each actor works alone. Modern plant breeding stands among the greatest scientific and human success stories of all time. Yet the fruits of major advances in agricultural science, such as those from the Green Revolution, have bypassed millions of farmers in developing countries, most of whom operate small farms under unstable and difficult growing conditions. The adoption of new plant varieties by this group has been low, an issue that has challenged scientists, development workers, governments, donors, and all others with a stake in agricultural progress and the fight against poverty. Their response has been the creation of a novel and promising set of research methods collectively known as participatory plant breeding (PPB). While the principal aims of PPB are to create more relevant technology and equitable access to it, there are often other objectives, depending on the organizations involved. For example, large-scale breeding programs by international or national research agencies may wish to cut research costs. Other organizations, such as farmer's groups and NGOs, may wish to affirm local people's rights over genetic resources, produce seed, build farmers' technical expertise, or develop new products for niche markets, like organically grown food.

Conventional breeding has tended to focus heavily on "broad adaptability"- the capacity of a plant to produce a high average yield over a range of growing environments and years. Unfortunately, candidate genetic material that produces very good yields in one growing zone, but poor yields in another, tends to be quickly eliminated from the breeder's gene pool. Yet, this may be exactly what small farmers in some areas need. And the resulting "improved" varieties often require heavy doses of fertilizer and other chemicals, which most poor farmers can't afford. Professional breeders, often working in relative isolation from farmers, have sometimes been unaware of the multitude of preferences - beyond yield, and resistance to diseases and pests — of their target farmers. Ease of harvest and storage, taste and cooking qualities, how fast a crop matures, and the suitability of crop residues as livestock feed are just a few of the dozens of plant traits of interest to small-scale farmers. Despite this wealth of knowledge, in many cases farmers' participation in conventional breeding programs has been limited to evaluating and commenting on a few advanced experimental varieties just prior to their official release. Such token participation means that few farmers feel a sense of ownership of the research or have been able to contribute their technical expertise. Many of the varieties reaching on-farm trials would have been eliminated from testing years earlier if farmers had been given the chance to critically assess them. Farms in many have been the chief engineers of crop and variety development for thousands of years, and continue today to actively select and breed most crops, including the so-called 'minor' or 'neglected ones, which are the key for family nutrition.

A goal of the PPB program is to build on farmers' knowledge, which involves clearly identifying farmers' needs and preferences and the reasoning behind them. So far, more than 80 participatory plant breeding efforts have been documented. The systematic recording of this knowledge, and its application in formal breeding programs, has been one of PPB's major achievements, but while the technical issues of PPB are moving ahead, the social, ethical and legal issues are lagging behind. People generally recognize the role of farmers in managing and improving germplasm, but there's little agreement yet on how to value the role and research contributions of both the farming community and the formal breeding system. It makes sense, therefore, to analyze PPBas a new approach to germplasm development, especially in the public sector. The CGIAR program on Participatory Research and Gender Analysis for Technology Development and Institutional Innovation (PRGA) currently has detailed documentation on 65 PPB programs and projects (Smith *et al.*, 2003,). Most of the cases, whether located in public sector or nongovernment (NGO) crop improvement programs, were begun within the last 10 years. A lack of consensus about terminology is common when a new science is in its early stages, and PPB is no exception. Terms commonly used interchangeably include: Collaborative Plant Breeding (CPB) (Soleri *et al.*, 1999) Farmer Participatory Breeding (FPB) (Courteois *et al.*, 2001), and Participatory Crop Improvement (PCI) (Witcombe *et al.*, 1996). The latter is sometimes subdivided into two areas, one for work with stabilized materials, termed Participatory Varietal Selection (PVS) and another referring exclusively to work with variable or segregating materials, confusingly also sometimes termed PPB. All labels presently used describe broadly the same activities of what is a multifaceted technical and organizational collaboration in plant breeding by scientists and users of their results (Smith *et al.*, 2003). Most important reason for differentiating among approaches to participatory plant breeding is to understand how each approach can lead to a different outcome, and so to be able to make informed choices among approaches. Participatory plant breeding (PPB) also supports the preservation of existing genetic diversity and the maintenance of evolutionary processes that form diversity, and simultaneously improve cultivar performance? In recent years, participatory approaches have contributed significantly to the understanding of evolutionary processes (biological and social) that form, and affect, diversity in agricultural systems. By studying and understanding the local farming and seed management system PPB strategies can be developed that respect farmers' diversity needs and that fit into the local farming systems and that help to overcome the limitation of the local seed system.

This chapter sets up a framework for relating different participatory plant breeding approaches to breeding outcomes and impacts. It is only when these variables are clearly described that current and potential practitioners can start to link the 'type of PPB' employed (method and organizational forms) with the type of impacts achieved. Such clarity is essential if PPB is to have the scientific and organizational foundations to judge its utility for a given objective. It is also essential for choosing the appropriate PPB approach.

Advantages of PPB

Local approaches to plant breeding based on participatory methods offer a number of potential advantages compared to the traditional global approach to plant breeding. Improved local adaptation breeding- PPB methods are well suited for niche breeding, or development of varieties that perform well in specialized environments. Niches can be defined not only by biophysical variables, but also by human preferences and needs. The advantage of PPB methods derives from the strong links that they generate between scientists and end users. By making selection criteria more relevant to end user needs, PPB can reach poor households that have not yet benefited from MVs. Promotion of genetic diversity- Unlike the current global breeding model, which for the most part has concentrated on developing a limited number of varieties that are stable over time and adapted to a wide range of environments, the breeding model based on PPB methods encourages the maintenance of more diverse, locally adapted plant populations (Joshi and Witcombe, 1996). To the extent that diverse populations are taken up and grown by farmers, in-situ conservation of crop genetic resources is encouraged, and genetic diversity is enhanced (Witcombe *et al.*, 2001). PPB methods could, however, lead to loss of genetic diversity if only a few genetically similar plant populations are taken up and grown by farmers, displacing an array of more diverse populations. Increased breeding efficiency- Returns to investment in plant breeding research will increase if use of PPB methods increase MV adoption levels. Similarly, economic benefits will be created if use of PPB methods accelerates adoption of MVs by reducing the time required to develop new varieties. No matter how excellent the science, if improved germplasm is never adopted, or if it is adopted only after a long lag, the breeding process must be considered inefficient. Although relatively little empirical work has been done to document the speed of PPB compared to conventional breeding, recently evidence has started to emerge suggesting that PPB can lead to earlier adoption of MVs, with no major additional costs (Witcombe *et al.*, 2003). Empowerment of rural communities- PPB allows rural communities to maintain germplasm they value and enables them to participate in the development of new varieties that suit their needs. PPB methods thus can empower groups that traditionally have been left out of the development process (McGuire *et al.*, 1999).

Goals of PPB

Over the last decade, PPB has been applied as a crop improvement strategy primarily in response to the need for impact in non-commercial crops and in very unpredictable, stressed production environments. Its successes in reaching highly diversified, very specialized and segmented markets have been less well publicized. A range of other goals have also been defined within PPB programs: for instance, enhancing biodiversity and germplasm conservation; developing adapted germplasm for especially disadvantaged user groups (e.g., women, poor farmers); making breeding programs more cost-efficient, particularly through decentralization of programs which target more niches. Close analysis of the set of PPB cases shows that some goals are explicit and often attained (for instance, production increase) while others are poorly articulated and usually not addressed, unless they are explicitly built into the research design (for instance, reaching specialized interest groups). Case study analysis also suggests that many of the goals are not obviously compatible (for instance, biodiversity enhancement and reaching the poorest farmers). The trade-offs among goals is one of the areas where a good deal more structured or focused work needs to be pursued within the PPB field. As partners usually have to accept trade-offs in reaching certain goals, it is important at the very beginning of a PPB collaboration for those concerned- scientists, farmers, development/NGO personnel - to discuss explicitly primary and secondary goals, and the minimal agreed-upon outcomes for which collaborators are aiming.

Farmer contribution found in formal-led PPB programs

There are four main thrusts where farmers contribute:

1. Contribution of knowledge and experiences,
2. Contribute genetic materials,
3. Conduct trials,
4. Select and evaluate germplasm.

Farmer's knowledge and information

In many PPB programs farmers' knowledge was often integral to achieving a better understanding of the varieties that farmers grow. Farmers' descriptions of their cropping system, major constraints, key aspects of the farming system, and social institutions are all information that makes an important contribution to the orientation of a breeding program. Another important requirement for a breeding program is to understand farmers' needs and preferences for the specific traits of a crop. This kind of understanding often leads to descriptions of appropriate type of varieties, a reorientation of the breeding program in terms of selection strategy, and germplasm used for breeding. Farmers' knowledge is also essential to better understand a local seed system, and opportunities for increasing its efficiency and its ability to serve a wide range of farmers.

In some of the cases, the description of locally-grown varieties in the target areas allowed breeders to become familiar with the varieties that farmers grow and helps breeders to choose appropriate materials for testing and breeding that could fulfill some of the farmers' basic requirements; or it allowed the breeder to choose materials for testing that differ sufficiently from the locally grown varieties for key traits (Witcombe *et al.*, 1996). The description of the local process of seed production and management is another example where farmers' information was key to a program's success. Based on a careful analysis of farmers' strategies for seed potato, production technologies for improved seed health and storage conditions were developed in the Andean region and in eastern Africa (Haugerud and Collinson, 1990). Another example where farmers' information played a key role in orientating the breeding program is the work on pearl millet in the desert region of Rajasthan. Farmers in this region do not differentiate particularly between different varieties, but rather between different plant types. They associate strongly between adaptation to specific growing conditions and specific plant types. Farmers who regularly produce their own seed often select panicles representing very different plant types for their seed lots. For a breeding program targeting this region, adoption of any new variety will depend on its capacity for contributing to the success of these diverse seed lots (Dhamotharan *et al.*, 1997). This type of farmer contribution thus provides breeders with necessary information for identifying the appropriate type of variety, both in terms of genetic make-up and in terms of key plant traits. Similarly, research efforts aimed at improving institutional support for seed production and distribution can derive key interventions from a sound analysis of farmers' information. Better understanding key Farmer Participation and Formal-Led Participatory Plant Breeding Programs production constraints may allow breeders to modify the testing environments on the research station to test materials more effectively for key adaptations. This understanding, especially when combined with an understanding for the major uses of a crop or varieties, will also help breeders reorient their programs and be better able to meet farmers' key needs.

A number of varieties have been developed and released through PPB in Guangxi, southwest China from 2000 to 2011. Here the researchers again utilized the farmer's knowledge and experience to develop these varieties. The 1st variety is Xin Mo 1, which derived from the cross of farmer improved Tuxpeno 1 (as female line, from Wenteng village) and Jiahe white (as male line, from Zicheng village) in 2002. It is an open-pollinated variety (OPV). The 2nd one is Zhong Mo 1, which derived from a cross of Xin Mo 1, Suwan 1 and Amarinto 966 (as male line) in 2004. Since Xin Mo1 is in white color, PPB farmers discussed with breeders and would like to improve it into a yellow color, which has high commercial value. This was the motivation for breeding out Zhong Mo 1 and the 3rd one is Zhong Mo 2, which derived from a cross between Xin Mo 1 and Amarinto 9 in 2006 (Yiching, 2011).

These examples indicate that the types of impact that a PPB program can expect from letting farmers contribute their information to the joint program are largely related to making research more effective and improving the dissemination of new technologies. Farmers' information may have some impact on the level of biodiversity maintained in the target farming system if the breeders start targeting a wider range of plant types, or if more variable types of varieties are available for dissemination as a result of these discussions with farmers. An important advantage of explicitly searching for and including farmers' knowledge for developing and implementing a PPB program is that it is easy to involve many farmers and, with appropriate care, stakeholders can also be involved who are often overlooked or excluded (e.g., women, poor farmers, and minorities). Discussions with farmers on specific topics related to the breeding effort do not have to be conducted during the growing season when farmers' time is often limited and valuable.

Farmers contribute genetic materials

Farmers often contribute genetic materials to breeding programs, especially participatory breeding programs focusing on adaptation to specific stresses, production systems, or niches. Specific quality traits or crops, for which little breeding has been done so far, rely strongly on farmers' contributing their own genetic materials to a joint breeding effort. In such cases, farmers' genetic materials are commonly key to success. Farmers' genetic materials can be used in different ways. In a few cases with cross pollinated crops, the farmers deliberately create new variability by facilitating out crossing between highly diverse types of varieties, often their landrace and an introduced modern variety. These outcrosses often reveal enormous genetic variation and many new combinations of traits. In the case of pearl millet in Namibia, one such population was used as the base for creating a new breeding population (Bidinger, 1998). In several cases, farmers have contributed landraces as parents for crossing, or for targeted improvement efforts. A good example for this is the project on breeding for chilling tolerance in rice in Nepal. Breeders identified the local landrace in a series of tests as highly tolerant of chilling temperatures during the early growth phase and during grain filling. This variety was crossed with a high yielding, chilling-susceptible variety. Farmers used the progenies from this cross for selection of a new variety (Sthapit *et al.*, 1996). Farmers' varieties usually form the basis for further improvement in projects that mainly aim at improving farmers own skills in improving the genetic composition of their varieties and seed stocks. An example is a project, led by the national program of Honduras and Cornell University, with a component of teaching farmers techniques for pollination control for maize in a region with a high degree of local varietal diversity (Gomez, 1996). Thus the base material for the efforts for improvement is the farmers' varieties or possibly other materials that are available to them.

Farmers' genetic materials often broaden the genetic base of the participating breeding program considerably. The material derived and disseminated from these efforts can also represent a wider range of diversity than the products of previous efforts. The use of farmers' genetic materials sometimes contributes to conserving landrace materials in the local farming system, particularly in cases that focus on building farmers' skills. The types of impact that were observed or expected from using the farmers' genetic materials have been mostly related to enhancing biodiversity or conserving local germplasm in the farming community and enhanced productivity in specific production systems.

Farmer's conduct trials

In most of the cases of PPB, farmers manage trials on their own land as part of the PPB program. Farmers normally decide how to experiment, that is, they choose the field for growing the trial(s), manage the nutrients and other aspects of crop husbandry, choose the control variety, and contribute to the trial design. In many of the cases where farmers evaluate stable varieties, this contribution is key to the breeding program's success. Farmers provide the appropriate testing conditions, which enhance the selection program's efficiency.

An example where farmers' trials are key to the joint breeding effort is the case of pearl millet in southern Africa, where farmers grow nurseries of diverse pearl millet varieties and germplasm accessions usually in community plots for selecting material for further testing on their own farms. In this case, the farmers' fields also provide more appropriate testing conditions and better access to farmers (Monyo *et al.*, 1996). More farmers can come and see and assess the new breeding materials. The key impact from farmers' contributing trials to a participatory breeding effort is an increase in research efficiency through providing appropriate testing conditions. Through on farm trials, farmers get early access to new genetic materials and have the option to adopt these varieties. Increases in productivity and initial adoption can be immediate results from farmer managed trials. This would normally lead to further dissemination and adoption of the new varieties. Farmers may also decide to use them in their own breeding work and thus influence the level of genetic diversity in their farming system. Farmers' experimentation with new germplasm is a powerful tool to create the basis for a range of impacts on-farm. However, among the cases examined, several report that poor farmers or women can experience greater difficulty in growing trials. Better-off farmers often have more time to devote to farming or have a stronger role in decision making about land allocation on their farm. If poor or women farmers are among the targeted group, specific planning is needed so that they can benefit from the planned activities.

Farmer's select and evaluate

This contribution of decision making among a set of varietal choices goes beyond providing information alone. Farmers in these cases make judgments based on multiple criteria; they decide which trade-offs to make and which combination of traits to favor. These are often complex decisions because numerous traits can be considered and because differences between individuals may be small for individual traits. To identify materials that contribute new opportunities for enhanced productivity or stability of production requires an intimate knowledge of the target farming system, production system, and social system. In many PPB programs, farmers make the final selection decisions, for example, among entries in an on-farm trial, or among entries in larger nurseries grown on-farm or on-station. Selection usually involves evaluating a variety for a number of traits in combination simultaneously. Trade-offs are assessed between the different traits, for example, yield potential and earliness, or panicle size and tillering behaviour. A farmer is often the better judge in predicting which combination of traits may have potential use for specific growing conditions and production goals on his farm or in his area of cultivation. Several well-documented cases show that farmers' selections indeed performed well, better than breeders' selections, in the conditions for which they were selected (Sperling *et al.*, 1993). Breeders often find it difficult to have a broad enough understanding of the diversity of growing conditions and the corresponding relative weaknesses of the local cultivars to make appropriate selection decisions that anticipate the full range of potential adoption possibilities. Thus, farmers' involvement in decision making can have far reaching impacts and often leads directly to more adoptable varieties especially in areas where none were before. Researchers usually gain a better understanding of farmers' assessment processes and can thus target their program better towards meeting these needs. Most programs use several types of contributions from farmers and to different degrees. They also change their organizational set-up to adapt to the needs of the farming communities and to improve the program's impact. The fast and comprehensive learning experience for the participating scientists is a basis for these rapid changes in the operational forms of PPB projects. The other basis is the rapidly increasing role of farmers in the decision-making process and in guiding the project toward other areas of work (e.g., seed system support) to improve the scope and scale for impact.

Examples were farmers selected and evaluated genetic materials is the improvement of bean varieties **through participation** of Honduran hillside farmers and regional scientists.

The focus was at the field level where teams of farmer-researchers (CIALs), supported by a local non-governmental organization (NGO), the Foundation for Participatory Research with Honduran Farmers (FIPAH), collaborated in a participatory plant breeding (PPB) programme with the Pan-American Agricultural School, Zamorano. Scientists at the school crossed a popular local bean variety with improved materials, and CIAL members, trained in participatory research by FIPAH, conducted successive selections in their fields. Farmers learnt formal selection techniques, including selection from early generations, in a process that has taken more than four years to complete (Sulaco and Victoria, 2005).

Another example of this type of contribution is the Krishak Bharati Cooperative Indo British Rainfed Farming Project (KRIBP) in India, were a number of released and pre-released cultivars of rice, maize, chickpea and black gram were introduced. The three steps of PVS were carefully followed: 1) identification of farmers' needs in a cultivar, 2) a search for suitable material to test with farmers, and 3) experimentation on farmers' fields. Some of the adopted cultivars had been cultivated in other areas of India for many years, but had never been released in the project's working area. This was done after in participatory rural appraisals (PRA) local landrace characteristics were assessed with techniques such as ranking multiple characters of local landraces. That was done in focused group discussions before and after harvest on many crop aspects such as agronomic characteristics, taste, market value, threshing characteristics and storability. This method was used to ensure that only potentially useful cultivars were given to farmers for testing. Introductory trials with nine released and seven pre-released rice cultivars were conducted in 1993 with 128 farmers in six villages. In each village, five cultivars were grown and every cultivar was replicated across three to five farmers (each farmer representing one replicate). Farmers used a number of characters that had not been used by breeders, and the degree of agreement on farmers' perceptions was very high between group discussions. Farmers scored performance in three classes: better, similar or worse than the local cultivar. Data were analyzed with a one-way anova with farms as replicates and villages as replicate blocks within a class. Two rice cultivars, Kalinga III and Sathi-34-36 were identified as giving greatest production. Unsubsidized seed sales of Kalinga III have increased afterwards. The selection of Kalinga III is remarkable, as the cultivar had been released only in Orissa, reported for chickpea, maize and black gram (Witcombe and Joshi, 1996a, 1996b).

PPB and *in situ* germplasm conservation

The assumption that PCI provides more genetic diversity that better matches farmers' needs provides the rationale for PCI as a tool in *in situ* conservation. *In situ* conservation of crop genetic diversity is only possible through its use by farmers as it are precisely the farmers' practices in combination with the local growing conditions that shaped the varieties and crops and that are considered valuable for conservation *in situ* (Brush, 1999). *In situ* conservation is seen as a conservation strategy that is complementary to *ex situ* conservation (Almekinders *et al.*, 2000). *In situ* conservation of many local varieties and varieties of minor crops is considered desirable because they may have genes or gene combinations that are not stored in gene banks but can be of value in the future. Maintaining the evolutionary process in farmers' fields is another important argument for *in situ* conservation: it may provide new genes and gene combinations and plays a role in the adaptive potential of farmers' agricultural production systems. Since farmers' practices and environment are not strictly controlled, and evolution of crop genetic diversity is a desired element of *in situ* conservation, it is not conservation in a strict sense. *In situ* conservation may therefore be described or defined as an *in situ* or on farm management of crop genetic diversity (Almekinders *et al.*, 2000). Particularly local materials and evolution in marginal areas that are characterized by high environmental variation, complex stresses and presence of wild crop relatives are considered valuable for *in situ* conservation. Potatoes in the Andes, maize in Central America, sorghum and other crops in Ethiopia are examples of such situations (although valuable diversity can also develop outside the primary centers of diversity of these crops) the diversity in such areas has only been influenced to a limited extend by modern plant breeding. In these marginal areas there seem to be more cases in which modern varieties have been added to the portfolio of planted varieties rather than that they displaced them (Brush *et al.*, 1999). However, productivity of crop varieties in these situations is often low and threatens the sustainability of farmers' livelihood. Poverty is an important explanation for farmers' loss of seeds due to crop failure, urgent cash needs or because consumption needs of the household exceed production. Various surveys indicate that in rural communities the wealthiest farmers plant most crop genetic diversity. Poor farmers tend to have restricted access to seed from other sources than their own planting. Improving farmers' seed management and access to crop genetic diversity will therefore contribute to possibilities of farmers to maintain *in situ* those materials which are of value to them. This also points to the fact that agricultural development and conservation of genetic diversity are not conflicting objectives but are to be combined. Improving farmers' seed production capacities is particular relevant for local varieties since the formal sector does not provide seed of these materials and they are considered relevant materials for *in situ* conservation (Almekinders and Louwaars, 1999).

Another way to support the *in situ* conservation of valuable crop genetic diversity is to provide farmers with improved versions of local varieties that are obtained through crossing local varieties with modern varieties. Improved performance makes the materials more competitive with improved varieties and there with reduce the likeliness that farmers lose interest in planting local materials. This approach does not maintain farmers' use of the local varieties *in situ*, but the genes or gene combinations for characteristics of those varieties that are relevant to the farmers (Bellon, 1996). Directed crossing of local varieties with exotic materials in formal breeding programmes is however a relatively costly genetic improvement as it usually involves various back-crossings and the products may only be suitable for relatively small target areas. PCI offers interesting opportunities in this respect, in particular in cross-pollinating crops. In Indonesia, for example, farmers plant small amounts of hybrids around their fields with landrace maize, and as those two cross-fertilize, the landrace will receive some of the hybrid's genes. Such examples show that, following PCI strategies, an important part of the selection process after the back-crossing may be carried out by the farmers. The contribution of PCI to *in situ* conservation of genetic diversity is based on the assumption that the diversity of farmers' preferences and environment, will result in the local selection and use of a wider diversity of materials than when one or two broadly adapted varieties would have been introduced (Witcombe *et al.*, 2001). Furthermore, local selection of the genetically heterogeneous population will develop an adaptation and genetic composition that may differ from location to location. An increased selection pressure on genetically heterogeneous local varieties through rigorous farmers' seed selection may however lead to a narrowing of the genetic basis of these materials. If indeed this narrows the genetic basis of varieties, than this may reduce their yield stability. However, there exists little consistent information of the effect of farmers' selection on genetic diversity within varieties.

This dynamic concept of *in situ* conservation in which varieties may be lost or modified and in which new materials may be introduced does also emphasis the importance and complementarity of *ex situ* conservation: materials or genetic characteristics for which farmers lose interest with changing agro-ecologically or socio-economic environment can be collected and stored in gene banks. And, when materials are lost because of crop failures and when interest in old materials revives, such materials can be re-introduced from *ex situ* germplasm collections.

Impacts of PPB

Impact of PPB projects led by formal sector institutions was analyzed for both groups of partners in this process: the farmers, their families, their communities, and associations; and the researchers and their institutions. The farm-level impacts have in many cases exceeded researchers' expectations.

Changes in orientation of breeding program

Many participatory breeding projects were initiated in regions that are marginal in terms of the environmental conditions for crop growth and productivity increase. These regions are often where plant breeding programs have had limited prior impact. The PPB research targeting these types of environment has examined the suitability of specific breeding strategies in achieving better genetic gains under such condition through insights gained from farmers, and farmer expertise in selecting among plants, progenies, or varieties. Such close interactions can profoundly influence the overall strategy of an evolving breeding program.

An example is the pearl millet breeding program of the International Crops Research Institute for the Semi-Arid Tropics (ICRISAT). The program is for the dry zones of the state of Rajasthan, and was initiated in 1989 in response to findings indicating that local landraces tended to show higher grain yields under severe stress conditions. The approach taken was to improve testing sites in the target region and develop breeding populations using the local germplasm. Farmers were involved in the project implicitly to identify their needs and preferences and thus contribute to refining the program's goals and objectives. The initial findings of this work indicated that farmers in different regions and representing different social strata and gender had widely differing preferences and needs (Weltzien *et al.*, 1998).

A more detailed understanding of farmers' seed management practices revealed that farmers often use small quantities of modern varieties as components of seed mixtures and thus considerably broaden the genetic base of their own seed lots. Farmers could thus enhance their chances of obtaining increased yields under better growing conditions. At the same time, poor farmers, especially women, expressed concern over this practice of mixing improved varieties with local ones, because the seed produced from these mixtures shows less adaptation to the predominantly poor soil conditions. Poor farmers are often dependent on seed from better-off farmers, but the seed these provide does not serve the poor farmers' needs as well as it used to. Because poor farmers dominate in these regions and better-off farmers also own field with poor soils, the breeding program decided to fill this gap by focusing its efforts on developing materials specifically targeted towards poor fertility soil conditions and a higher stability of grain yield. This constituted a profound change, because it required changes in the management of research station fields and in the germplasm base for breeding, and further research into the methodology for achieving these goals, including sharing responsibilities with farmers. The program also helped a local nongovernment organization (NGO) identify farmer-preferred sources of landrace-type seed for increase and distribution to those farmers who are facing difficulties with seed production. In collaboration with the national authorities, steps were initiated to change release procedures for pearl millet so that varieties specifically adapted to these marginal conditions would have a higher chance of being identified as superior by the national variety testing scheme, and thus a chance to be released.

Adoption of PPB products

Generating new varieties for farmers who now cultivate 'old' varieties in marginal environments is the reason for success of PPB. Although PPB is still in its early days, practitioners need to show that PPB-generated cultivars are adopted by farmers. Productivity levels in low rainfall environments fluctuate sharply from year-to-year. These fluctuations will swamp the effects of varietal change, giving the impression that nothing significant is happening in the region when, in fact, adoption is robust. Secondary data cannot be relied on to assess the effects of Final PPB. Early acceptance and later adoption surveys should feature prominently in the PPB toolkit. In general, the evidence for adoption is variable with several dry holes and undocumented outcomes, but the generation of PPB-related varieties provides grounds for optimism.

In India, CAZS-NR and their partners have generated improved rice varieties for the rainfed uplands of eastern India and improved maize varieties for the poverty-ridden Chhotanagpur Plateau of eastern India and for the hill areas of the western state of Gujarat (Virk *et al.*, 2005). In Nepal, CAZS-NR and their partners have produced improved rice varieties for the high-altitude hills, several mid-hill districts, and for all 21 terai districts. Early acceptance studies of adoption suggest that all these varieties enjoy bright prospects. Very thorough adoption research also documents the rapid spread of Kalinga III, a PVS-related product of CAZS-NR collaboration (Witcombe *et al.*, 2003).

The ICARDA Barley Team started its work a few years after CAZS-NR, and it has yet to reach the level of adoption performance that the CAZS-NR group has obtained in South Asia. Nonetheless, in the first cycle of crop improvement with PPB, farmers have selected the following number of PPB-generated barley varieties: 12 in Syria, 1 in Jordan, 5 in Egypt, 3 in Eritrea, and 2 in Yemen where 2 lentil varieties have also been selected. Several of the selections in Syria are already on several thousand hectares. One elite line, Zanbaka, was submitted to the official system of variety release in the early eighties and was rejected. It was then included in the PPB trials and was adopted because it matched farmers' preferences.

Zanbaka has now spread from the 'participatory-breeding' villages to about five thousand hectares in Syria in several provinces (Mustafa *et al.,* 2006). Recently, a second variety, also rejected by the variety-release committee several years ago is repeating the Zanbaka story of adoption in the PPB trials and is expected to reach 5,000 hectares next year. An ex-ante assessment of the ICARDA Barley Team's PPB investment in Syria shows high economic potential of that work given that barley is cultivated on 800,000 hectares and that results should be rapidly forthcoming (Lilja and Aw-Hasaan, 2002). A very early ex-post assessment has been carried out (Mustafa *et al.,* 2006).

Conspicuous for its absence is any hard evidence on adoption outcomes in Latin America of PPB-related work. Northeast Brazil of Brazil is one of the earliest PPB projects in Latin America (Fukuda and Saad, 2000). Clones that are resistant to root rot and that are highly acceptable to farmers have been released, but a recent adoption study concluded that it was too early to mount a survey to assess the level of adoption of those materials (Saad *et al.,* 2005). In Ecuador in 2005, a PVS-related good-processing potato variety I-Fripapa-99 occupied 5,000 hectares in the central part of the country (Montesdeoca *et al.,* 2006). That variety was selected over a three-year period in farmers' fields with producers, consumers, processors, and traders. PROINPA, the semi-autonomous national potato program of Bolivia, has conducted PPB for several years, but PPB-related cultivar adoption as yet does not exceed a few hundred hectares (Thiele, 2006).

Improved varieties for marginal areas

Because farmers were involved in the selection in segregating generations, a rice variety (*Oryza sativa*) was bred that combined the high level of Farmer Participation and Formal-Led Participatory Plant Breeding Programs chilling tolerance of landrace rice varieties of the Nepal mountains with the increased productivity from modern varieties. So for the first time the farmers in these hilly regions could benefit from a plant breeding effort targeting their region (Sthapit *et al.,* 1996). Initial adoption of this variety was high and, more interestingly, the success of farmers' selection encouraged them to continue further selection in this and other material. They were thus carrying the process further even with reduced support from researchers. Another example is the case of Bean (*Phaseolus vulgaris* L.) breeding in Brazil which became successful through involving farmers in breeding a variety that combined disease resistance (*Macrophomina and Fusarium wilt*) with the preferred seed coat color, drought adaptation, and yield characteristics for the dry zone of northeastern Brazil. Farmers were involved in selection in segregating materials on-station and later in testing experimental varieties on their own farms. Before the variety could be formally recommended for cultivation in this region, farmers were growing it and demanding more seed (Zimmermann, 1995).

Immediate adoption and yield increases

Farmers in a remote area of India, where soils are highly degraded and production is subsistence oriented, could identify varieties of several major crops that provided them with new options, more food, and greater stability of production (Witcombe and Joshi, 1996). The documentation of adoption of the rice variety "Kalinga III" is a good example of the spread of a variety identified through farmers' participation in variety testing. This study shows that adoption of the variety began in the study villages the year following the first trials conducted by farmers. Seed of the preferred variety began spreading to other villages in the second year following farmers' trials. Although the project helped spread the seed in several ways, much of the found spread resulted from farmers' own initiatives in selling and giving away seeds to others, inside and outside the village. The spread of this variety continues, but official release in the state of Rajasthan depends and thus government subsidies for producing and distributing seed of this variety are not available to farmers. A study of the impact of this one rice variety, identified through farmer participation, estimates rates of return for the overall project between 47% and 70%. This figure is higher than for regular successful, non-participatory agricultural research. This indicates that participatory research with a comparatively local focus, and thus a smaller target region, can have similar rates of returns. In this case, it was mainly because this variety provided farmers with great yield benefits and more production stability. The advantage of participatory variety testing is precisely that these attributes can be rapidly identified because varieties are being tested under a wide range of conditions that farmers manage themselves.

Benefits for resource poor farmers

Breeders have often worked under the assumption that benefits of new varieties are scale- and user neutral, that poor and rich farmers, or men and women can achieve the same type of benefit from these new varieties. In many cases of widespread adoption, differences in preferences and potential benefits were observed, but the conventional approach to breeding (of variety release and of extension) has often been unable to offer effective solutions to meet these diverse needs. Participatory approaches are often designed to involve as partners in their program a wide range of farmers representing different social groups, gender, or otherwise differentiated groups. These approaches potentially have the in-built capacity to respond to specific groups' needs and to make use of different user talents. Women in particular are often plant breeders in small-scale farmer production systems, responsible for domesticating wild species, selecting germplasm, and saving seed.

In many instances, women's criteria may be significantly different from men's—dual involvement in breeding/selection is necessary to meet each partner's needs. In Mali, maize evaluations showed men putting production and early maturity as the main criteria, with women focusing on organoleptic and processing aspects (Kamara *et al.*, 1996). Rice work in West Africa had a similar gender division, with scientists from the West Africa Rice Development Association (WARDA) reporting that men focused on yield and yield-related traits, such as plant vigor, and women concentrated on quality attributes, such as bold grains. Involving women in PPB can confer benefits for all the community not just poor women. Over a 3-year period, Rwandan experts selected a pool of varieties (21 separate bean types) that outdid breeders' choices in their own production terms by up to 33%—and met quality characteristics of interest to diverse community groups (Sperling *et al.*, 1993).

Varietal diversity increased

Many of the examined cases of PPB involve farmers in the testing of more varieties than traditionally takes place during the adaptive or extension phase of formal research. It is commonly observed that farmers' selections differ as their needs differ, for example, farmers who own poor land tend to prefer different varieties than farmers who have better fields. Women farmers often select more explicitly on grain quality characteristics and make different choices than do men farmers. The outcome of PPB programs is usually that different farmers in different communities select different varieties and thus PPB in most cases has contributed to increased varietal diversity.

A clear example is the "Informal Research and Development" (IRD) program initiated in Nepal with the explicit objective of increasing the varietal diversity of rice varieties suitable for early planting under irrigated conditions (chaite rice) (Joshi *et al.*, 1995). Potentially useful rice varieties identified by researchers were distributed to about 1800 households in the target area for this project. Farmers were given some written information about the variety they received and were encouraged to compare it with their own variety in their own fields. Researchers did not participate in these evaluations, but returned 2 years later to a sample of the recipient farmers to assess the effect of the approach. The follow-up revealed that about 35% of farmers who had received a seed packet had adopted from one to four varieties of the six distributed for testing. In this area, a single cultivar was dominating cultivation prior to the participatory work. Most of the farmers who decided to adopt a new variety belonged to the group of farmers who regularly produce sufficient or surplus food for their families. Only 15 % of the adopters belonged to the food deficit group. Interestingly, the preferences for specific cultivars seemed to differ between farmers belonging to these different groups, indicating that they have different needs for cultivars and their key characteristics. A conclusion of this report is that not any one single variety could meet all these needs. This IRD approach proved to be a powerful tool for offering a wide range of diversity to a wide range of farmers.

Similarly impressive PPB results relating to enhanced varietal diversity occurred in two major cassava-growing regions of Colombia: a seasonally dry ecosystem in the north (an area with poor soils and 800-1000 mm rainfall annually, bimodally distributed), and more recently the highlands of southwest Colombia. Researchers from the International Center Tropical Agriculture (CIAT) initiated a participatory crop improvement effort in 1996. The aims were to learn more about farmer criteria for choosing cassava varieties for consumption, marketing, and processing, and to evaluate traits within a genetic base that had not been preselected, hopefully to generate varieties that were both more acceptable and more "biodiverse". An average of 28 communities per year was involved in the clonal evaluation effort in northern Colombia, with community participation organized via chip-drying cooperatives. Researchers quickly realized that effective varietal comparisons could be made only if the planting material for local varieties and new breeders' clones was produced under similar conditions, to avoid bias caused simply by its health and vigor. This concern was addressed by producing all planting material in a common location under conditions approximating those of the farmers. It also became clear that farmers and researchers often used different terms for variety evaluation; a glossary of farmer evaluation terms was compiled. Farmer evaluations of advanced clones from cassava breeding programs resulted in the release of three new varieties in northern Colombia (a significant addition to the two varieties in use). Through this process, researchers acquired a better understanding of farmers' selection criteria and were able to quantify certain of them in ways that would facilitate researchers' selections (Iglesias *et al.*, 2000)

Changes in variety release procedure and seed production system

Centralized forum or committees for any individual country usually enforce variety release procedures. Needs of specific regions, especially in marginal areas, and those specific consumers or users are not easily considered in such centralized procedures. As PPB tends to reveal such differentiated needs and those for diversity, recommendations on how to change the existing release procedures are often a direct result of working more closely with farmers. Participatory research also contributes to making farmers more aware of these procedures and can spur ways for farmers to initiate policy changes. A PPB program in India is working to encourage formal committees to give greater official weight to farmer evaluations. The program argues that data synthesized from farmer varietal evaluations (i.e., qualitative assessments) should be used as a base for varietal release decisions. In many cases, such data may be more predictive of future adoption than the standard yield measurements, which form the core of most release decisions (Witcombe *et al.*, 1998).

Often PPB goes hand in hand with recommendations to develop or build on local, more decentralized seed systems, which can provide location-specific varieties those farmers themselves effectively, multiply and distribute. For instance, prior to the civil strife in the early 1990s, both Rwanda and Burundi Ministries of Agriculture and Rural Development tabled plans for decentralizing seed services—partly to accommodate more decentralized breeding (Sperling, 2001). Programs of PPB are also increasingly including seed production components, which are integrated and innovative, to quickly deliver the positive impacts that PPB can achieve. Good examples of this come from the PPB and seed work with cassava in Colombia (Iglesias *et al.*, 2000).

Impact on Specific adaptation

Specific adaptation is foremost in participatory plant breeding. In principle, plant breeders should always be able to make progress in a specific locale. But the generated product may not be transferable to other locales and therefore may only be adopted locally. The CAZS-NR group has provided evidence that varieties bred for specific adaptation may be more transferable than initially thought. One of the varieties bred in the terai has given excellent results in the High Barind Tract of Bangladesh in aus, aman, and even in the irrigated boro seasons (Joshi *et al.*, 2007). In Nepal, two rice varieties bred for the high hills actually performed relatively better in the mid-hills where rice cultivation is more common (Joshi and Witcombe, 2003). Part of the reason for wider than expected adaptation is the shared preferences for traits: aside from yield, farmers usually prize earliness, market quality, and fodder quality and quantity in many cereals.

Net benefits per hectare

Comparing the selected variety and the check variety yields in farmer fields suggest higher net benefits than expected. For the three success stories of the CAZS-NR group in India, the average yield difference of the selected variety was 40% higher than the check

(Virk *et al.*, 2005). These relative differences are higher in farmer fields than on research stations. For cereals, these gains are equivalent to about 50-60 U.S. dollars per hectare, assuming a base yield of about one ton.

Future prospects

After analyzing the impacts the prospects are bright for participatory plant breeding to make a positive contribution to varietal change in marginal environments. The next ten years are critical to the development of PPB, and they will define the size of that contribution. Some pleasant surprises have emerged during the last years. Research has shown that varieties bred and selected for specific adaptation can perform well in larger niches. Several products of PPB have wider adaptation than expected. Yield gains of 30-50% in farmers' fields, compared to the farmer's variety under the same management, are larger than expected.

The target for PPB should still be in geographic poverty traps characterized by low and/ or unexploited production potential. Although one success has taken place in one heterogeneous irrigated tract, the PPB varieties, in this case, were competing against cultivars that farmers had adopted in the late 1950s. Very high weighted mean varietal age suggested that PPB could be successful. High varietal age signals the potential demand for PPB. Crops with multiple uses seem to be particularly attractive for PPB projects as conventional breeding tends to focus only on the dominant use. As with farming systems research, the initial institutionalization of PPB has been slow and perhaps even disappointing to some practitioners but progress has been made. PPB is starting to figure as a sub-discipline of plant breeding and is beginning to be taught in workshops and university curricula. PPB-related varieties are entering public-sector varietal testing systems. In the next ten years, we should see examples of induced change on formal seed systems and varietal testing and release procedures precipitated by the accommodation of PPB products. (Already institutional change precipitated by these approaches has occurred in Nepal (Joshi *et al.*, 2006).

We are also beginning to see what production-oriented PPB is and is not. As practiced today, PPB respects the comparative advantage of what a farmer does best and what a breeder does best. Farmers' information is critical on the choice of parents, but farmers are not crossing nor are they directly choosing parents. In the next ten years, we will also have a better appreciation of what works when, where, and why as experience allows researchers to gradually approximate Morris and Bellon's (2004) ideal of efficient participatory plant breeding. Unlike to conventional breeding, PPB requires good plant breeders. The standard for human capital in PPB is as high or even higher for capacity in conventional breeding. Breeders working in PPB have to be creative to come up with flexible yet workable designs tailored to a specific context and to deal with day-to-day tasks that are not routine. (The CAZS-NR group in several of their projects employed an international consultant to perform very specialized work in plant breeding). Given that training in genetics and plant breeding today is severely skewed towards biotechnology, the human capital requirements for PPB could be a cause for concern underscoring the importance for simple generalized designs. Accumulating more contextual information on the spatial, temporal, and commodity adaptation of PPB models and breeding strategies of the CAZS-NR team and the ICARDA barley program is a priority. In this unexpectedly long and increasingly technologically uncertain interim, both conventional plant breeding and highly client-oriented approaches such as PVS and PPB require our support if the aim is to effect varietal change for this generation of poor people and their children.

Conclusion

A wide range of impacts were achieved through the use of participatory approaches, despite the short time that most of these projects have been in operation. Through the close collaboration of farmers and formal researchers, breeding progress was made in regions where variety improvement efforts of centralized programs with large uniform target regions have been less than successful. Farmer participation opened new opportunities for reducing the lag-time between variety testing, release, and adoption; and, through making new germplasm available to farmers in experimental quantities; the basis for adoption was repeatedly laid. The improved understanding of local seed production and dissemination systems allowed researchers and project managers to overcome specific seed system constraints and to make targeted interventions to overcome seed supply problems quickly. This helped farmers gain access to the desired seeds rapidly, from known and reliable seed sources. Farmer participatory breeding programs also led to new alliances between research and development organizations. As a result, changes occurred in the orientation of breeding programs and in variety release procedures, and changes in formal seed regulatory policies were recommended and in some cases (in Latin America, Africa, and Asia) implemented.

Achieving specific benefits for specific user groups, such as poor or women farmers, really only became possible through participatory research programs. Enhancing their capacity to voice their concerns and needs was a major outcome of programs that explicitly targeted this type of goal. Such user-driven programs were also often those that had a complementary strong capacity building component—to strengthen farmers' own breeding, selection, and seed management skills. In the cases examined, such skill building proved integral to the scaling-up of the specific participatory breeding effort.

Although the impact of PPB to date has been impressive, this review of programs also highlights that many options for farmer participation in the plant breeding process are yet to be explored. Having more widespread goals, focusing on a broader range of crops, working with explicitly more diverse user groups, and experimenting with organizational options may identify wider and more varied benefits that PPB can achieve.

In short we can say that, Participatory plant breeding reduces the chances of rejection of a variety by the farmers. PPB may be the only possible type of breeding for crops grown in remote regions, for crops for which a high level of diversity is required within the same farm, or for those crops considered as minor crops and therefore neglected by formal breeding. Combined farmers and breeders efforts might lead to more suitable varieties that would ultimately benefit farmers. Farmer participation opens new opportunities for reducing the lag-time between variety testing, release, and adoption; and, through making new germplasm available to farmers. The main conclusion of PPB research is that formal plant breeding has nothing to lose and much to gain from farmers' participation.

REFERENCES

Almekinders C, Louwaars N (1999) Farmers' seed production. New approaches and practices. *IT Publications, London.* p. 292

Almekinders CJM, Köhler-Rollefson I (2000) Similarities and differences between plant and animal genetic resource management and conclusions for technical co-operation. *Report prepared for the GTZ.* 17

Bidinger FR (1998) Farmer participation in pearl millet research in Namibia. In: Proc. participatory plant improvement. MS Swaminathan Research Foundation (MSSRF) - International Crops Research Institute for the Semi-Arid Tropics (ICRISAT) workshop. *MS. Swaminathan Research Foundation, Chennai, India.* p 21-30

Brush SB (1999) The issues of *in situ* conservation of crop genetic resources. In: S.B. Brush (Ed), Genes in the field. *Lewis Publishers, IPGRI & IDRC.* p. 3-26

Ceccarelli S, Grando S (2006) Decentralized-Participatory Plant Breeding: An Example of Demand Driven Research. *Euphytica* 20: 126-132

Courtois B, Bartholome B, Chaudhary D, McLaren G, Misra CH, Mandal NP, Pandey S, Paris T, Piggin C, Prasad K, Roy AT, Sahu RK, Sahu VN, Sarkarung S, Sharma S, Sigh A, Singh HN, Sinh ON, Singh NK, Singh RK, Singh S, Sinha PK, Sisodia BVS, Takhur R (2001) Comparing farmers' and breeders' rankings in varietal selection for low-input environments: a case study of rainfed rice in eastern India. *Euphytica* 122: 537–550

Dhamotharan M, Weltzien RE, Whitaker ML, RattundeWHF, Anders MM, Tiagi LC (1997) Seed management strategies of farmers in western Rajasthan in their social and environmental contexts: Results from a workshop using new communication techniques for a dialogue between farmers and scientists, 5-8 February 1996, Digadi village, Jodhpur district, Rajasthan, India. Integrated Systems Project Progress Report No. 9, pp. 47. *ICRISAT, Patancheru 502324 India*

Fukuda W, Saad N (2000) Participatory research in cassava breeding with farmers in Northeastern Brazil. Centro Nacional de Pesquisa de Mandioca e Fruicultura Tropical, *EMBRAPA, Cruz de Almas, Brazil and the PRGA, Cali, Colombia*

Gomez F (1996) Conservation and enhancement of maize with small farmers in Honduras. 1995 Annual Report of collaborative project between the Escuela Agricola Panamericana zamorano, and cornell international institute for food, Agriculture and development (CIIFAD). *Project report.* p 15

Haugeurd A, Collinson (1999) Plants, genes and people; improving the relevance of plant breeding in Africa. *Experimental Agriculture* 26: 341-362

Iglesias CA, R-Hernandez LA, Lopez A (2000) Participatory selection with farmers in northern Colombia. Internal mss. *Centro Internacional de Agricultura Tropical (CIAT), Cali, Colombia*

Joshi KD, Rana RB, Subedi M, Kadayat KB, Sthapit BR (1995) Addressing diversity through farmer participatory variety testing and dissemination approach: A case study of *chaite* rice in the western hills of Nepal. In: Sperling L; Loevinsohn M, eds. Using diversity –

enhancing and maintaining genetic resources on-farm. *International Development Research Centre (IDRC), New Delhi, India.* p 158-175

Joshi KD, Witcombe JR (2003) The impact of participatory plant breeding (PPB) on landrace diversity: A case study for high-altitude rice in Nepal. *Euphytica* 134: 117-125

Joshi KD, Musa AM, Johansen C, Gyawali S, Harris D, Witcombe JR (2007) Highly client-oriented breeding, using local preferences and selection, produces widely adapted rice varieties. *Field Crops Research* 100: 107-116

Kamara A, Defoer T, Groote HD (1996) Selection of new varieties through participatory research, the case of corn in south Mali. *Tropicultura* 114: 100-105

Lilja N, Ashby JA (1999) Types of participatory research based on locus of decision-making. Focus on who decides, who participates and when? Consultative Group on International Agricultural Research (CGIAR), Participatory Research and Gender Analysis (PRGA), *PRGA Working Document,* No. 6. 10 p

Lilja N, Aw-Hassan A (2003) Benefits and costs of participatory plant breeding in Syria. A paper submitted to the 25th *International Conference of IAAE, Durban, South Africa* 16-22

McGuire S, Manicad G, Sperling L (1999) Technical and institutional issues in participatory plant breeding done from a perspective of farmer plant breeding. CGIAR system wide program on Participatory Research and Gender Analysis for Technology Development and Institutional Innovation Working Department 2.

Montesdeoca F, Jimenez JM, Reinoso I (2006) Generacion de riqueza a partir de la innovacion tecnologica: *Impacto economico de I-Fripapa-99.* Un estudio de caso. Programa Nacional de Raices y Tuberculos del INIAP, INIAP, Quito, Ecuador. 26p

Monyo ES, Ipinge SA, Mndolwa SI, Mangombe N, Chintu EM (1996) The potential of local land races in pearl millet improvement. In: Leuschner K; Manthe CS, eds. Drought tolerant crops for southern Africa: Proc. Southern Africa Development Community (SADC)-International Crops Research Institute for the Semi-Arid Tropics (ICRISAT), *Regional Sorghum and Pearl Millet Workshop, 25-29 Jul, 1994, Gaborone, Botswana. ICRISAT, India.* p 153-162

Morris ML, Bellon MR (2004) Participatory plant breeding research: Opportunities and challenges for the international crop improvement system. *Euphytica* 136: 21-35

Mustafa Y, Grando S, Ceccarelli S (2006) Assessing the benefits and costs of participatory conventional barley breeding programs in Syria. *ICARDA report for a study supported by the International Development Research Centre.* 54 pp

Saad N, Lilja N, Fukuda W (2005). Participatory cassava breeding in Northeast Brazil: Who adopts and why? Working Document No. 24. Cali,Colombia: *CGIAR PRGA program.* 28 p

Soleri D, Cleveland CA, Smith SE (1999) Creating common ground in collaborative crop improvement. *ILEIA Newsletter* 3(4): 20-22

Sperling L, Loevinsohn M (1995) Proceedings of a workshop, 19-21 June 1995, New Delhi. IDRC, New Delhi

Sperling L, Loevinsohn M, Ntambovura B (1993) Rethinking the farmer's role in plant breeding: Local bean experts and on-station selection in Rwanda. *Exp Agric* 29: 509-519

Sperling L, Ashby JA, Smith ME, Weltzien E, McGuire S (2001) A framework for analyzing participatory plant breeding approaches and results. *Euphytica* 122: 439-450

Sthapit B, Joshi KD Witcombe JR (1996) Farmer participatory crop improvement III. Participatory plant breeding, a case-study from Nepal. *Exp Agric* 32: 479-496

Sulaco V (2005) Linking small farmers to the formal research sector: lessons from a participatory bean breeding programme in honduras. *Agreen 142*

Thiele G (2006) Informal potato seed systems in the Andes: Why they are important and what we should do with them. *World Develop* 27: 83-99

Virk DS, Chakraborty M, Ghosh J, Prasad SC, Witcombe JR (2005) Increasing the client orientation of maize breeding using farmer participation in Eastern India. *Expl Agric* 41: 413-426

Weltzien E, Smith ME, Meitzner, LS, Sperling L (2003) Technical and institutional issues in participatory plant breeding-from the perspective of formal plant breeding: A global analysis of issues, results, and current experience. *PPB Monograph No. 1. Cali, Colombia: PRGA Program.* 226 p

Weltzien-RE, Whitaker ML, Rattunde HFW, Dhamotharan M, Anders MM (1998) Participatory approaches in pearl millet breeding. In: Witcombe JR; Virk DS; Farrington J, eds. Seeds of choice. Making the most of new varieties for small farmers. *Oxford and IBH Publ Co., New Delhi, Calcutta.* p. 143-170

Witcombe JR, Joshi A (1996b) The impact of farmer participatory research on biodiversity in crops. In: Using Diversity. p. 87-101 in: *Enhancing and Maintaining Genetic Resources On-Farm, Eds.* L.

Witcombe JR, Joshi A, Goyal SN (2003) Participatory plant breeding in maize: A case study from Gujarat, India. *Euphytica* 120: 413-422

Witcombe JR, Joshi L, Joshi KD Sthapit BR (1996) Farmer participatory crop improvement. I. Varietal selection and breeding methods and their impact on biodiversity. *Expl Agric* 32: 445-460

Witcombe JR, Joshi KD, Rana RB, Virk DS (2001) Increasing genetic diversity of participatory varietal selection in high potential production systems in Nepal and India. *Euphytica.* 122: 575-588

Witcombe JR, Joshi KD, Gyawali S, Musa SL, Johansen C, Virk DS, Sthapit BR (2005) Participatory plant breeding is better described as highly client-oriented plant breeding. I. Four indicators of client-orientation in plant breeding. *Expl Agric* 41: 299-319

Witcombe JR, Packwood AJ, Raj AGB, Virk DS (1998) The extent and rate of adoption of modern cultivars in India. pp. 53-68. in Seeds of Choice. Making the most of new varieties for small farmers. J.R. Witcombe, D.S. Virk and J. Farrington (Eds). *Published by Oxford IBH, New Delhi and Intermediate Technology Publications, London*

Yiching S, Jingsong L (2011) The role of biodiversity, traditional knowledge and participatory plant breeding in climate change adaptation in Karst mountain areas in SW China. *Biocultural Heritage*

Zimmermann MJDO (1995) Breeding for marginal/ drought-prone areas in northeastern Brazil. In: Eyzaguirre P; Iwanaga M, eds. Participatory plant breeding. *International Plant Genetic Resources Institute (IPGRI), Wageningen, Neths.* p 117-122

Index

F

G

H

I

J

K

L

M

N

O

P

Q

R

S

T

U

V

W

X

Y

Z